I0051591

Multifunctional and Multiband Planar Antennas for Emerging Wireless Applications

This work focuses on designing multiband-printed single/Multiple Input Multiple Output (MIMO) CP antennas for WLAN/V2X and NR Sub-6GHz 5G applications. It also delves into the design and implementation of a Four-Port MIMO antenna for wireless applications, addressing theoretical foundations and challenges. Additionally, the book explores critical aspects of software-defined radios (SDRs), including modulation, signal processing, radio systems, TX/RX blocks, SDR-enabled phased arrays, and beam hopping techniques, with relevance to 5G, 6G, and IoT applications.

Features:

- Explores advancements in planar monopole antennas, including bandwidth enhancement techniques.
- Analyzes innovative antenna design structures, like miniaturized and conformal monopole antennas; and discusses modeling and implementation.
- Spotlights WLAN and Wi-Fi 6/6E antenna design for next-gen laptops with practical insights.
- Addresses the use of triple-band antenna arrays for MIMO applications in laptops.
- Focuses on planar antenna advancements for diverse wireless bands and applications.
- Explores multiband-printed single/MIMO CP antennas for WLAN/V2X and NR Sub-6GHz 5G.
- Covers the design and implementation of a Four-Port MIMO antenna for wireless applications, including theoretical foundations and challenges.
- Explores SDR, modulation, signal processing, radio systems, TX/RX blocks, SDR-enabled phased arrays, and beam hopping techniques for 5G, 6G, and IoT applications.

This book is aimed at graduate students and researchers in electrical and electronic engineering, antennas, and wireless communication systems.

Multifunctional and Multiband Planar Antennas for Emerging Wireless Applications

Jayshri Kulkarni, Chow-Yen-Desmond Sim,
Jawad Yaseen Siddiqui, Anisha M. Apte,
Ajay Kumar Poddar, and Ulrich L. Rohde

CRC Press
Taylor & Francis Group
Boca Raton London New York

CRC Press is an imprint of the
Taylor & Francis Group, an informa business

Designed cover image: © Jayshri Kulkarni, Chow-Yen-Desmond Sim, Jawad Yaseen Siddiqui, Anisha M. Apte, Ajay Kumar Poddar, and Ulrich L. Rohde

First edition published 2024
by CRC Press
2385 NW Executive Center Drive, Suite 320, Boca Raton FL 33431

and by CRC Press
4 Park Square, Milton Park, Abingdon, Oxon, OX14 4RN

CRC Press is an imprint of Taylor & Francis Group, LLC

© 2024 Jayshri Kulkarni, Chow-Yen-Desmond Sim, Jawad Yaseen Siddiqui, Anisha M. Apte, Ajay Kumar Poddar, and Ulrich L. Rohde

Reasonable efforts have been made to publish reliable data and information, but the author and publisher cannot assume responsibility for the validity of all materials or the consequences of their use. The authors and publishers have attempted to trace the copyright holders of all material reproduced in this publication and apologize to copyright holders if permission to publish in this form has not been obtained. If any copyright material has not been acknowledged please write and let us know so we may rectify in any future reprint.

Except as permitted under U.S. Copyright Law, no part of this book may be reprinted, reproduced, transmitted, or utilized in any form by any electronic, mechanical, or other means, now known or hereafter invented, including photocopying, microfilming, and recording, or in any information storage or retrieval system, without written permission from the publishers.

For permission to photocopy or use material electronically from this work, access www.copyright.com or contact the Copyright Clearance Center, Inc. (CCC), 222 Rosewood Drive, Danvers, MA 01923, 978-750-8400. For works that are not available on CCC please contact mpkbookspermissions@tandf.co.uk

Trademark notice: Product or corporate names may be trademarks or registered trademarks and are used only for identification and explanation without intent to infringe.

ISBN: 9781032362588 (hbk)
ISBN: 9781032362595 (pbk)
ISBN: 9781003331018 (ebk)

DOI: 10.1201/9781003331018

Typeset in Times
by KnowledgeWorks Global Ltd.

Contents

Preface

The presented book focuses on the specific topic of innovative antenna designs and structures that are integrated into laptop computers and modern electronic devices. This book also discusses the software-defined radio (SDR) technology that offers flexibility, adaptability, and advanced signal processing capabilities that greatly benefit antenna designs. Antennas play a critical role in enabling wireless communication for laptop computers and modern electronic devices. With the increasing demand for wireless connectivity, compactness, and sleek designs, antennas need to be carefully designed to meet these requirements while maintaining optimal performance.

Laptop antennas are typically integrated into the device, with some common types including monopole antennas, planar antennas, and microstrip antennas. These antennas are designed to operate in specific frequency bands, such as the 2.4 GHz and 5 GHz bands for WLAN/Wi-Fi standards. The challenge lies in designing miniaturized antennas that can cover multiple wireless standards and fit within the limited space available inside laptops.

Modern electronic devices, beyond laptops, also rely on antennas for wireless connectivity. Smartphones, tablets, smartwatches, and Internet of Things (IoT) devices all require antennas to establish wireless connections. These antennas are designed to support various wireless standards, such as 4G LTE, 5G sub-6 GHz wireless LAN, and Wi-Fi 6/6E. Furthermore, antennas for modern electronic devices need to be compatible with the device's form factor and aesthetics. Integration of antennas within the device's chassis, display bezels, or other components is often required to maintain a sleek and appealing design while ensuring efficient wireless communication.

Advancements in antenna technology, such as multiple input multiple output (MIMO) techniques and adaptive tuning, are being explored to enhance the performance and reliability of wireless connections in modern electronic devices. These technologies enable better signal reception, increased data rates, improved coverage, and reduced interference.

The primary objective of this book is to provide insights into the design of miniaturized monopole antennas for laptops and modern electronics devices with dual/triple-band and MIMO operations. These antennas aim to enhance performance, achieve wider bandwidth, and increase data rates without the need for additional hardware, lumped elements, or holes/vias. Finally, it discussed how SDR technology enables the selection of optimal antennas, agile adjustment of parameters, digital beamforming, software optimization, interference mitigation, and efficient prototyping. By leveraging the power of software control, SDR enhances antenna performance, adaptability, and overall system capabilities.

By addressing these various topics, the book aims to contribute to the understanding and advancement of antenna design for laptop computers and modern electronic devices, fostering innovation and improved wireless communication capabilities.

Authors:
Jayshri Kulkarni
Chow-Yen-Desmond Sim
Jawad Yaseen Siddiqui
Anisha M. Apte
Ajay Kumar Poddar
Ulrich L. Rohde

About the Authors

Jayshri Kulkarni received a Bachelor of Engineering degree in Electronics and Telecommunications (E&TC) from Shivaji University, Master of Engineering degree from PICT College, and a PhD in Information and Communication Engineering from Anna University, Chennai. From 2006 to 2009, she was a Lecturer at the GSM College of Engineering, Pune, India. From 2010 to 2022, she was an Assistant Professor at the E&TC Department, Vishwakarma Institute of Information Technology (VIIT), Pune, India. She is currently working as a postdoctoral research associate with Baylor University, Texas, USA. She has authored more than 10 books, and more than 50 research papers in reputed journals and conferences. Her research interests include antennas, microwave engineering, wireless communications, and wireless sensors. She received several awards, including the Desmond Sim Award for Best Antenna Design Paper at IEEE InCAP 2019, the Outstanding Oral Presentation Award at ICRAMET-2020, and the Best Paper of the Session Award at IEEE ESCI-2021. She was one of the recipients of the reputed Mojgan Daneshmand Grant awarded by IEEE (AP-S)-2022.

Chow-Yen-Desmond Sim was born in Singapore in 1971. He received a Bacherlor of Science degree from the Engineering Department, University of Leicester, UK, in 1998, and a PhD from the Radio System Group, Engineering Department, University of Leicester, UK, in 2003. In 2007, he joined the Department of Electrical Engineering, Feng Chia University, Taiwan, where he became a Full Professor in 2012 and a Distinguished Professor in 2017. He has authored or coauthored more than 200 SCI papers, and his current research interests include 5G antenna design and RFID applications. He is a Fellow of the IET, Senior Member of the IEEE AP Society, and a Life Member of the IAET. Since 2016, he has served as the technical consultant of Securitag Assembly Group (SAG), which is one of the largest RFID tag manufacturers in Taiwan. He is also the technical consultant of Avary Holding (the largest PCB manufacturer) since 2018.

Jawad Yaseen Siddiqui is currently working as an Associate Professor in the Institute of Radio Physics and Electronics, University of Calcutta, India. He received his Doctor of Philosophy degree in Radio Physics and Electronics, University of Calcutta, in 2005. He worked as a postdoctoral fellow at the Royal Military College of Canada and Visiting Researcher at Queen's University, Canada, at different periods during 2008–2021. He has more than 150 publications in peer-reviewed journals and conferences. His research areas include printed circuits and antennas, radar, and nano-photonics. He is Co-Principal Investigator of the Stratosphere Troposphere (ST) Radar Project at the University of Calcutta. He is the recipient of the 2022 Institution of Electronics and Telecommunication Engineers (IETE)-S.N. Mitra Memorial Award for contribution and leadership role in radio broadcast science and technology. He is currently the IEEE AP-S Committee Chair for AP-S SIGHT Committee.

Anisha M. Apte (Senior Member, IEEE) graduated in Instrumentation and Control Engineering from The Savitribai Phule University of Pune, India; received a Master of Science in Electrical Engineering from the New Jersey Institute of Technology (NJIT), New Jersey, USA; and a Dr.-Ing. from Brandenburg Technical University, Cottbus, Germany. She worked at Hindusthan Instruments and Ultraline Instruments, in Pune, India; Discovery Semiconductors, in New Jersey, USA, and then joined Synergy Microwave in 2003, and has been working on signal-generation and signal-processing electronics, reference-frequency-standards, and metamaterial resonators for signal-source applications.

Dr. Apte has published many papers in IEEE conferences and journals and received awards for her scientific contributions. She is a dedicated IEEE volunteer, currently serving as the Vice-Chair-2, and various other positions of the IEEE North Jersey Section, and also the IEEE AP-S society, where she is an elected AdCom member for the term 2022–2024. She is the recipient of the 2021 and 2012 IEEE Region-1 Support for IEEE Mission Award for Outstanding Contributions to the IEEE North Jersey Section.

Ajay Kumar Poddar (SM 2005, Fellow 2015) is working as a Chief Scientist at Synergy Microwave, New Jersey, USA, responsible for the design and development of signal generation and signal processing electronics, RF-MEMS, and metamaterial-sensors/electronics for industrial, medical, and space applications. He is also serving as a visiting professor in the University of Oradea, Romania, and the Indian Institute of Technology Jammu, India. He has received more than a dozen awards, such as the 2015 IEEE IFCS Cady Award in recognition of his outstanding scientific contributions. Dr. Poddar published more than 350 scientific papers in journals, magazines, and conference proceedings, co-authored four technical books/chapters, and more than 40 patents for scientific and technological innovations. For the past 30 years, he has supervised many PhD scholars. He is currently serving on several scientific committees, professional societies, and voluntary organizations. He has been actively involved with IEEE SIGHT/HAC, IEEE sister societies, and IEEE MGA activities, including charitable services.

Ulrich L. Rohde (Life-Fellow IEEE, 2005), is a Partner of Rohde & Schwarz, Munich, Germany; Chairman of Synergy Microwave Corp., Paterson, New Jersey, USA; President of Communications Consulting Corporation; serves as an honorary member of the Senate of the University of the Armed Forces Munich, Germany, and is a Professor of Microwave Systems, faculty of Informatics, and of the Senate of the Brandenburg University of Technology Cottbus–Senftenberg, Germany. Dr. Rohde is serving as Professor of Radio-Microwave Frequency Theory and Techniques at several universities worldwide, to name a few: Honorary Professor Indian Institute of Technology (IIT), Delhi, India, and Honorary Chair Professor IIT Jammu, India. Member, Indian National Academy of Engineering (INAE)

Rohde has published more than 400 scientific papers, co-authored more than a dozen books, and more than four dozen patents; received several awards, recently including: recipient of 2022 IEEE Photonics Society Engineering Achievement Award, 2021 Cross of Merit of the Federal Republic of Germany, and the 2017 IEEE AP-S Distinguished Achievement Award.

1 Introduction to Wireless Technologies

1.1 WIRELESS TECHNOLOGY

Today, wireless technology is gaining popularity rapidly due to the numerous advantages it offers. In essence, wireless is a blessing for humanity. Take a moment to consider what would happen if all the wireless connections worldwide were replaced with cables. Devices would be cluttered with wires, creating a tangled mess. This is where wireless technology works wonders by eliminating the need for cables, making communication simple and easy. In wireless technology, devices communicate with each other without the use of physical connections or cables. The key question here is how do these devices communicate without physical connections or cables? In wireless communication, the connection between devices is established using antennas. Antennas are electrical devices that convert electrical signals into radio signals in the form of electromagnetic (EM) waves and vice versa. Wireless technology is the fastest-growing communication technology industry. Antenna engineering plays a crucial role in the global adoption of wireless technologies by providing aesthetic design, improving the overall frequency-link budget, and enabling multiple users to utilize a single interface, thereby increasing capacity. It is indisputable that good antenna design facilitates better wireless devices. Therefore, antennas are the most critical component, whether it be for Wireless Local Area Network (WLAN), Worldwide Interoperability for Microwave Access (WiMAX), sub-6 GHz 5G, or any other wireless device. Consequently, antennas must be designed with utmost care, as they can make or break any wireless system.

Over the past decade, laptop computers have experienced exponential growth and have become an indispensable part of everyday life worldwide. Laptop computers utilize WLAN, WiMAX, and sub-6 GHz 5G bands for wireless communication. For WLAN and WiMAX to function effectively and efficiently, the antenna must be designed to operate flawlessly within these networks. Therefore, to design an antenna, one must first understand WLAN, WiMAX, and sub-6 GHz 5G wireless networks.

1.2 WLAN

WLAN, or Wi-Fi, is a data transmission system designed to enable network access between computing devices without the need for a physical connection. It achieves this by utilizing EM waves. WLAN serves as a wireless computer network, connecting multiple devices through wireless technology to establish a Local Area Network (LAN) in small areas like offices, colleges, and homes. Additionally, it provides internet access through an Access Point (AP), also known as a gateway. The popularity of WLAN has grown significantly in homes and offices due to its affordability,

DOI: 10.1201/9781003331018-1

1

flexibility, reliability, ease of use, and installation. It has particularly revolutionized the business world, enhancing productivity and accessibility. Unlike wired LANs, WLANs enable devices to communicate without wires, providing instant information exchange. This allows employees the freedom to work from any location within their office, instead of being tied to a specific area. Furthermore, WLAN facilitates prompt responses to time-sensitive or urgent situations.

1.2.1   EVOLUTION OF WLAN (WI-FI)

The evolution of Wi-Fi can be traced through a series of advancements and standards. Initially, WLAN technology emerged as a cost-effective solution for areas where traditional cabled LAN connections were challenging or impractical. Early developments in WLANs involved industry-specific solutions and proprietary protocols. However, by the late 1990s, standardized protocols began to replace these proprietary approaches. The most significant standard that revolutionized Wi-Fi was the IEEE 802.11 series. This series defined the foundation for modern WLANs and introduced various versions of the 802.11 standard.

In 1997, the 802.11 committee was established, leading to the introduction of Wi-Fi as a popular wireless communication technology. IEEE 802.11 provided a comprehensive set of standards that governed communication protocols for WLANs. In parallel, a European alternative called High-Performance Radio Local Area Network 1 (HiperLAN 1) emerged. It was declared by the European Telecommunications Standards Institute (ETSI) and saw its first version released in 1996. HiperLAN 1 was followed by the development of HiperLAN/2, which incorporated Asynchronous Transfer Mode (ATM) influences and was finalized in February 2000.

Over time, Wireless Fidelity (Wi-Fi) technology continued to evolve, with new versions and enhancements being introduced. These advancements have brought about significant improvements in terms of data transfer speeds, range, security, and compatibility. Today, Wi-Fi is widely adopted and has become an integral part of our daily lives, enabling wireless connectivity in various settings such as homes, offices, public spaces, and beyond.

The progression of Wi-Fi standards is illustrated in Figure 1.1 and is further elaborated in the following paragraphs.

1.2.1.1   802.11 (Year: 1997)

The initial version of the IEEE 802.11 standard, which laid the foundation for Wi-Fi technology, was released in 1997. This version introduced the basic principles and specifications for WLANs, defining the protocols and mechanisms necessary for wireless communication between devices. It provided a framework for wireless networking and served as a starting point for subsequent versions and advancements in Wi-Fi technology. It defined two-bit rates, namely 1 Mbps and 2 Mbps, along with the inclusion of forward error correction code. Additionally, it specified three alternative physical layer technologies: diffuse infrared, which operated at 1 Mbps; frequency-hopping spread spectrum (FHSS), which operated at either 1 Mbps or 2 Mbps; and direct-sequence spread spectrum, which operated at either 1 Mbps or 2 Mbps.

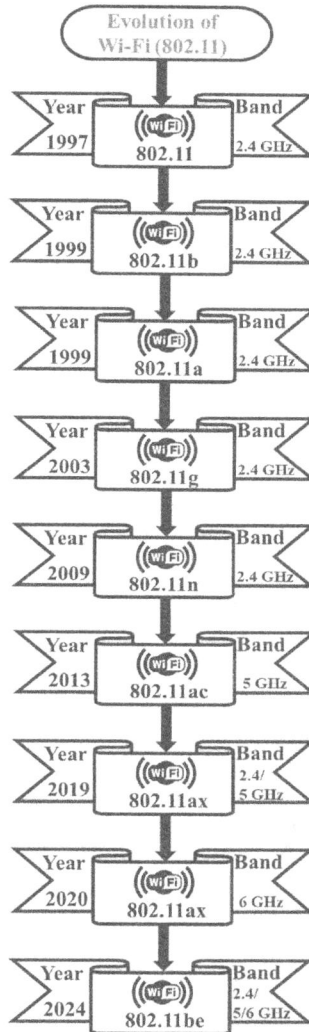

FIGURE 1.1 Evolution of Wi-Fi 802.11 (in modern communication Wi-Fi is also known as a WLAN].

1.2.1.2 802.11b (Year: 1999)

A subsequent version of the IEEE 802.11 standard, 802.11b was introduced as an enhancement to Wi-Fi technology. It was released in 1999 and brought significant improvements compared to the initial version. One of the key advancements of 802.11b was the increase in data-transmission rates. It defined a maximum data rate of up to 11 Mbps, which was a substantial improvement over the 1 Mbps or 2 Mbps rates of the earlier version. This higher data rate allowed for faster and more efficient wireless communication.

The 802.11b version utilized the same 2.4 GHz frequency band as the initial version, which enabled compatibility with existing hardware. However, it employed a different modulation technique called Complementary Code Keying (CCK) to achieve higher data rates. This modulation scheme provided better robustness and improved signal quality.

Another advantage of 802.11b was its backward compatibility with the original 802.11 standard. This meant that devices supporting 802.11b could communicate with devices using the earlier version, ensuring interoperability and a smooth transition. The 802.11b version had a significant impact on the adoption of Wi-Fi technology, as it offered faster speeds and better performance compared to its predecessors. It became widely deployed in various settings, including homes, offices, and public spaces, contributing to the proliferation of wireless networking. The Apple iBook was the first laptop sold with optional 802.11b networking.

The only disadvantage of the 802.11b standard is that it faced interference from other devices operating in the 2.4 GHz band.

1.2.1.3   802.11a (Year: 1999)

Another version of the IEEE 802.11 standard, 802.11a was introduced as a complementary technology to 802.11b. It was released around the same time as 802.11b in 1999 and brought its own set of advancements and features. One significant difference between 802.11a and 802.11b lies in the frequency band on which they operate. While 802.11b utilizes the 2.4 GHz frequency band, 802.11a operates on the 5 GHz frequency band. This higher frequency allowed for more available channels and reduced interference, resulting in improved overall performance and higher data rates.

Also, 802.11a provided faster data transmission speeds compared to 802.11b. It supported data rates of up to 54 Mbps, offering a significant boost in bandwidth. This increase in speed made 802.11a suitable for applications that required more intensive data transfer, such as multimedia streaming and large file transfers. However, due to the higher frequency band, the range of 802.11a was generally shorter than that of 802.11b. The 5 GHz signal had more difficulty penetrating obstacles like walls and had a reduced range compared to the 2.4 GHz band. As a result, 802.11a was often deployed in settings where shorter range coverage was sufficient, such as office environments.

Despite its shorter range, 802.11a provided advantages in terms of higher data rates and reduced interference. Over time, subsequent versions of the IEEE 802.11 standard, such as 802.11n and 802.11ac, incorporated the best features of both 802.11a and 802.11b to provide improved performance, range, and compatibility, leading to the widespread adoption of Wi-Fi technology.

1.2.1.4   802.11g (Year: 2003)

The 802.11g version of the IEEE 802.11 standard was introduced as an evolution of Wi-Fi technology. It was released in 2003 and aimed to combine the advantages of both 802.11a and 802.11b into a single standard. One of the key features of 802.11g is its backward compatibility with 802.11b. This means that devices supporting 802.11g can seamlessly communicate with devices using the earlier 802.11b standard. This compatibility ensured a smooth transition and allowed users to upgrade their networks without rendering existing devices obsolete.

The 802.11g version operates in the same 2.4 GHz frequency band as 802.11b, providing a wider coverage area compared to 802.11a. It offered data-transmission rates of up to 54 Mbps, matching the speeds of 802.11a. This higher data rate enabled faster and more efficient wireless communication, supporting applications such as video streaming and online gaming. The modulation scheme used by 802.11g is orthogonal frequency division multiplexing (OFDM), which provides improved signal quality and resistance to interference. This modulation technique allows for better performance in environments with potential sources of interference, such as other wireless devices or appliances.

The 802.11g version brought enhanced performance and compatibility to Wi-Fi networks, making it a popular choice for both home and business use. It provided faster speeds and a wider coverage area compared to 802.11b, while maintaining compatibility with older devices. This widespread adoption of 802.11g contributed to the continued growth and prevalence of Wi-Fi technology in various settings.

1.2.1.5 802.11n (Year: 2009)

A significant advancement in the IEEE 802.11 standard, 802.11n was designed to provide faster speeds, improved range, and better overall performance for Wi-Fi networks. It was released in 2009 and quickly became a widely adopted standard. One of the key features of 802.11n is its ability to utilize multiple input multiple output (MIMO) technology. MIMO employs multiple antennas, both on the transmitter and receiver sides, to transmit and receive multiple data streams simultaneously. This technology significantly enhances the data throughput and overall network capacity, allowing for higher data rates and better reliability.

The 802.11n version supports both the 2.4 GHz and 5 GHz frequency bands, offering greater flexibility and compatibility with different devices and network setups. The dual-band capability allows for better channel selection and reduces potential interference from other devices operating in the same frequency range. The maximum data rate of 802.11n can reach up to 600 Mbps or even higher with the use of channel bonding and higher end equipment. This increased bandwidth enables smooth streaming of high-definition video, faster file transfers, and improved performance for bandwidth-intensive applications.

To further enhance the range and coverage, 802.11n introduced features like beamforming. Beamforming focuses the Wi-Fi signal directly toward the client device, improving signal strength and reducing dead zones. This technology allows for better coverage throughout a larger area, even in environments with obstacles or interference. Ensuring compatibility with devices that support earlier standards such as 802.11a/b/g, 802.11n is backward compatible with previous versions of Wi-Fi. This compatibility allows for a seamless transition and integration of 802.11n into existing Wi-Fi networks.

The adoption of 802.11n brought significant advancements to Wi-Fi technology, offering faster speeds, improved range, and better overall performance. Its widespread use in homes, businesses, and public spaces has transformed the way we connect and interact in the digital world.

1.2.1.6 802.11ac (Year: 2013)

A major advancement in the IEEE 802.11 standard, 802.11ac is introduced to provide even faster speeds, increased capacity, and improved performance for Wi-Fi networks. It was released in 2013 and quickly gained popularity as the next generation of Wi-Fi technology. One of the key features of 802.11ac is its ability to utilize wider channel bandwidths. It supports channel widths of up to 160 MHz, compared to the maximum of 40 MHz in the previous 802.11n standard. This wider channel bandwidth allows for greater data-transmission rates and increased network capacity.

The 802.11ac version operates exclusively in the 5 GHz frequency band, which offers several advantages. The 5 GHz band provides more available channels and less interference compared to the crowded 2.4 GHz band, resulting in improved performance and reliability. Additionally, 802.11ac introduces MultiUser MIMO (MU-MIMO) technology, which enables simultaneous data transmission to multiple devices. This technology optimizes network efficiency, allowing for faster and more efficient communication in environments with multiple connected devices.

The maximum data rate of 802.11ac can reach up to several gigabits per second (Gbps). It achieves these high speeds through the use of advanced modulation techniques and more efficient data encoding. This makes 802.11ac ideal for bandwidth-intensive activities such as streaming 4K video, online gaming, and large file transfers. To enhance range and coverage, 802.11ac also employs beamforming technology, similar to 802.11n. Beamforming focuses the Wi-Fi signal toward connected devices, improving signal strength, and reducing signal degradation over longer distances.

Also, 802.11ac is backward compatible with previous Wi-Fi standards, allowing devices supporting older standards to connect to an 802.11ac network. However, to fully benefit from the enhanced features and performance of 802.11ac, compatible devices are recommended. The adoption of 802.11ac has significantly improved wireless networking capabilities, providing faster speeds, higher capacity, and better overall performance. Its deployment in homes, businesses, and public spaces has enabled seamless connectivity and enhanced user experiences in today's increasingly connected world (Table 1.1).

Since 2012, the 802.11ac standard has been implemented in numerous laptop computers. In June 2012, Asus introduced the ROG G75VX gaming notebook, becoming the first consumer-oriented notebook to fully comply with the 802.11ac standard. In

TABLE 1.1

Comparison of the 802.11ac and 802.11n Standards

S. No.	Description	802.11ac	802.11n
1.	Channel bandwidth	80 MHz or 160 MHz	40 MHz
2.	Spatial streams	8	4
3.	Modulation	256 QAM	64 QAM
4.	Data rate	Max 6.9 Gbps	Max. 600 Mbps
5.	Operating bands	5 GHz	2.4 GHz and 5 GHz

June 2013, Apple unveiled the new MacBook Air, which featured 802.11ac wireless networking capabilities. They later announced that the MacBook Pro and Mac Pro would also incorporate the 802.11ac standard. As of December 2013, HP started incorporating the 802.11ac standard in their laptop computers.

1.2.1.7   802.11ax, Wi-Fi 6 (Year: 2019)

Commonly referred to as Wi-Fi 6, 802.11ax is the latest iteration of the IEEE 802.11 standard for wireless networking. It was introduced in 2019 and builds upon the foundation laid by its predecessor, 802.11ac (Wi-Fi 5). One of the key features of 802.11ax is its ability to handle high-density environments with numerous connected devices more efficiently. It employs advanced techniques such as Orthogonal Frequency Division Multiple Access (OFDMA) and MU-MIMO to enable simultaneous data transmission to multiple devices, improving overall network capacity and performance.

Wi-Fi 6 introduces a significant boost in data rates compared to previous standards. It supports peak data rates of up to 9.6 Gbps, achieved through increased channel bandwidth and more efficient data encoding schemes. This enhanced speed makes Wi-Fi 6 ideal for bandwidth-intensive activities like 4K video streaming, online gaming, and large file transfers. Another notable feature of Wi-Fi 6 is Target Wake Time (TWT), which enhances power efficiency for connected devices. TWT allows devices to schedule their wake-up times, reducing power consumption and extending battery life. This feature is especially beneficial for Internet of Things (IoT) devices and other battery-powered devices.

Wi-Fi 6 operates in both the 2.4 GHz and 5 GHz frequency bands, maintaining compatibility with older devices while delivering improved performance and reduced interference. It also includes other enhancements such as improved security protocols and better coexistence with neighboring networks. As Wi-Fi 6 continues to gain adoption, an increasing number of devices, including smartphones, laptops, routers, and other networking equipment, are being manufactured with support for this standard. Upgrading to Wi-Fi 6 provides users with faster speeds, increased capacity, and a more seamless wireless experience in today's connected world.

1.2.1.8   802.11ax - Wi-Fi 6E (Year: 2020)

Wi-Fi 6E is an extension of Wi-Fi 6 that operates in the newly allocated 6 GHz frequency band, in addition to the 2.4 GHz and 5 GHz bands used by Wi-Fi 6. The introduction of the 6 GHz band in Wi-Fi 6E provides several advantages. First, it offers a significant increase in available spectrum, allowing for the deployment of wider channels and higher data rates. This results in improved performance and reduced congestion, especially in dense environments where many devices are competing for bandwidth.

By utilizing the 6 GHz band, Wi-Fi 6E is able to provide more channels with less interference from other devices operating in the 2.4 GHz and 5 GHz bands. Wi-Fi 6E allows for 14 additional 80 MHz channels and 7 additional 160 MHz channels. This now makes it practical to achieve the benefits of the wider channels while still maintaining channel diversity for adjacent APs. This improves the overall quality of the Wi-Fi connection and ensures a more reliable and consistent experience for users.

Wi-Fi 6E also inherits the key features and advancements of Wi-Fi 6, including OFDMA, MU-MIMO, and TWT. These technologies enhance efficiency, capacity, and power management, allowing for better utilization of the available spectrum.

As Wi-Fi 6E is relatively new, the adoption of devices supporting this standard is gradually increasing. Smartphones, laptops, routers, and other networking equipment with Wi-Fi 6E capabilities are being introduced to take advantage of the expanded frequency range. Overall, Wi-Fi 6E offers an exciting extension to the Wi-Fi 6 standard, providing additional spectrum and improved performance for wireless networks, particularly in high-density environments.

1.2.1.9   802.11be - Wi-Fi 7 (Year: 2024, Upcoming)

Wi-Fi 7, also known as IEEE 802.11be extremely high throughput (EHT), is an upcoming generation of the 802.11 standard for the implementation of WLAN computer communication.

Features and benefits of Wi-Fi 7:

- Wi-Fi 7 increases maximum speeds from 9.6 Gbps to a stunning 46 Gbps, which enables today's ultra-fast internet plans with gigabit-plus performance on devices.
- 100 times improvement in lower latency significantly upgrades today's interactive online experiences and opens the door for immersive next-gen Augmented Reality (AR) and Virtual Reality (VR).
- 20% more data transmission with 4096 QAM modulation.
- Maximum bandwidth of 320 MHz with high-capacity channels.

Table 1.2 concludes the evolution of the 802.11 Wi-Fi standards by giving a small comparison of all the standards.

TABLE 1.2
Comparison of 802.11 Standards

S. No.	Year	Standard	Operating Band	Modulation	Bit Rate	MIMO	Bandwidth (MHz)
1.	1997	802.11	2.4 GHz	DSSS/FHSS	2 Mbps	–	20
2.	1999	802.11b	2.4 GHz	QPSK	11 Mbps	–	20
3.	1999	802.11a	5 GHz	64 QAM	54 Mbps	–	20
4.	2003	802.11g	2.4 GHz	64 QAM	54 Mbps	–	20
5.	2009	802.11n	2.4/5 GHz	64 QAM	600 Mbps	4×4 MIMO	20–40
6.	2013	802.11ac	5 GHz	256 QAM	6.8 Gbps	4×4UL/DL MU-MIMO	20/40/80/160
7.	2019	802.11ax	2.4/5 GHz	1024 QAM	9.6 Gbps	8×8 UL/DL	20/40/80/160
8.	2020	802.11ax	6 GHz	1024 QAM	9.6 Gbps	MU-MIMO	20/40/80/160
9.	2024 (Upcoming)	802.11be	2.4/5/6 GHz	4096 QAM	46 Gbps	16×16 UL/DL MU-MIMO	Up to 320

1.2.1.10   HiperLAN 1 (Year: 1996)

HiperLAN 1 was a WLAN standard developed by the ETSI. It was introduced as an alternative to the IEEE 802.11 standard. HiperLAN 1 aimed to provide high-speed wireless connectivity for LANs. The standard was declared by ETSI in the early 1990s, with the first version released in 1996. HiperLAN 1 utilized an Frequency Hopping Spread Spectrum FHSS technique operating in the 5 GHz frequency band. It supported data rates of up to 24 Mbps.

The development of HiperLAN 1 was driven by European telecommunications companies and focused on meeting the requirements of European regulatory environments.

"HiperLAN1 covers the Physical layer and the Media Access Control (MAC) part of the Data link layer, similar to 802.11. It introduces a new sublayer called the Channel Access and Control sublayer (CAC), which handles access requests to the channels based on channel usage and request priority. The CAC layer enables hierarchical independence through the Elimination-Yield Non-Preemptive Multiple Access (EY-NPMA) mechanism. This mechanism consolidates priority choices and other functions into a variable-length radio pulse before the packet data, reducing collisions and enabling the network to accommodate a large number of users. The EY-NPMA priority mechanism is particularly beneficial for multimedia applications in HiperLAN. The MAC layer defines protocols for routing, security, and power saving, facilitating data transfer to the upper layers. On the physical layer, HiperLAN 1 utilizes Frequency-Shift Keying (FSK) and Gaussian Minimum Shift Keying (GMSK) modulations."

The features of HiperLAN 1 are:

*   It covers a range of 50 meters.
*   It provides slow mobility (1.4 m/s).
*   It supports both asynchronous and synchronous traffic.
*   It has a bit rate of 23.2 Mbps.
*   It operates in the 5 GHz frequency band.

One advantage of HiperLAN 1 is its ability to operate without interfering with devices like microwaves and other appliances that operate at 2.4 GHz. What sets HiperLAN apart from other wireless networks is its innovative feature of forwarding data packets through multiple relays.

However, despite its initial development, HiperLAN 1 did not achieve widespread adoption compared to the IEEE 802.11 standard. Over time, HiperLAN 1 was superseded by subsequent WLAN standards, including HiperLAN 2 and the widely adopted IEEE 802.11 family of standards. These later standards offered improved performance, increased compatibility, and broader industry support.

1.2.1.11   HiperLAN 2 (Year: 2000)

HiperLAN 2 was a WLAN standard developed by the ETSI as an evolution of HiperLAN 1. It aimed to provide improved performance, higher data rates, and increased functionality compared to its predecessor. HiperLAN 2 utilized the same frequency band as HiperLAN 1, operating in the 5 GHz range, delivering data rates of up to 54 Mbps. It adopted a more-advanced OFDM modulation scheme, which allowed for higher data rates and improved robustness against interference.

One of the key features of HiperLAN 2 was its support for quality of service (QoS) mechanisms, which enabled the prioritization of different types of traffic. This made it suitable for multimedia applications that required low latency and guaranteed bandwidth. HiperLAN 2 also introduced advanced security features, including encryption and authentication protocols, to ensure secure communication over the wireless network. It distinguishes itself with its media access control, utilizing a dynamic Timie Division Multiple Access (TDMA) multiple access protocol, in contrast to the Carrier Sense Multiple Access with Collision Avoidance (CSMA/CA) protocol used in 802.11a/n.

1.2.2 WLAN TOPOLOGIES

WLAN topologies refer to the ways in which wireless devices are interconnected to form a network. Here are some commonly used WLAN topologies:

Infrastructure mode: This is the most common WLAN topology. In infrastructure mode, wireless devices such as laptops, smartphones, and tablets connect to a central network device, the AP. The AP acts as a hub and facilitates communication between the wireless devices and the wired network.

Ad-hoc mode: In ad-hoc mode, also known as peer-to-peer mode, wireless devices communicate directly with each other without the need for an AP. Ad-hoc networks are often temporary and formed on the fly. This topology is useful in situations where there is no existing network infrastructure or when direct device-to-device communication is desired.

Mesh mode: In a mesh network, multiple APs are interconnected to create a self-configuring and self-healing network. Each AP communicates with nearby APs, forming multiple paths for data transmission. Mesh networks are robust and provide better coverage and reliability, as they can automatically route traffic around failures or congestion.

Extended service set (ESS): An ESS is created by connecting multiple APs together to extend the coverage area of a WLAN. All APs in the ESS share the same service set identifier (SSID), allowing wireless devices to seamlessly roam between APs without losing connectivity.

Point-to-multipoint: In point-to-multipoint topology, a central AP serves as a hub, and multiple wireless devices connect to it. This topology is commonly used in scenarios where a single AP needs to provide connectivity to multiple wireless devices in a specific area.

The WLAN topologies are explained in detail in the following section:

1.2.2.1 Infrastructure or Star Topology

The infrastructure or star topology is a common and widely used WLAN topology. In this topology, wireless devices (such as laptops, notebooks, tablets, smartphones, etc.) connect to a central network device, the AP. The AP serves as a central hub for communication between wireless devices and the wired network infrastructure.

The infrastructure topology offers several advantages:

- **Centralized management:** The AP acts as a central control point, allowing for easier network management, configuration, and security implementation. It simplifies the administration of the WLAN network.
- **Scalability**: The infrastructure topology is scalable, making it suitable for networks of various sizes. Additional APs can be added to extend the coverage area or accommodate more wireless devices.
- **Enhanced performance:** Since the AP controls the communication between devices, it can optimize data flow and manage network traffic efficiently. This can result in improved performance and reduced interference.
- **Seamless roaming:** In an infrastructure topology, wireless devices can move between APs without interruption or loss of connectivity. This enables seamless roaming within the WLAN coverage area.
- **Integration with wired network:** The AP serves as a bridge between wireless devices and the wired network infrastructure, allowing wireless devices to access resources and services available on the wired network.

The infrastructure or star topology is commonly used in various settings, such as offices, homes, public spaces, and enterprise environments. It provides a reliable and manageable wireless network infrastructure for connecting multiple devices and facilitating communication between them.

Figure 1.2 shows an example of Infrastructure topology.

FIGURE 1.2 Infrastructure or star network topology.

1.2.2.2   Ad-hoc or Peer-to-Peer Topology

The ad-hoc or peer-to-peer topology is another type of WLAN configuration. In this topology, wireless devices communicate directly with each other, without the need for an AP or infrastructure. Each device functions as both a transmitter and receiver, allowing direct device-to-device communication. Here are some key characteristics of the ad-hoc topology:

- **Direct device communication:** Devices in an ad-hoc network can communicate with each other without relying on a central AP. This enables wireless devices to establish a network on the fly, without the need for an existing network infrastructure.
- **Decentralized structure:** Ad-hoc networks have a decentralized structure, with each device having equal importance and capability. Devices can directly transmit and receive data from other devices in the network.
- **Flexibility and mobility:** Ad-hoc networks are highly flexible and mobile. Devices can join or leave the network easily, and the network adapts dynamically to changes in device availability or movement.
- **Limited coverage area:** Ad-hoc networks typically have a limited coverage area since the range of direct device-to-device communication is shorter compared to infrastructure-based networks. Devices need to be within each other's transmission range to establish a connection.
- **Temporary networks**: Ad-hoc networks are often temporary and formed for specific purposes or situations. They are commonly used in scenarios where there is no existing network infrastructure or where direct communication between devices is desired, such as in peer-to-peer file sharing or multiplayer gaming.

Ad-hoc networks offer advantages such as simplicity, flexibility, and independence from a central infrastructure. However, they may have limitations in terms of coverage range and scalability compared to infrastructure-based networks. Ad-hoc topology is suitable for small-scale, localized networks where direct device-to-device communication is desired and where a central AP is not available or necessary. Figure 1.3 shows the ad-hoc, or peer-to-peer, network topology.

1.2.2.3   Mesh Topology

The mesh topology is a WLAN configuration where multiple APs are interconnected to create a self-configuring and self-healing network. In a mesh network, each AP communicates with nearby APs, forming multiple paths for data transmission. This redundancy and interconnectivity enhances the network's reliability, coverage, and performance. Here are some key characteristics of the mesh topology:

- **Multiple APs:** A mesh network consists of multiple APs strategically placed to provide coverage across a wide area. Each AP functions as a relay, transmitting and receiving data from other APs within its range.

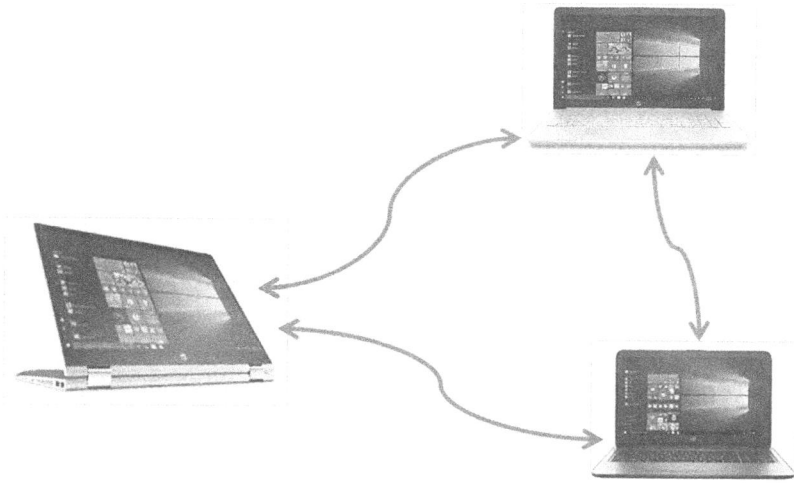

FIGURE 1.3   Ad-hoc, or peer-to-peer, network topology.

- **Self-configuring:** Mesh networks are self-configuring, meaning the APs automatically discover and connect to neighboring APs. This self-configuring capability simplifies the deployment and expansion of the network.
- **Self-healing:** Mesh networks are self-healing, which means they can dynamically reroute data packets in case of AP failure or network congestion. If one AP becomes unavailable, the network can automatically reroute traffic through alternative paths, ensuring continuous connectivity.
- **Improved coverage and scalability:** The interconnection of multiple APs in a mesh network extends the coverage area compared to a single AP. Mesh networks can be easily expanded by adding more APs to provide coverage in larger areas or to accommodate a larger number of devices.
- **Enhanced reliability and performance:** The redundancy in mesh networks improves reliability and performance. If one AP fails or experiences interference, the network can dynamically redirect traffic through other APs, minimizing downtime and maintaining optimal performance.

Mesh topologies are commonly used in environments where reliable and widespread wireless coverage is required, such as large-scale deployments in enterprises, campuses, or smart cities. They offer robust and scalable solutions for ensuring seamless connectivity and efficient data transmission throughout the network.

1.2.2.4   Extended Service Set (ESS)

The ESS is a WLAN configuration that extends the coverage area of a network by interconnecting multiple APs. In an ESS, all the APs share the same SSID, allowing wireless devices to seamlessly roam between APs without losing connectivity. Here are some key characteristics of the ESS:

- **Seamless roaming:** The ESS enables wireless devices to move between different APs within the coverage area without experiencing any interruption in connectivity. As a user moves from one area to another covered by a different AP, the device automatically associates with the new AP while maintaining a consistent network connection.
- **Single network identifier:** The APs within an ESS share the same SSID. This allows wireless devices to perceive the entire ESS as a single network entity, making it easy for users to connect and move between APs without manually switching networks.
- **Enhanced coverage area:** By deploying multiple APs within an ESS, the coverage area of the WLAN is expanded. This ensures that wireless devices can connect and stay connected to the network regardless of their location within the coverage area.
- **Load balancing:** In an ESS, APs can distribute the network load by dynamically assigning wireless devices to different APs based on factors such as signal strength, network congestion, or device capacity. Load balancing helps optimize the performance and efficiency of the WLAN.
- **Centralized management:** The ESS can be centrally managed, allowing administrators to configure, monitor, and secure the entire network through a centralized management system. This simplifies network administration and enables consistent configuration across multiple APs.

The ESS is widely used in environments where seamless mobility and extended coverage are crucial, such as large offices, educational institutions, airports, and public venues. By leveraging multiple APs working together as a unified network, the ESS provides a reliable and scalable solution for wireless connectivity.

1.2.2.5 Point-to-Multipoint

Point-to-Multipoint (P2MP) is a wireless communication topology where a single transmitting device (point) communicates with multiple receiving devices (multipoint). In this configuration, the transmitting device broadcasts data to multiple receiving devices simultaneously, allowing for efficient distribution of information to a group of recipients. Here are some key characteristics of the P2MP topology:

- **One-to-many communication:** P2MP enables one transmitting device to communicate with multiple receiving devices simultaneously. The transmitting device serves as a central hub or base station, while the receiving devices can be located at different positions within the coverage area.
- **Efficient data distribution:** With P2MP, the transmitting device broadcasts data to all the receiving devices at once. This eliminates the need for individual point-to-point connections, making it an efficient solution for delivering information to a group of recipients simultaneously.

- **Shared resources:** The receiving devices within the P2MP configuration share the available resources, such as bandwidth or network capacity, provided by the transmitting device. This enables efficient utilization of resources and ensures equitable distribution among the recipients.
- **Simplified network architecture:** P2MP reduces the complexity of the network architecture compared to point-to-point connections. Instead of establishing individual connections between the transmitting device and each receiving device, a single connection can serve multiple recipients.
- **Suitable for broadcasting or multicast applications:** P2MP is particularly useful for applications that involve broadcasting or multicasting, such as video streaming, teleconferencing, or wireless distribution of data to multiple users simultaneously.

P2MP topologies are commonly used in various scenarios, including wireless broadband access, wireless surveillance systems, public safety networks, and outdoor wireless deployments. They offer an efficient and scalable solution for delivering data to multiple recipients simultaneously, simplifying network architecture and optimizing resource utilization.

1.2.3 WLAN OR WI-FI SIGNALS

Wi-Fi signals refer to the radio frequency signals used to transmit data wirelessly in a Wi-Fi network. These signals are EM waves that carry information between Wi-Fi-enabled devices, such as smartphones, laptops, routers, and IoT devices. Wi-Fi signals operate in specific frequency bands allocated for wireless communication. The two primary frequency bands used for Wi-Fi are 2.4 GHz and 5 GHz. The 2.4 GHz band is more crowded and susceptible to interference from other devices like microwaves, Bluetooth devices, and cordless phones. The 5 GHz band offers higher data-transfer rates and less interference due to its wider available channels.

Wi-Fi signals use various modulation techniques to encode and transmit data. The most commonly used modulation schemes are OFDM and its variants. OFDM divides the available frequency spectrum into multiple subcarriers to transmit data simultaneously, enhancing the signal's efficiency and robustness (Figure 1.4).

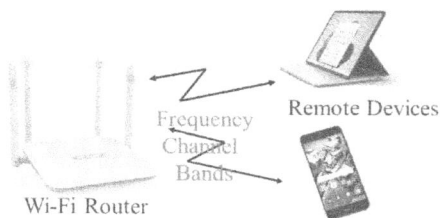

Remote Devices

Frequency
Channel
Bands

Wi-Fi Router

FIGURE 1.4 Wi-Fi communications.

1.2.3.1   2.4 GHz Frequency Band

The 2.4 GHz frequency band is one of the primary frequency bands used for Wi-Fi communication. The 2.4 GHz band provides a longer range but transmits data at slower speeds. The 2.4 GHz frequency band is the most common Wi-Fi technology in use today. Here are some key points about the 2.4 GHz frequency band:

- **Frequency range:** The 2.4 GHz frequency band spans from 2.400 GHz to 2.4835 GHz. It is divided into multiple channels, typically with a channel width of 20 MHz.
- **Widely used:** The 2.4 GHz band is widely used for various wireless communication technologies, including Wi-Fi, Bluetooth, cordless phones, and other wireless devices. This widespread use can lead to potential interference issues, as multiple devices may operate in the same frequency range.
- **Limited channels:** The 2.4 GHz band provides a limited number of non-overlapping channels for Wi-Fi networks. In most countries, it offers 14 channels with only three channels (1, 6, and 11) being non-overlapping. The limited number of non-overlapping channels can result in interference between neighboring Wi-Fi networks.
- **Interference:** The 2.4 GHz band is susceptible to interference from various sources, including other Wi-Fi networks, microwave ovens, cordless phones, Bluetooth devices, and certain types of wireless cameras. These devices can cause signal degradation and impact the performance of Wi-Fi networks operating in the same frequency band.
- **Longer range:** One advantage of the 2.4 GHz band is its longer range compared to higher frequency bands, which can enable Wi-Fi signals to travel through walls and other obstacles more effectively. However, this longer range comes at the cost of potential interference and slower data transfer rates compared to the higher frequency bands.
- **Slower data-transfer rates:** The 2.4 GHz band has a lower maximum data transfer rate compared to the higher frequency bands, such as the 5 GHz band. The maximum data rates achievable in the 2.4 GHz band depend on the Wi-Fi standard being used (e.g., 802.11b, 802.11g, 802.11n) and the specific modulation techniques employed.
- **Legacy support:** The 2.4 GHz band is backward compatible with older Wi-Fi standards, allowing devices that support older standards to connect to Wi-Fi networks operating in this frequency band. This compatibility is beneficial for maintaining connectivity with older devices but can limit the overall network performance.

When setting up Wi-Fi networks, it is essential to consider the potential interference and congestion issues associated with the 2.4 GHz band. Channel selection, positioning of APs, and proper network configuration can help optimize the performance and mitigate interference-related problems in the 2.4 GHz frequency band.

The 2.4 GHz frequency band consists of a total of 14 channels, as depicted in Figure 1.5. However, it is important to note that not all of these channels are

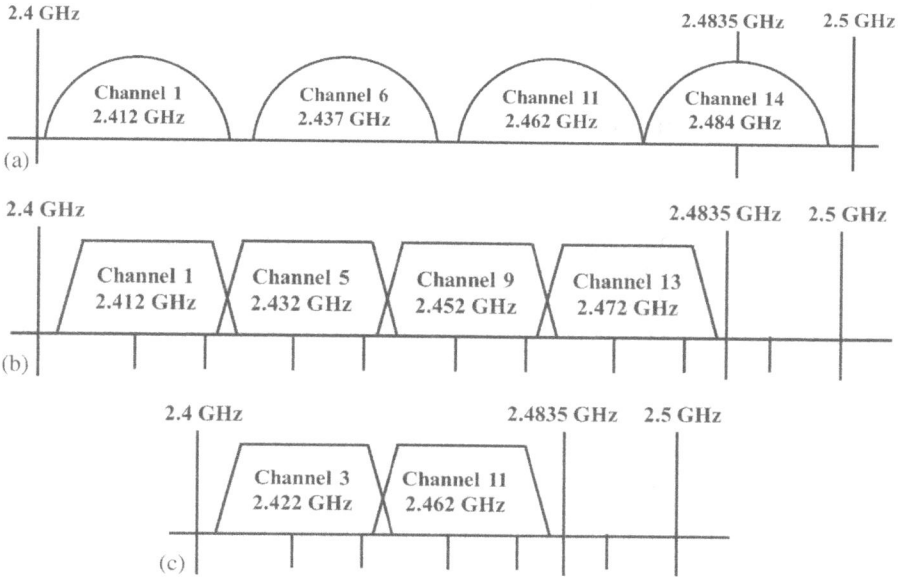

FIGURE 1.5 WLAN channels in the 2.4 GHz band: (a) 802.11b Direct Sequence Spread Spectrum width of 22 MHz, (b) 802.11g/n OFDM 20 MHz channel width with 16.25 MHz used by subcarriers, and (c) 802.11n OFDM 40 MHz channel width with 33.75 MHz used by subcarriers.

universally available for use. The channels within this band are spaced 5 MHz apart, except for the last two channels, which have a spacing of 12 MHz.

Due to the 22 MHz bandwidth and the 5 MHz channel spacing, adjacent channels overlap with each other, as illustrated in Figure 1.6. As a result, signals transmitted on these neighboring channels can interfere with one another, leading to potential disruptions in wireless communication.

For a comprehensive overview of the frequencies corresponding to these 14 Wi-Fi channels available globally, refer to Table 1.3.

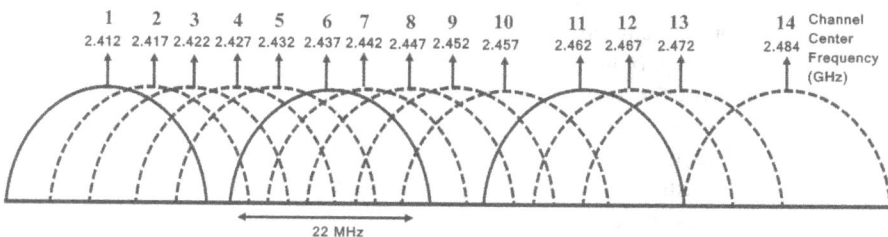

FIGURE 1.6 Channel overlapping of the 2.4 GHz band.

TABLE 1.3
2.4 GHz Channels and Frequencies

Channels	Lower Frequency (GHz)	Center Frequency (GHz)	Upper Frequency (GHz)
1	2.401	2.412	2.423
2	2.406	2.417	2.428
3	2.411	2.422	2.433
4	2.416	2.427	2.438
5	2.421	2.432	2.443
6	2.426	2.437	2.448
7	2.431	2.442	2.453
8	2.436	2.447	2.458
9	2.441	2.452	2.463
10	2.446	2.457	2.468
11	2.451	2.462	2.473
12	2.456	2.467	2.478
13	2.461	2.472	2.483
14	2.473	2.484	2.495

1.2.3.2 5 GHz Frequency Band

The 5 GHz frequency band is one of the primary frequency bands used for wireless communication, including Wi-Fi networks. Here are some key points about the 5 GHz frequency band:

- **Frequency range:** The 5 GHz frequency band spans from 5.150 GHz to 5.815 GHz. It is divided into multiple channels, typically with a channel width of 20 MHz, 40 MHz, 80 MHz, or 160 MHz.
- **Wi-Fi compatibility:** The 5 GHz band is used for Wi-Fi networks operating under various Wi-Fi standards, such as 802.11a, 802.11n, 802.11ac, and 802.11ax (Wi-Fi 6 and Wi-Fi 6E). Wi-Fi 6E specifically refers to the usage of the 5 GHz band in conjunction with the newly allocated 6 GHz band, providing even more available channels for high-speed Wi-Fi communication.
- **Higher data transfer rates:** The 5 GHz band offers higher potential data-transfer rates compared to the 2.4 GHz band. This increased bandwidth allows for faster and more efficient wireless communication, making it suitable for applications that require high-speed data transmission, such as video streaming, online gaming, and large file transfers.
- **Reduced interference:** Compared to the 2.4 GHz band, the 5 GHz band is less crowded and generally experiences less interference from other devices. This is because many common household devices, such as microwaves and cordless phones, typically operate in the 2.4 GHz band, while the 5 GHz band is less congested.
- **Shorter range:** One consideration with the 5 GHz band is that it generally has a shorter range compared to the 2.4 GHz band. Higher frequency signals tend to have more difficulty penetrating obstacles like walls and

furniture, resulting in a reduced coverage area. However, the shorter range can also help minimize interference between neighboring Wi-Fi networks.

- **More non-overlapping channels:** The 5 GHz band provides more non-overlapping channels compared to the 2.4 GHz band, which allows for better coexistence of multiple Wi-Fi networks in close proximity. This helps reduce interference between neighboring networks and improves overall network performance.

When setting up Wi-Fi networks, the 5 GHz band can be advantageous for high-performance applications and environments with heavy network traffic. It offers faster data-transfer rates, reduced interference, and more available channels, making it suitable for demanding wireless applications that require reliable and high-speed connections.

The 5 GHz frequency band offers a total of 25 channels when using a 20 MHz channel width. However, the specific channels that can be used vary depending on the regulations implemented in each country. Compared to the 2.4 GHz band, the 5 GHz band experiences significantly less interference due to its larger number of channels. In the 802.11n standard, it is possible to bond two 20 MHz channels together to create a wider 40 MHz channel within the 5 GHz band. Furthermore, the 802.11ac standard allows for the bonding of even more channels. In this case, four channels can be combined to form an 80 MHz channel, and eight channels can be combined to form an even larger 160 MHz channel. Figure 1.7 provides a visual representation of the WLAN channels available in the 5 GHz band.

1.2.3.3   6 GHz Frequency Band

The 6 GHz frequency band refers to a range of radio frequencies that have been recently allocated for wireless communication. Here are some key points about the 6 GHz frequency band:

- **Frequency range:** The 6 GHz frequency band spans from 5.925 GHz to 7.125 GHz. It provides additional spectrum for wireless communication, allowing for increased capacity and improved performance.

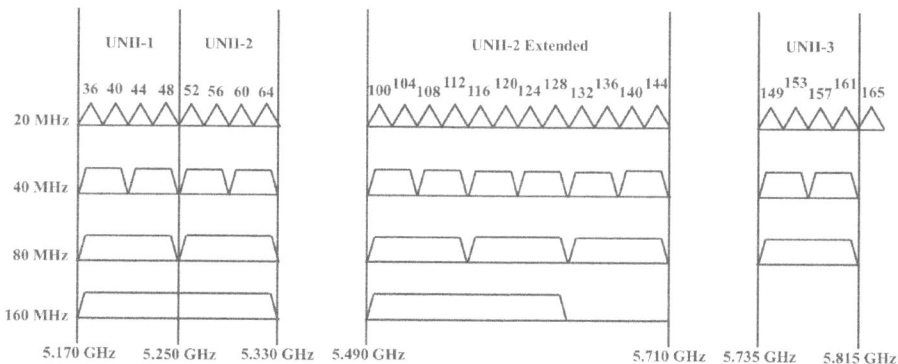

FIGURE 1.7   WLAN channels in 5 GHz band.

- **Wi-Fi 6E:** The 6 GHz band is commonly associated with the introduction of Wi-Fi 6E, an extension of Wi-Fi 6 (802.11ax) technology. Wi-Fi 6E utilizes the 6 GHz band to deliver enhanced speeds, reduced latency, and increased capacity for wireless networks. It offers a significant expansion of available channels and bandwidth, enabling higher quality connections and improved user experiences.
- **Reduced interference:** The 6 GHz band is relatively free from interference compared to the crowded 2.4 GHz and 5 GHz bands. With more available spectrum, Wi-Fi devices operating in the 6 GHz band can benefit from cleaner and less congested wireless environments.
- **Compatibility:** Devices supporting Wi-Fi 6E are designed to operate in the 6 GHz band, offering backward compatibility with previous Wi-Fi standards. However, older devices that do not support Wi-Fi 6E will not be able to utilize the 6 GHz band.
- **Benefits and applications:** The 6 GHz band provides a significant increase in available spectrum, enabling higher speed and more reliable wireless connections. It is particularly beneficial for applications that require high bandwidth and low latency, such as VR, AR, high-definition video streaming, online gaming, and other bandwidth-intensive tasks.

As the 6 GHz band continues to be deployed and adopted, it holds promise for unlocking new possibilities and expanding the capabilities of wireless communication systems, offering improved performance and experiences for a wide range of wireless applications. Figure 1.8 provides a visual representation of the WLAN channels available in the 6 GHz band.

Initially, the 2.4 GHz band was the recommended frequency range for Wi-Fi networks. However, with the introduction of technology supporting the 5 GHz and 6 GHz band, it gained significant popularity due to its wider channel bandwidth capability and higher transmission speeds. Today, many devices are equipped to operate in the 2.4 GHz, 5 GHz, and 6 GHz bands, offering users the flexibility to choose between the two.

Considering the information provided about WLAN, it is essential to design antennas that can operate in the 2.4 GHz, 5 GHz, and 6 GHz bands. By doing so, the antennas can achieve long-range coverage, high transmission speeds, and experience minimal interference. This ensures optimal performance and reliable wireless connectivity for various applications and environments.

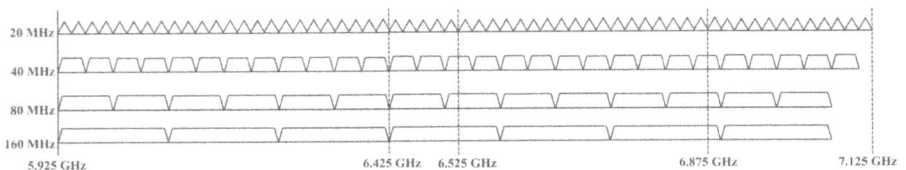

FIGURE 1.8 A visual representation of the wireless local area network channels available in the 5 GHz band.

1.3 WiMAX

WiMAX is a wireless communication technology that provides high-speed internet access over long distances. Here are some key points about WiMAX:

- **Technology overview:** WiMAX is based on the IEEE 802.16 standard and operates in both licensed and unlicensed frequency bands. It utilizes a P2MP architecture, allowing a base station to communicate with multiple subscriber stations over a wide area.
- **Broadband connectivity:** WiMAX is designed to deliver broadband connectivity in areas where it may be challenging or impractical to lay traditional wired infrastructure. It provides wireless broadband access to both fixed and mobile devices, enabling users to access the internet, make voice calls, and stream multimedia content.
- **Coverage and range:** WiMAX offers a greater coverage range compared to Wi-Fi, making it suitable for providing internet access in rural and remote areas. Depending on the deployment, WiMAX can cover distances ranging from a few kilometers to several tens of kilometers.
- **Speed and capacity:** WiMAX supports high data-transfer rates, with theoretical speeds ranging from a few Mbps to several tens of Mbps. It provides sufficient bandwidth to support various applications, including video streaming, online gaming, and Voice Over-Internet Protocol (VoIP) services.
- **Backhaul connectivity:** WiMAX can also serve as a backhaul solution, connecting cellular base stations or other network nodes to the core network. It enables the transportation of data between the access network and the core network, ensuring reliable and efficient connectivity.
- **Standards and evolution:** The initial version of WiMAX, based on the IEEE 802.16-2004 standard, provided fixed wireless access. Subsequent enhancements were made with the introduction of the IEEE 802.16e-2005 standard, which added support for mobility, allowing WiMAX to be used in mobile applications.
- **Deployment and adoption:** WiMAX has been deployed in various regions globally, particularly in areas where wired infrastructure is limited. It has found applications in both developed and developing countries, offering an alternative means of broadband connectivity.
- **Transition to 4G and beyond:** With the evolution of cellular networks, WiMAX faced competition from long-term evolution (LTE) technology, which became the dominant standard for 4G networks. As a result, the deployment of WiMAX networks has slowed, and the focus has shifted to 4G LTE and subsequent generations of cellular technology.

Despite the transition to newer technologies, WiMAX has contributed to expanding internet access and bridging the digital divide in certain areas. It has played a role in extending broadband connectivity to underserved regions and has served as

TABLE 1.4
Comparison of IEEE 802.16 WiMAX Standards

S. No.	Specifications	802.16	802.16a	802.16d	802.16e
1.	Frequency of operation	10–66 GHz	2–11 GHz	Less than 6 GHz	Less than 6 GHz
2.	Data rate	32–134 Mbps, (28 MHz BW channel)	Less than 70 or 100 Mbps, (20 MHz BW channel)	Up to 70 Mbps with no mobility	60–70 Mbps with mobility
3.	Mobility	Fixed	Fixed	Fixed	Less than and equal to 75 MPH
4.	Modulation	QPSK, 16 QAM, 64 QAM	256-point FFT with QPSK, 16 QAM, 64 QAM, 256 QAM	256-point FFT with QPSK, 16 QAM, 64 QAM, 256 QAM	256-point FFT with QPSK, 16 QAM, 64 QAM, 256 QAM and OFDMA modulation scheme with variable FFT sizes
5.	Mode of configuration	Line-of-sight	Non-line-of-sight	Non-line-of-sight	Non-line-of-sight
6.	Channel bandwidth	20/25/28 MHz	Selectable 1.25–20 MHz	3.5 MHz, 7 MHz	1.25–28 MHz, OFDMA modulation scheme
7.	Cell radius	1–3 miles	3–5 miles	3–5 miles	1–3 miles

an alternative to wired solutions in specific deployments. Table 1.4 shows various IEEE 802.16 WiMAX standards

From Table 1.4, it can be concluded that the 802.16d/e standards can be used for wireless communication in laptop computers. The advantages of WiMAX are that it provides non-line-of-sight (NLOS) connectivity within a radius up to 5 miles. It also provided flexibility, high-speed internet, data transfer, and video applications at a very low cost. As WiMAX provides high-speed internet over long distances, even in remote and scarcely populated areas and does not need direct line-of-sight to operate, it is of great use in the laptop computers.

1.4   SUB-6 GHz 5G

5G technology is at the forefront of advanced technology, expected to revolutionize connectivity. It promises to increase capacity, provide low latency, offer incredibly high data speeds, and ensure reliable service quality. By overcoming the limitations of earlier technologies, 5G is poised to drive progress and meet the growing demand for faster and more sophisticated wireless applications. Additionally, 5G technology plays a vital role in supporting the IoT, enabling the seamless connection of numerous devices within a robust network. The sub-6 GHz band in 5G refers to the frequency range below 6 GHz that is allocated for the deployment of 5G networks. Here are some key points about the sub-6 GHz 5G band:

- **Frequency range:** The sub-6 GHz band covers a range of frequencies below 6 GHz, typically including frequency bands such as 600 MHz, 700 MHz, 2.4 GHz, 3.5 GHz, and 4.9 GHz, among others. These frequencies provide a balance between coverage and capacity in 5G networks.
- **Wide coverage:** One of the advantages of the sub-6 GHz band is its ability to provide wider coverage compared to higher frequency bands. The lower frequencies can propagate over longer distances and penetrate buildings and obstacles more effectively, making them suitable for delivering 5G services in both urban and rural areas.
- **Enhanced capacity:** While the sub-6 GHz band offers wider coverage, it has relatively lower bandwidth compared to higher frequency bands like millimeter wave (mmWave). However, advancements in 5G technology, such as the use of advanced antenna systems and spectrum aggregation, help maximize the capacity and performance of the sub-6 GHz band.
- **Interference and congestion:** The sub-6 GHz band is less prone to interference and congestion compared to higher frequency bands. It can support a larger number of connected devices and provide a more consistent and reliable user experience, particularly in densely populated areas.
- **Network deployment:** Many countries and regions around the world have allocated portions of the sub-6 GHz spectrum for 5G deployment. Network operators utilize this band to provide broad coverage and lay the foundation for 5G connectivity, allowing for widespread adoption of 5G services.
- **Compatibility and transition:** The sub-6 GHz band is backward compatible with existing 4G LTE networks, enabling a smooth transition from 4G to 5G. This compatibility ensures that devices can seamlessly connect to both 4G and 5G networks, allowing for gradual adoption of 5G technology.
- **Use cases:** The sub-6 GHz 5G band supports a wide range of use cases, including Enhanced Mobile Broadband (eMBB) services for faster data speeds, massive machine-type communications (mMTC) for connecting a large number of IoT devices, and ultra-reliable low-latency communications (URLLC) for applications that require highly reliable and real-time connectivity.

Effective antenna design and development are essential for optimal performance of any 5G equipment. These antennas must be designed with a small physical footprint and cost-effectiveness in mind. It is crucial to select the appropriate 5G band that offers increased bandwidth, gain, and minimal radiation losses. Striking the right balance between aesthetics and performance is paramount when designing antennas for 5G devices. This ensures that both functionality and visual appeal are optimized.

In conclusion, the advancements brought about by 5G technology are set to satisfy the increasing need for efficient communication. Through its superior capabilities, such as low latency, high-speed data transmission, and support for the IoT, 5G will

TABLE 1.5
5G NR Bands

Band	Operating Bandwidth (GHz)	Uplink (GHz)	Downlink (GHz)	Total Bandwidth (MHz)
n77	3.300–4.200	3.300–4.200	3.300–4.200	900
n78	3.300–3.800	3.300–3.800	3.300–3.800	500
n79	4.400–5.000	4.400–5.000	4.400–5.500	600
n80	1.710–1.785	1.710–1.785	NA	75
n81	0.880–0.915	0.880–0.915	NA	35
n82	0.832–0.862	0.832–0.862	NA	30
n83	0.703–0.748	0.703–0.748	NA	45
n84	1.920–1.980	1.920–1.980	NA	60
n86	1.710–1.780	1.710–1.780	NA	70
n89	0.824–0.849	0.824–0.849	NA	25
n90	2.496–2.690	2.496–2.690	2.496–2.690	194
n91	0.832–1.432	0.832–0.862	1.427–1.432	600
n92	0.832–1.517	0.832–0.862	1.432–1.517	685
n93	0.880–1.432	0.880–0.915	1.427–1.432	552
n94	0.880–1.517	0.880–0.915	1.432–1.517	637
n95	2.010–2.025	2.010–2.025	NA	15

shape the future of connectivity and enable a wide range of innovative applications. Antenna design, playing a critical role in 5G equipment, should prioritize efficiency, cost-effectiveness, and aesthetic integration to achieve optimal performance.

Table 1.5 provides a categorization of frequency ranges, including the 5G new radio (NR) frequency band FR-1. The FR-1 band encompasses frequencies ranging from 0.703 GHz to 5.00 GHz. This band offers a wider spectrum and enhanced capacity, enabling a greater number of users, high-speed data transmission, and faster connections. As the usage of legacy networks decreases, there are plans to utilize the current low band spectrum for 5G to accommodate potential future use cases. Deploying 5G requires a range of equipment, including small-cell and macro-coverage antenna packages, as well as various types of terminal equipment.

The sub-6 GHz 5G band can be further divided into two segments: the low-band and the mid-band. The low-band operates within the frequency range of 470 MHz to 2.4 GHz, which closely resembles the operational range of 4G networks. On the other hand, the mid-band of the sub-6 GHz band operates between 3 GHz and 5 GHz. Currently, many countries have chosen the mid-band as the preferred band for deploying 5G on a larger scale, aiming to make 5G accessible to a wider audience.

1.5 EVOLUTION OF ANTENNAS FOR LAPTOPS

The evolution of antennas for laptops has witnessed significant advancements over the years. Initially, wireless devices were integrated into laptops using personal cards (PC) that were inserted into PC slots. However, as wireless technology gained

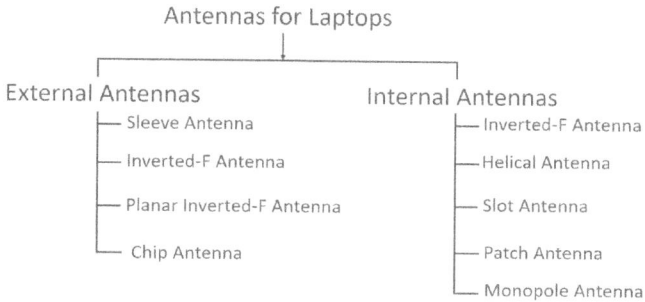

```
                        Antennas for Laptops
       ┌──────────────────────────┴──────────────────────────┐
   External Antennas                              Internal Antennas
       ├── Sleeve Antenna                            ├── Inverted-F Antenna
       ├── Inverted-F Antenna                        ├── Helical Antenna
       ├── Planar Inverted-F Antenna                 ├── Slot Antenna
       └── Chip Antenna                              ├── Patch Antenna
                                                     └── Monopole Antenna
```

FIGURE 1.9 Antennas for laptop computers.

popularity and became more affordable, manufacturers began incorporating antennas directly inside laptops. This shift not only eliminated the need for external PC
cards, but also contributed to making laptop computers more compact. Figure 1.9
illustrates the various types of external and internal antennas, which are further
explained next.

1.5.1 EXTERNAL ANTENNA

External antennas for laptops provide an alternative option for enhancing wireless
connectivity. These antennas are designed to be attached externally to the laptop,
typically through a USB port or a dedicated antenna port.

One type of external antenna commonly used is the USB adapter as depicted in
Figure 1.10 [1]. The card consisted of the controller, the radio frequency (RF) front-
end, and an integrated antenna. This small device plugs into a USB port on the laptop
and contains its own antenna, which can be positioned for optimal signal reception.
USB adapters often offer better range and signal strength compared to the internal
antennas of laptops, especially in situations where the laptop's built-in Wi-Fi capabilities may be limited.

Another type of external antenna is the antenna extension cable. This cable allows
for the placement of the laptop's internal antenna in a location where it can receive

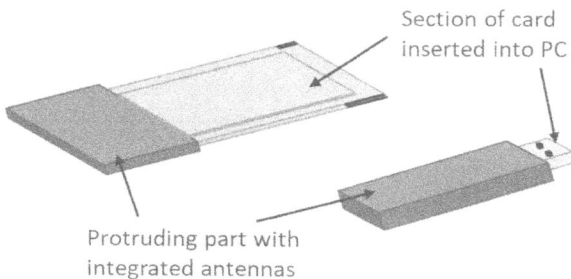

Section of card
inserted into PC

Protruding part with
integrated antennas

FIGURE 1.10 Antenna mounted inside the PC card.

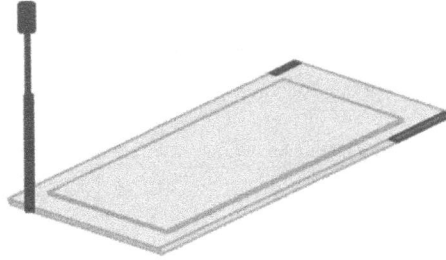

FIGURE 1.11   Sleeve antenna integrated in a PC card.

a stronger signal. By extending the reach of the antenna, these cables can improve Wi-Fi performance in areas with poor signal quality or when the laptop is located far from the wireless router.

Some laptops also feature dedicated antenna ports that allow for the connection of high-gain external antennas. These antennas are designed to provide even greater range and signal strength compared to USB Wi-Fi adapters or internal antennas. They are often used in professional or specialized scenarios where reliable and robust wireless connectivity is essential, such as in large offices or remote locations.

It is worth noting that while external antennas can enhance wireless performance, they do add bulk and require additional hardware. Therefore, their usage may vary depending on the specific needs and preferences of laptop users.

Discussed next are the various types of external antennas specifically designed for laptop computers:

1.5.1.1   Sleeve Antenna

The sleeve dipole [1], depicted in Figure 1.11, was initially utilized in the early implementations of wireless cards for laptops. It represents a modified version of the dipole antenna, where the center feed has been replaced with an end-feed approach. The cross section of the coaxial sleeve dipole is illustrated in Figure 1.12(a). This type of

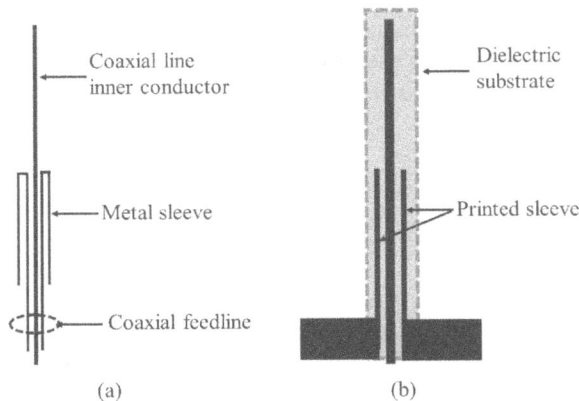

FIGURE 1.12   Sleeve dipole: (a) cross section of coaxial version and (b) printed version.

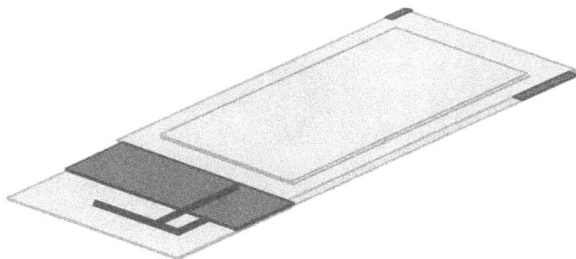

FIGURE 1.13   Inverted-F antenna integrated in a PC card.

antenna consists of an asymmetric dipole constructed with conductors of different diameters and slightly varying lengths. The thinner conductor, typically an extension of the inner conductor in the coaxial feed, needs to be of an appropriate length to ensure proper antenna matching within the operational frequency range. On the other hand, the larger diameter conductor sleeve serves the purpose of effectively suppressing RF currents at its open end and halfway along the radiating portion. It is also possible to implement a planar version of the sleeve dipole as a strip, as shown in Figure 1.12(b).

1.5.1.2   Inverted-F Antenna

Figure 1.13 illustrates an inverted-F antenna (IFA) printed on a PC card, while its structure is depicted in Figure 1.14. The IFA's design incorporates a quarter-wavelength arm aligned parallel to the edge of the ground plane, facilitating its integration within confined spaces. Essentially, it functions as a half of the traditional $\lambda/2$ slot antenna, operating on similar principles. By relocating the feeding stub from the shorting stub to the open slot end, the IFA's input impedance transitions from very low to very high values. The feeding point is carefully chosen to achieve impedance matching with the 50-Ω line. The IFA can be printed on the extended section of the PC card [2] and fed using a coplanar waveguide.

1.5.1.3   Planar Inverted-F Antenna

A planar inverted-F antenna (PIFA) is derived from a microstrip patch antenna with a half-wavelength length. Its size is reduced by incorporating shorting walls

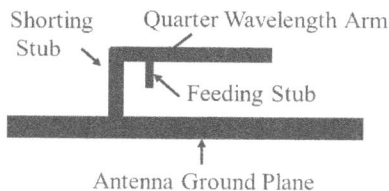

FIGURE 1.14   Structure of inverted-F antenna.

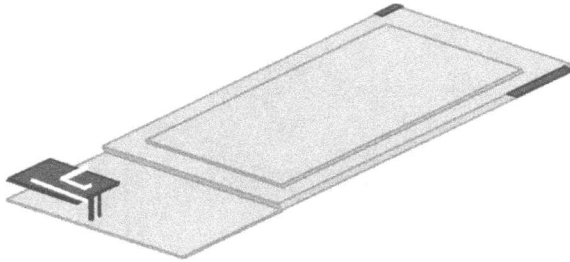

FIGURE 1.15   Triple-band planar inverted-F antenna integrated in a PC card.

or shorting pins. In a typical single-band PIFA, the length of the patch element is approximately one-quarter of a wavelength. By introducing slots in the radiating patch, the current resonant path can be altered, enabling further miniaturization. Additionally, careful design of the patch shape, feed location, and the use of shorting pins can facilitate multiple resonant paths, resulting in operation across multiple frequency bands. Figure 1.15 showcases a compact, triple-band PIFA [3] integrated into a PC card as an example.

Due to its low-profile structure, a PIFA exhibits robustness against the influence of nearby components or antennas [3]. Consequently, it can be easily integrated into the extended section of a PC card.

1.5.1.4   Chip Antenna

The chip antenna is a highly compact device that can be surface-mounted (Figure 1.16). It consists of a dielectric body with a high permittivity ($\varepsilon_r >7$), incorporating a meandering metal line embedded within it. In practical implementations, the low-temperature co-fired ceramic (LTCC) technology is often employed. This technique involves printing conducting strips on different ceramic layers, which are interconnected by metal via metal posts, creating a continuous three-dimensional (3D) path. The shape of the path varies depending on the application and desired

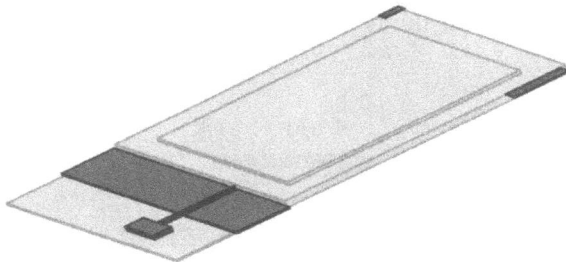

FIGURE 1.16   Chip antenna integrated in a PC card.

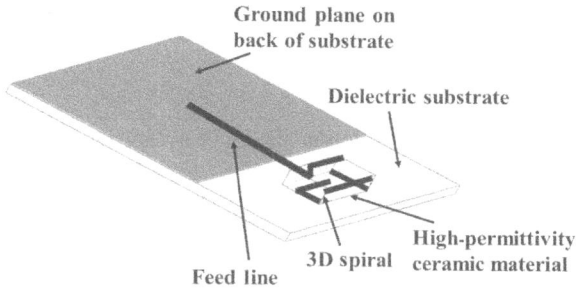

FIGURE 1.17   A 2.4/5.2 GHz dual-band chip antenna integrated on a dielectric board.

miniaturization, with spiral configurations [4] being commonly used, as depicted in Figure 1.17.

There are two main types of ceramic chip antennas. One features a ground plane printed on the bottom of the ceramic base, while the other does not have a ground plane. The former type, more versatile for practical applications, exhibits a narrower impedance bandwidth and lower radiation efficiency. Therefore, in most modern designs, the chip antenna is mounted on a circuit board section without an underlying ground plane, as illustrated in Figure 1.16. The distance between the antenna elements and the ground plane plays a role in determining the impedance bandwidth [5].

Due to their small size and surface-mountable nature, chip antennas find application in laptop wireless interfaces, both in external plug-in devices and built-in antenna systems [5]. Despite being mounted on a potentially high-loss circuit board, the chip antenna can achieve acceptable radiation efficiency owing to the concentration of the EM field in the low-loss ceramic substrate.

1.5.2   INTERNAL ANTENNA

Internal antennas for laptops are designed to be integrated directly into the laptop's chassis, eliminating the need for external protrusions. These antennas are typically smaller in size and are strategically placed within the laptop to optimize wireless signal reception. The majority of modern laptop computers come equipped with built-in wireless interfaces, eliminating the need for external WLAN/PAN plug-in cards. The location of the antenna element is often inconspicuous to the user, cleverly integrated within a narrow space between the laptop's shielding layers and plastic case. This integration of the wireless solution within the laptop design eliminates concerns related to antenna breakage and physical limitations associated with external antennas [6]. As a result, almost all laptops available in the market today are equipped with integrated WLAN devices. However, the size limitations for internal antennas differ from those of PC cards and USB dongles and are primarily influenced by their placement within the laptop. The most critical design constraint for internal antennas is typically their height and thickness, which must be carefully considered during the laptop's development process.

1.5.2.1   IFA

The IFA is a commonly used antenna design for laptops [7]. It is specifically designed to fit within the limited space available inside a laptop. The IFA consists of a printed metal trace on a circuit board, typically located near the edge of the laptop's display.

The shape of the IFA resembles an inverted "F", with a vertical arm and a horizontal arm. The vertical arm is usually a quarter-wavelength long, while the horizontal arm extends parallel to the laptop's ground plane. This configuration allows the IFA to be easily integrated into the laptop's internal structure.

By adjusting the position of the feeding stub, the impedance of the IFA can be matched to the 50-Ω transmission line. This ensures efficient transfer of RF signals between the antenna and the wireless module. The compact size and design of the IFA make it an ideal choice for laptops, as it can provide reliable wireless connectivity while occupying minimal space. The integration of the IFA into laptops allows for wireless communication without the need for external antennas or protruding components. This design approach helps maintain the sleek and compact form factor of modern laptops while providing efficient wireless performance. The IFA can be cut from a thin metal sheet and fed by a miniaturized coaxial cable, as shown in Figure 1.18.

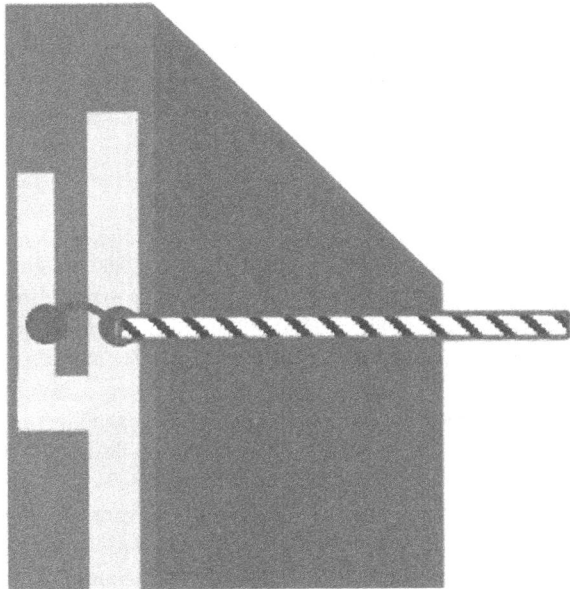

FIGURE 1.18   A 2.4 GHz inverted-F antenna integrated into a laptop.

When the IFA is mounted on the edge of a laptop screen, it exhibits excellent omnidirectional properties, which are crucial for mobile terminal applications. Additionally, the IFA offers a wide impedance bandwidth, ensuring compatibility with various frequency bands. Over time, several modifications have been introduced to enhance the functionality of the IFA. These advancements have resulted in the development of dual-band designs [8–10]. Furthermore, there have been notable achievements in creating triple-band IFA designs [11]. These innovations have allowed for improved wireless performance and expanded frequency coverage in laptop antennas.

1.5.2.2 Helical Antenna

The helical antenna is another type of antenna used in laptop applications. It features a conducting wire wound in a helical shape and connected to a ground plate using a feeder line, as illustrated in Figure 1.19. The helical antenna design offers several advantages for laptop use. It provides circular polarization, allowing for better reception and transmission in different orientations. The helix design also enables compactness, making it suitable for integration within the limited space of a laptop. The number of turns in the helix and the spacing between the turns can be adjusted to achieve the desired performance characteristics. Helical antennas have been employed in laptop systems to enhance wireless connectivity and ensure reliable communication. For optimal performance, helical antennas, along with monopole antennas, should be positioned at the top of the display.

The helical antenna offers a compact physical size, but its bandwidth is narrower compared to a monopole antenna. This narrower bandwidth can present challenges in achieving a good match across the relatively wide industrial, scientific, and medical bands. Additionally, the helical antenna can be affected by nearby objects, leading to detuning, which may limit its suitability for handheld wireless applications.

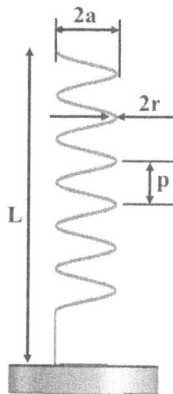

FIGURE 1.19 Helical antenna structure.

1.5.2.3 Slot Antenna

The slot antenna for laptops is a design that integrates a narrow slot within the laptop's casing. This slot serves as the radiating element of the antenna, allowing for wireless communication. The slot is typically positioned along the edges of the laptop or in a specific location within the device.

The slot antenna design offers several advantages for laptops due to their wider bandwidth characteristics. It provides a compact, simple design and low-profile solution, allowing for easy integration into the laptop's form factor. The slot antenna is designed with a slot length that is typically a half- or quarter-wavelength long. It primarily exhibits one polarization, and its radiation pattern is less omnidirectional compared to the inverted-F antenna (IFA) antenna. However, the slot antenna tends to radiate more energy in the horizontal direction, making it advantageous for WLAN applications. It is worth noting that the slot antenna has a high impedance, usually several hundred Ohms. The antenna impedance can be adjusted by varying the feed point location. If the slot antenna is cut in half, it transforms into an open-slot antenna with propagation characteristics similar to a monopole antenna. This transformation allows for further flexibility in antenna design. The structure of slot antennas is depicted in Figure 1.20.

Additionally, the slot antenna can offer good radiation patterns and performance, particularly when properly positioned and designed. The slot antenna's performance can be influenced by factors such as the size, shape, and position of the slot, as well as the surrounding components and materials within the laptop. Proper design considerations are crucial to ensure optimal radiation efficiency and signal reception. Overall, the slot antenna is a viable choice for laptops, providing a wireless communication solution while maintaining the device's sleek and portable design. They are a reliable choice for achieving wireless connectivity while maintaining the desired performance and cost requirements.

1.5.2.4 Patch Antenna

Patch antennas are widely used in laptop applications due to their compact size, ease of integration, and good performance characteristics. They are commonly found in devices such as Wi-Fi cards and Bluetooth modules embedded within laptops. A patch antenna consists of a flat metallic patch, typically made of copper or another conductive material, mounted on a dielectric substrate. The patch is usually square,

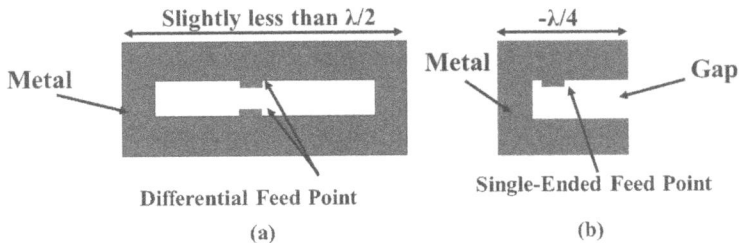

FIGURE 1.20 Structure of a slot antenna.

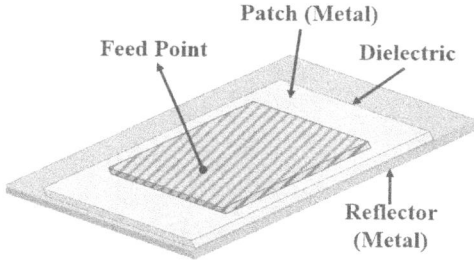

FIGURE 1.21 Structure of patch antenna.

rectangular, or circular in shape and is designed to resonate at a specific frequency. The size of the patch and the substrate material determine the operating frequency of the antenna.

The patch antenna is typically placed on the surface of the laptop's circuit board or integrated within the laptop's display lid. The ground plane of the circuit board or the metal casing of the laptop acts as the bottom conducting surface, forming a half-wavelength resonant structure. The patch is fed with a microstrip transmission line, which provides the necessary RF signal to excite the antenna and radiate EM waves.

Patch antennas offer several advantages for laptop applications. They have a low profile, allowing for easy integration and placement within the laptop's design. They can be manufactured using printed circuit board technology, making them cost-effective. Patch antennas also provide a relatively wide bandwidth, allowing for reliable wireless communication.

However, patch antennas have a limited radiation pattern, typically providing a directional radiation pattern with maximum radiation perpendicular to the patch surface. This characteristic can affect the antenna's performance in terms of signal coverage and range. To overcome this limitation, multiple patch antennas or additional diversity techniques may be employed to improve the overall wireless performance of the laptop.

Overall, patch antennas are a popular choice for laptop applications due to their compact size, cost-effectiveness, and reasonable performance characteristics. They enable reliable wireless connectivity while maintaining the sleek and compact design of modern laptops. Figure 1.21 shows an example of a patch antenna.

1.5.2.5 Monopole Antenna

Monopole antennas are commonly used in laptop applications for wireless communication. They are compact and efficient, making them suitable for integration within the limited space available in laptops. A monopole antenna consists of a single conducting element, typically a vertical rod or wire, that acts as the radiating element, as illustrated in Figure 1.22. One end of the monopole is connected to the ground plane or the laptop's circuitry, while the other end serves as the radiating element. The length of the monopole is typically a quarter-wavelength or a half-wavelength at the desired operating frequency.

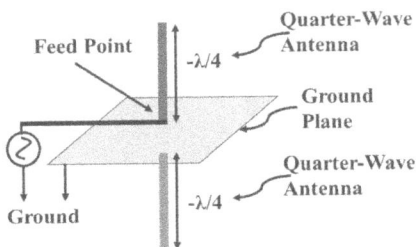

FIGURE 1.22 Quarter-wave monopole antenna.

In laptops, the monopole antenna is often integrated into the laptop's display or housed within the laptop's chassis. It can be positioned on the top edge of the display or embedded in the bezel area to achieve optimal performance. The ground plane for the monopole antenna is provided by the laptop's circuit board or the metal casing, ensuring a sufficient ground reference for radiation.

Monopole antennas offer several advantages for laptop applications. They provide omnidirectional radiation, meaning they radiate EM waves in all directions around the antenna. This characteristic allows for better signal coverage and reception from various directions. Monopole antennas also offer good efficiency and can operate over a wide frequency range, making them suitable for multiple wireless communication standards.

However, the simple quarter-wave monopole is not commonly used as an internal antenna in laptops due to its relatively large size. To overcome this limitation, several modifications have been proposed to adapt monopole antennas for laptop applications. One such modification is the compact printed tab monopole, which offers a broad 50% bandwidth and can be integrated onto a dielectric board by protruding over a laptop's display panel (Figure 1.23). This design allows for reduced size while maintaining acceptable performance characteristics. In cases where lower bandwidth requirements are sufficient, the compact printed tab monopole can be replaced with a single-band meander monopole or a dual-band branched monopole [12]. These variations of monopole antennas provide flexibility in achieving the desired operating characteristics while considering the space constraints within a laptop.

Further, the performance of a monopole antenna can also be affected by its proximity to other components within the laptop, such as display cables or metallic

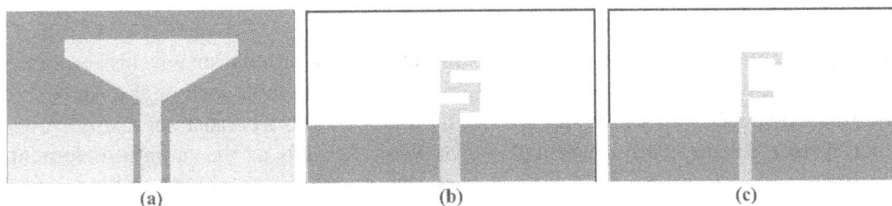

FIGURE 1.23 Printed monopole antenna: (a) wide-band monopole, (b) miniature meander monopole, and (c) dual-band branched monopole.

structures. These nearby objects can cause detuning and affect the antenna's radiation pattern and impedance. To mitigate these effects, careful placement and design considerations are considered during the laptop's development.

Overall, monopole antennas are a popular choice for laptops due to their compact size, omnidirectional radiation pattern, and wide frequency coverage. They enable reliable wireless communication while maintaining the portability and aesthetics of modern laptops.

1.6   IMPACT OF EMERGING WIRELESS STANDARDS

The landscape of wireless telecommunications is experiencing rapid evolution, driven by emerging technologies that propel wireless systems toward our envisioned future. The emergence of new wireless standards has had a profound impact on laptops. Some of the key impacts are as follows:

- **Enhanced connectivity:** Emerging wireless standards have significantly improved the connectivity capabilities of laptops. These standards offer faster data-transfer speeds, increased capacity, and reduced latency, resulting in a smoother and more reliable wireless connection for laptops.
- **Improved performance:** With the adoption of new wireless standards, laptops can take advantage of advanced features and technologies. For example, Wi-Fi 6 introduces features like OFDMA and MU-MIMO, which enable better performance in crowded environments and improved overall network efficiency.
- **Higher data-transfer speeds:** The latest wireless standards support higher data transfer speeds, allowing laptops to handle bandwidth-intensive tasks more effectively. This is particularly beneficial for activities such as streaming high-definition videos, online gaming, and large file transfers.
- **Increased battery efficiency:** Emerging wireless standards often come with power-saving mechanisms and optimizations. These technologies help reduce the power consumption of wireless modules in laptops, resulting in improved battery life.
- **Seamless integration with IoT devices:** Many emerging wireless standards prioritize compatibility with IoT devices. This allows laptops to easily connect and interact with a wide range of smart home devices, wearable technology, and other IoT devices, enhancing the overall user experience.
- **Enhanced security:** Newer wireless standards often incorporate improved security features to protect data transmitted over wireless networks. For instance, Wi-Fi 6 includes Wi-Fi protected access 3 (WPA3) encryption protocols, providing stronger security measures compared to previous standards.
- **Future-proofing:** By incorporating the latest wireless standards in laptops, manufacturers ensure compatibility with future network infrastructure and technologies. This future-proofing aspect enables laptops to remain relevant and compatible with evolving wireless networks for an extended period.

In summary, emerging wireless standards have transformed laptops by delivering faster, more reliable connectivity, improved performance, extended battery life, and enhanced integration with IoT devices. These advancements contribute to a more seamless and efficient user experience, empowering users to take full advantage of wireless capabilities in their laptops. Specifically, two prominent emerging standards, namely, 802.11ac (Wi-Fi) and 802.16e (WiMAX), along with the upcoming 802.11ax (Wi-Fi), have a significant impact on laptop computers. This section will delve into the implications of these emerging and upcoming standards on laptops.

1.6.1 802.11ac (Wi-Fi)

The new IEEE 802.11ac standard offers a robust platform to meet the growing demands of the future by implementing substantial enhancements to the RF spectrum utilization, surpassing the capabilities of the 802.11n standard. These technological advancements bring significant improvements to wireless speeds and throughput capabilities. Unlike its predecessors, the 802.11ac standard is specifically designed to deliver faster wireless data rates. It operates exclusively in the 5 GHz band, which grants it greater resilience by avoiding the interference commonly encountered in the 2.4 GHz band.

Features of 802.11ac are:

The 802.11ac standard, also known as Wi-Fi 5, introduces several notable features and improvements over its predecessors. Some key features of 802.11ac include:

- **Higher data rates:** 802.11ac offers significantly higher data rates compared to previous Wi-Fi standards. It achieves this through the use of wider channel bandwidths (up to 160 MHz) and advanced modulation techniques, such as 256 quadrature amplitude modulation (QAM).
- **MIMO:** 802.11ac utilizes MIMO technology, which allows for the use of multiple antennas to transmit and receive data simultaneously. This enables improved wireless performance and higher throughput by increasing spatial multiplexing.
- **Beamforming:** Beamforming is a technique used in 802.11ac to focus and direct wireless signals toward the intended recipients. By dynamically adjusting signal direction and strength, beamforming enhances signal quality and extends the range of Wi-Fi coverage.
- **Extended channel bonding:** 802.11ac supports channel bonding, which combines adjacent channels to create wider transmission channels. This provides increased bandwidth and allows for higher data rates.
- **Backward compatibility:** Although 802.11ac operates in the 5 GHz frequency band, it maintains backward compatibility with devices supporting older Wi-Fi standards (e.g., 802.11a/b/g/n). This ensures that devices using different Wi-Fi standards can coexist and communicate with each other.
- **Improved power efficiency:** 802.11ac includes power-saving mechanisms to improve energy efficiency in devices. This helps prolong battery life in laptops, smartphones, and other wireless devices.

- **Enhanced security:** The 802.11ac standard incorporates advanced security features, including Wi-Fi protected access 2 (WPA2) encryption and authentication protocols, to ensure secure wireless communications.

Overall, 802.11ac delivers faster and more reliable wireless connectivity, improved network capacity, and better performance for multimedia streaming, online gaming, and other bandwidth-intensive applications.

802.11ac promises to deliver several key benefits, including:
- Increased capacity and higher throughput gained by multiple clients being able to connect at the same time without reducing performance.
- Lower latency will make higher quality connections to increase the quality of service for real-time application including VoIP and video.
- More efficient power usage will mean less power consumption when data are transmitted.

1.6.2 802.11ax (Wi-Fi)

The 802.11ax standard is specifically designed to tackle connectivity challenges in high-density networks and enhance overall network performance. It introduces several new features that contribute to significant improvements. One of the key advancements is the ability for multiple clients to transmit data simultaneously, resulting in increased network speed by up to 30% compared to the previous 802.11ac standard. This enhanced speed enables faster data transfer and a smoother user experience. In addition to speed, 802.11ax also addresses the issue of latency by reducing delays in data transmission. This improvement is particularly beneficial for real-time applications, such as online gaming or video conferencing, where low latency is crucial for seamless interactions.

The features of 802.11ax are:
- **Increased capacity:** 802.11ax introduces OFDMA technology, which allows for more efficient utilization of the available spectrum. This enables multiple devices to transmit data simultaneously, significantly increasing the network's capacity.
- **Higher throughput:** With the introduction of MU-MIMO technology, 802.11ax enables simultaneous data transmission to multiple devices. This enhances the network's throughput and improves overall performance, especially in high-density environments.
- **Improved efficiency:** TWT is a new feature in 802.11ax that enhances power efficiency. It allows devices to schedule specific times for data transmission, resulting in reduced power consumption and longer battery life for connected devices.
- **Lower latency:** The introduction of OFDMA and other enhancements in 802.11ax helps reduce latency, providing a more responsive and lag-free network experience. This is particularly beneficial for real-time applications such as online gaming and video streaming.

- **Enhanced coverage:** 802.11ax introduces improved spatial reuse and beam-forming techniques, allowing for better coverage and signal strength. This ensures a more reliable and stable connection, even in challenging environments.
- **Backward compatibility:** While 802.11ax is a new standard, it is designed to be backward compatible with previous Wi-Fi standards, such as 802.11ac. This means that devices supporting 802.11ax can still connect and communicate with devices using older Wi-Fi standards.

Overall, 802.11ax offers increased capacity, higher throughput, improved efficiency, lower latency, enhanced coverage, and backward compatibility. These features make it well-suited for high-density environments and support the growing demands of modern wireless networks.

802.11ax promises to deliver several key benefits, including:
- 30% faster speeds, allowing users to get their data in less time.
- Increased throughput to allow more simultaneous users.
- Reduced latency, so that an increase in the number of users will not affect speed, and it will provide a more reliable connection even in crowded environments.
- Superior outdoor service and increased range, so that networks will have fewer dead spots.

1.6.3 802.16e (WiMAX)

The 802.16e standard offers exceptional performance in terms of distance and throughput. It supports speeds of up to 70 Mbps and provides a range of up to 50 kilometers. Similar to the widely used Wi-Fi technology, it can be utilized for wireless networking in laptop computers.

One of the key advantages of the 802.16e standard is its ability to achieve higher data rates over longer distances. It ensures the efficient utilization of available bandwidth and minimizes the chances of interference. Technological advancements envision a future where households have a WiMAX receiver installed, accompanied by a Wi-Fi transmitter for in-home connections. Additionally, laptops and personal devices are expected to have the capability to directly transmit data to WiMAX towers.

The 802.16e standard relies on microwave radio technology to establish connections between computers and the internet. Users located within a 3–5-mile radius of the base station can establish a link using non-line-of-sight (NLOS) technology, enjoying data rates as high as 75 Mbps. This seamless connectivity empowers users to utilize their wireless devices, such as laptops, from the comfort of their homes, offices, while traveling, or practically from any location worldwide.

1.7 CHALLENGES FOR ESTABLISHING WIRELESS IN LAPTOPS

With new technology, come new challenges. A few of the challenges for mounting wireless antennas in laptop computers are discussed next:

1.7.1   MINIATURIZATION

One of the key challenges in wireless technology for laptops is the miniaturization of antennas. As laptops continue to become smaller and more compact, there is limited physical space available for integrating antennas while maintaining optimal performance. This miniaturization issue poses several challenges:

- **Limited antenna size:** The available space inside a laptop chassis is limited, making it challenging to design and integrate an antenna that meets the desired performance specifications. Antennas typically require a certain physical size to achieve optimal radiation efficiency and coverage, which may be compromised in smaller laptops.
- **Antenna efficiency:** Miniaturized antennas often experience reduced efficiency compared to larger counterparts. The reduced physical size can result in lower radiation efficiency, weaker signal strength, and limited coverage range. This can impact the overall wireless performance of the laptop.
- **Frequency band support:** Modern laptops require support for multiple wireless frequency bands, such as 2.4 GHz and 5 GHz for Wi-Fi. Designing compact antennas that can operate effectively across multiple frequency bands can be technically challenging.
- **EM interference (EMI):** The proximity of other components and circuitry within a laptop can introduce EMI, which can negatively affect the antenna's performance. Shielding and isolation techniques are necessary to mitigate these interference effects.

To address the miniaturization challenge, engineers and researchers are continually developing innovative antenna designs and technologies. These include compact and multiband antennas, metamaterial-based antennas, and advanced antenna tuning techniques. Additionally, advancements in materials, fabrication techniques, and simulation tools aid in optimizing antenna performance within the constraints of miniaturized laptop designs.

1.7.2   PLACEMENT OF ANTENNA INSIDE THE LAPTOPS

The placement of the antenna within the laptop chassis is crucial for both performance and aesthetics. It is essential to design antennas that seamlessly integrate with the laptop's sleek design, ensuring they fit within the back cover without disrupting the overall appearance. Moreover, the antennas should be positioned in a way that keeps them hidden from the users' view when looking at the laptop from the outside. Further, limited space options may restrict the ideal placement, leading to suboptimal antenna positioning and potential signal blockage or interference from internal components. Placing an antenna inside a laptop presents several challenges due to the limited space and complex internal structure of the device. Some of the key challenges include:

- **Space constraints:** Laptops have limited internal space, making it challenging to find an optimal location for the antenna. Components such as the display, motherboard, battery, and other peripherals occupy most of the available space, leaving limited options for antenna placement.

- **Signal interference:** The internal components of a laptop, including the processor, memory modules, and hard drives, generate EMI. Placing an antenna near these components can result in signal degradation and reduced wireless performance.
- **Structural obstacles:** The internal structure of a laptop, including metal casings, shields, and circuitry, can obstruct the antenna's radiation pattern and impede signal propagation. These obstacles can lead to reduced signal strength and coverage.
- **Antenna-body interaction:** The presence of the laptop's body and other components close to the antenna can affect its performance. The EM properties of the laptop's materials can interact with the antenna, causing detuning, impedance mismatch, and altered radiation characteristics.
- **Human body effects:** When using a laptop, the user's body, hands, and arms come in close proximity to the antenna. This can cause signal blockage and detuning, affecting the antenna's radiation pattern and overall performance.

To address these challenges, engineers employ various techniques such as careful antenna design and placement optimization, shielding and isolation techniques to minimize EMI, antenna diversity and MIMO systems, and antenna tuning to compensate for impedance variations. Advanced simulation tools and testing methods are used to evaluate the antenna's performance in realistic laptop environments and mitigate the impact of these challenges.

1.7.3   USE OF COAXIAL CABLE AND ITS ROUTING INSIDE THE LAPTOPS

The routing of coaxial cables inside laptops poses several challenges. One primary concern is space limitation. Laptops are designed to be compact, and internal space is often limited. Therefore, routing the coaxial cables effectively becomes crucial to ensure proper functioning without interfering with other components or hindering the laptop's overall design.

Several antennas, like the monopole antenna, tend to deliver their best performance when mounted on the top of the display. In laptops, the radio card is typically situated in the base, while the antenna is positioned on the display's top. This arrangement results in a feeding cable with a length of a few centimeters in most implementations. Furthermore, the routing of the coaxial cable inside laptop computers is a critical consideration. Proper cable routing ensures minimal interference and signal degradation, contributing to optimal wireless performance.

To accommodate the integration of antennas, coaxial cables with small diameters are utilized. This enables the cables to be routed through hinges and other internal components, ensuring they are placed away from sensitive components and critical signal paths to minimize interference and signal degradation. Furthermore, cable management techniques such as clips, brackets, or adhesive routing guides must be employed to secure and organize the coaxial cables. These measures help maintain cable integrity, prevent tangling or strain, and ensure the cables remain in their designated positions throughout the laptop's lifespan. However, it's important to note that these small-diameter coaxial cables can introduce certain losses at higher frequencies. Such cable losses can significantly impact wireless performance, making it crucial to carefully select the appropriate RF cable.

Overall, the use of coaxial cables and their routing inside laptops requires meticulous planning and consideration. Designers should aim to strike a balance between functionality, signal integrity, space optimization, and maintaining the laptop's design aesthetics. In summary, selecting the right RF cable and implementing effective cable routing within laptop computers are essential for maintaining excellent wireless performance. These considerations help mitigate signal losses and ensure optimal communication between the antenna and the radio card, ultimately enhancing the overall wireless experience for laptop users.

REFERENCES

1. Balanis, C. A. (Ed.), *Modern Antenna Handbook*, Canada, John Wiley & Sons, Inc., 2008.
2. Soras, C., Karaboikis, M., Tsachtsiris, G. & Makios, V. "Analysis and design of an inverted-F antenna printed on a PCMCIA card for the 2.4 GHz ISM band." *IEEE Antennas and Propagation Magazine*, vol. 44, no. 1, pp. 37–44, 2002.
3. Manteghi, M. & Rahmat-Samii, Y. "Novel compact tri-band two-element and four-element MIMO antenna designs." *2006 IEEE Antennas and Propagation Society International Symposium*, pp. 4443–4446, 2006.
4. Tang, C. L. "2.4/5.2 GHz dual-band chip antenna for WLAN application." *2005 IEEE Antennas and Propagation Society International Symposium*, vol. 1A, pp. 454–457, 2006.
5. Dakeya, Y., Suesada, T., Asakura, K., Nakajima, N. & Mandai, H. "Chip multilayer antenna for 2.45 GHz-band application using LTCC technology." *IEEE MTT-S International Microwave Symposium Digest*, vol. 3, pp. 1693–1696, 2000.
6. Chen, Z. N., *Antennas for Portable Devices*, John Wiley & Sons Ltd., 2007.
7. Liu, D., Flint, E. & Gaucher, B., "Integrated laptop antennas - Design and evaluation." *Proceedings of the IEEE APS International Symposium and URSI Radio Science Meeting*, San Antonio, Texas, vol. 4, pp. 56–59, 2002.
8. Liu, D. & Gaucher, B. "A branched inverted-F antenna for dual band WLAN applications." *IEEE Antennas and Propagation Society Symposium*, vol. 3, pp. 2623–2626, 2004.
9. Liu, D., Gaucher, B. & Hildner, T., "A dual band antenna for WLAN applications." *Proceedings IEEE International Workshop on Antenna Technology: Small Antennas and Novel Metamaterials IWAT*, pp. 201–204, 2005.
10. Yeo, J., Lee, Y. J. & Mittra, R. "A novel dual-band WLAN antennas for notebook platforms." *IEEE Antennas and Propagation Society Symposium*, vol. 3, no. 2, pp. 1439–1442, 2004.
11. Liu, D., Gaucher, B. P., Flint, E. B., Studwell, T. W., Usui, H. & Beukema, T. J. "Developing integrated antenna subsystem for laptop computers." *IBM Journal of Research and Development*, vol. 47, no. 2.3, pp. 355–367, 2003.
12. Peixeiro, C., Fernandes, C. A., Moreira, A. A., Kemp, B., Capstick, M., Burr, A. & Hofstetter, H. *Design and Performance of Antenna Prototypes*, IST-2001-32125 FLOWS, Deliverable Number: D16, 2004.

2 Planar Monopole Antenna

2.1 INTRODUCTION

Broadband planar antennas have emerged as highly promising candidates for driving the rapid growth of wireless communications. They are in high demand due to their ability to cater to various applications, including Global Positioning System (GPS), vehicle mobile communication (VMC), wireless local area network (WLAN), and ultra-wide band (UWB) for wireless personal area network (WPAN), while utilizing a reduced number of antennas. The term "planar antennas" generally encompasses all antenna configurations featuring planar or curved surface radiators, along with at least one feed radiator [1]. The planar monopole antenna holds significant importance in the field of wireless communications and radio frequency (RF) engineering. Here are some key reasons for the significance of planar monopole antennas:

- **Compact and low-profile design:** Planar monopole antennas are known for their compact and low-profile design, making them suitable for integration into small electronic devices, such as mobile phones, tablets, laptops, and Internet of Things (IoT) devices. Their planar structure allows for easy fabrication and integration into various platforms.
- **Broadband operation:** Planar monopole antennas typically exhibit wide bandwidth characteristics, enabling them to operate over a broad frequency range. This wideband operation is advantageous for applications that require communication across multiple frequency bands, such as multiband wireless systems.
- **Omnidirectional radiation pattern:** The planar monopole antenna offers an omnidirectional radiation pattern in the horizontal plane. This means that it radiates electromagnetic waves uniformly in all directions around the antenna, making it suitable for applications that require coverage in all directions, such as wireless communication in urban environments.
- **Simple construction and manufacturing:** Planar monopole antennas can be easily fabricated using standard printed circuit board (PCB) manufacturing techniques. This simplicity in construction leads to cost-effective production and allows for mass production of antennas, making them commercially viable for various applications.
- **Flexible frequency tuning:** The length and shape of the planar monopole antenna can be adjusted to tune its resonant frequency. This flexibility in frequency tuning allows the antenna to be customized for specific frequency bands or wireless communication standards, enhancing its versatility and adaptability to different applications.
- **Low cost:** Planar monopole antennas can be manufactured using low-cost materials and processes, such as PCB technology. This cost-effectiveness

 DOI: 10.1201/9781003331018-2

makes them economically viable for mass production and deployment in consumer electronics and wireless communication systems.

- **Compatibility with planar substrates:** Planar monopole antennas can be easily integrated into planar substrates, such as PCBs or flexible substrates. This compatibility facilitates their integration into electronic devices, where space is limited, and planar structures are prevalent.
- **Integration with active devices:** Planar monopole antennas can be easily integrated with active devices, such as RF transceivers or power amplifiers, on the same planar substrate. This integration simplifies the overall system design and improves the overall efficiency and performance of the wireless communication system.

The significance of planar monopole antennas lies in their compactness, broadband operation, omnidirectional radiation pattern, simple construction, flexibility in frequency tuning, cost-effectiveness, compatibility with planar substrates, and ease of integration with active devices. These features make them highly desirable for various wireless communication applications, especially in small and portable devices.

Extensively used planar antennas in wireless communication systems are:

- Printed microstrip patch antennas.
- Slot antennas.
- Suspended-plate antennas (SPAs).
- Planar inverted-L antennas (PILAs).
- Planar inverted-F antennas (PIFAs).
- Planar monopole antennas, etc.

Since Guglielmo Marconi's successful transmission and reception of the first radio signal in Italy in 1895, monopole antennas have been widely utilized as the most fundamental type of antenna, particularly at the periphery of planar transmission-line antennas such as microstrip patch antennas, SPAs, and PIFAs or PILAs. However, the high manufacturing and material costs associated with laptop computer devices pose a challenge. Moreover, the directional radiation patterns of these antennas in the elevation (E-plane) and/or azimuth (H-plane) angles can lead to pulse distortion. To address these issues, planar monopole antennas have emerged as a preferable alternative. They exhibit excellent radiation performance, impedance-frequency characteristics, and offer the significant advantage of compact size and volume [2, 3].

2.2 EVOLUTION OF MONOPOLE ANTENNAS FOR WIRELESS APPLICATIONS

2.2.1 PLANAR BI-CONICAL MONOPOLE ANTENNA STRUCTURE

A planar bi-conical monopole antenna structure refers to an antenna design that incorporates two conical elements on a planar substrate. This configuration combines

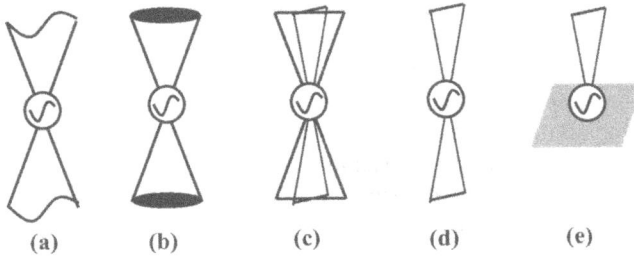

FIGURE 2.1 The evolution of planar finite triangle monopole from bi-conical structure.

the features of a monopole antenna, which typically consists of a single radiating element, with the bi-conical shape that resembles two back-to-back cones.

The planar bi-conical monopole antenna is a versatile and efficient antenna design that meets the demands of many wireless devices requiring broad impedance and radiation bandwidths. Traditional methods to increase the impedance bandwidth of a simple thin wire monopole antenna involve increasing the thickness of the radiating strip, which, unfortunately, leads to bulkier and heavier antennas, rendering them unsuitable for applications like laptop computers. To address this trade-off between bandwidth and size, the planar structure is widely adopted.

In Figure 2.1 Ref. [4], the evolution of the planar finite-size triangle monopole antenna is depicted, originating from the infinite bi-conical structure illustrated in Figure 2.1(a).

The latter refers to the principle of frequency independence, which occurs when the input impedance of the cone matches the characteristic impedance of the transmission line used to feed the bi-conical structure. The input impedance of the cone is determined by the cone's angle. In Figure 2.1(b), the finite structure exhibits inadequate bandwidth due to the inefficient radiation of the cone at lower frequencies, caused by the insufficient length of the cone. As a result, the lower edge of the bandwidth is determined by the cone's largest dimension. These bi-conical structures can be either solid or hollow.

To address the weight and potential wind-resistance concerns of the bi-conical design in Figure 2.1(c), a thin-wire structure, as depicted in Figure 2.1(c), was utilized. Alternatively, the planar structure shown in Figure 2.1(d) can be employed to reduce the antenna's installation space. Compared to three-dimensional designs illustrated in Figure 2.1(a–c), the two-dimensional triangle monopole shown in Figure 2.1(d) can be easily etched onto a PCB and integrated with other RF circuits on the same PCB, where the planar monopole can be co-planar with the ground plane as shown in Figure 2.1(e) [5].

2.2.2 PLANAR MONOPOLE

Several planar radiator designs have been extensively discussed and utilized [6], as depicted in Figure 2.2. Most planar monopoles exhibit remarkably broad impedance bandwidths exceeding 70% for Voltage Standing Wave Ratio (VSWR) <2 [7].

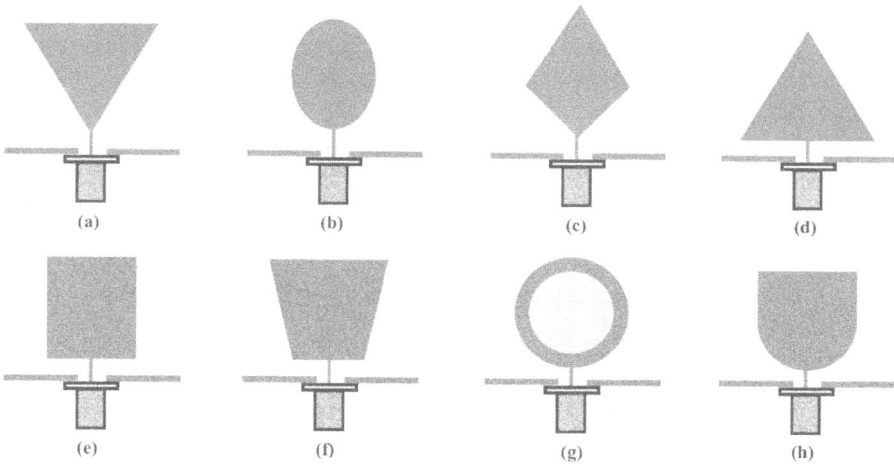

FIGURE 2.2   Different shapes of the planar monopole antenna.

For instance, circular planar monopoles display high-pass impedance characteristics [8]. The impedance transition is formed by the feeding probe, the bottom of the planar radiator, and the ground plane. When good impedance matching is achieved at multiple adjacent resonances simultaneously, the bandwidth is expanded. Furthermore, unlike the design of straight-wire monopoles, the development of planar monopoles necessitates specific considerations, such as the evaluation of the lower-edge frequency of bandwidth (LEFBW) and the achievement of omnidirectivity in radiation.

On one hand, evaluating the LEFBW for a broadband planar monopole becomes challenging when the dominant resonant frequency significantly deviates from the center frequency of the operating bandwidth due to the presence of multiple resonances [9]. On the other hand, as the operating frequency increases, the lateral dimension of the planar monopole becomes electrically large, causing the radiation from the two vertical edges to become directional in horizontal planes.

2.2.3   ROLL MONOPOLES

In the design of planar monopoles, an important consideration is the variation in omnidirectional radiation patterns across the wide bandwidth. The presence of non-axisymmetric structures in monopoles influences the radiation patterns, particularly in horizontal planes, at higher frequencies within the bandwidth. This can result in a compromised radiation performance that somewhat offsets the advantage of reduced antenna volume. Additionally, in vertical planes, the directed beam's maximum value shifts, and the gain decreases as the operating frequency increases.

However, for cylindrical monopoles operating over a wide bandwidth, the radiation performance can be enhanced by incorporating a sleeve around the stem [10, 11]. This technique, although effective in cylindrical monopoles, has not been widely utilized in planar designs.

A roll monopole antenna is a type of monopole antenna that is designed in a rolled or spiral shape. It consists of a conducting element, typically in the form of a thin strip or wire, wound in a spiral configuration. The structure of the antenna resembles a cylindrical coil, with the radiating element rolled or spiraled around a central axis.

The roll monopole antenna is a compact and space-efficient design that offers several advantages. Its spiral shape allows for a larger physical length of the radiating element to be accommodated within a limited space, enabling it to achieve improved radiation efficiency and bandwidth compared to straight monopole antennas of similar size. The spiral configuration also helps in reducing the overall physical size of the antenna, making it suitable for applications with space constraints.

Roll monopole antennas are commonly used in various wireless communication systems, including mobile devices, radio frequency identification (RFID) tags, and wireless sensor networks. They can operate over a wide frequency range and exhibit omnidirectional radiation patterns in the azimuthal plane, making them suitable for applications that require coverage in multiple directions.

The roll monopole antenna's compact size, improved radiation efficiency, and wide bandwidth make it a popular choice for applications where space is limited, such as portable electronics, wearables, and compact wireless devices.

To enhance the radiation capabilities of a broadband planar monopole while preserving its compactness, the introduction of a roll monopole concept has been proposed [12]. The geometry of the roll monopole is illustrated in Figure 2.3, where a perfectly electrically conducting (PEC) sheet with dimensions W × H = 75 mm × 50 mm is uniformly rolled to form the monopole.

The roll monopole configuration features the bottom of the monopole being parallel to the ground plane, with a feed gap of g = 1 mm. The feed is achieved using a 1.2 mm thick probe, an extension of the inner conductor of a 50 Ω coaxial line, connected to the bottom of the roll at the point (r, $\theta = 0°$) through the ground plane. Optimal positioning of the feed point can be determined to ensure a broad bandwidth.

Thanks to the symmetrical structure of the roll monopole, the measured radiation patterns in the x-y plane for the E_θ components exhibit nearly omnidirectional characteristics across the impedance bandwidth. Consequently, the radiation performance of the roll monopole surpasses that of a planar monopole.

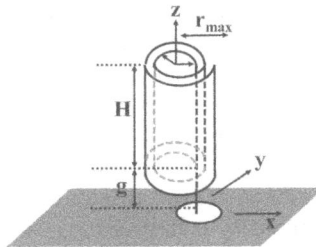

FIGURE 2.3   Geometry of the roll monopole.

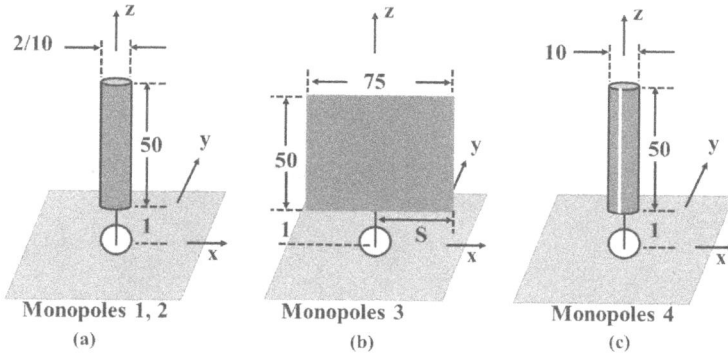

FIGURE 2.4 Geometry of monopoles for comparison, dimensions are in millimeters. (a) Thin-wire monopole and cylindrical monopole; (b) planar monopole; (c) roll monopole.

2.2.4 COMPARISON WITH OTHER MONOPOLES

In this section, the characteristics of four different monopole designs, namely, the thin-wire monopole, cylindrical monopole, planar monopole, and roll monopole, are explained and compared in terms of their impedance bandwidth. Figure 2.4 illustrates the four selected monopoles chosen for this comparison.

All monopoles were mounted vertically at the center of a 320 mm × 320 mm ground plane and connected to a 50 Ω Sub Miniature version A (SMA) connector located beneath the ground plane. The bottom of each monopole was parallel to the ground plane, with a feed gap of 1 mm. All monopoles had a height of 50 mm. Monopole 1 had a diameter of 2 mm, while Monopole 2 had a diameter of 10 mm. Monopole 3 was a planar monopole consisting of a rectangular copper sheet with a thickness of 0.2 mm and a width of 75 mm. The feed point for Monopole 3 was positioned at S = 30 mm to achieve maximum impedance bandwidth. Monopole 4 was created by rolling Monopole 3 uniformly.

The measured -10 dB impedance bandwidth for VSWR = 2 revealed that Monopole 4 (the roll monopole) exhibited the widest bandwidth of 71%, surpassing the 25%, 40%, and 53% bandwidths of Monopoles 1–3, respectively. Additionally, the lower edge frequencies of the bandwidth ranged from 1.12 GHz to 1.38 GHz. The coupling between the layers of the roll monopole generated parasitic capacitance, while the spiral structure produced parasitic inductance. The balance between these parasitic capacitance and inductance led to additional resonances, enabling the compact Monopole 4 to achieve a broader impedance bandwidth compared to Monopole 3.

2.2.5 ELECTROMAGNETIC COUPLING FEEDING METHODS

Planar radiators can be directly fed using probes, as discussed earlier. By adjusting the feed gap and the location of the feed point, excellent impedance matching can be achieved. Moreover, planar monopoles can also be excited electromagnetically through electromagnetic coupling (EMC) between the planar radiator and the

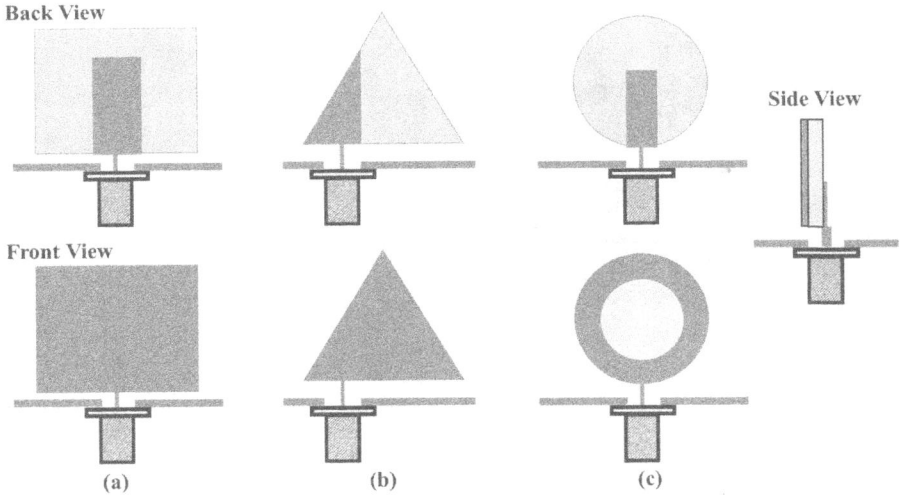

FIGURE 2.5   Planar monopoles with electromagnetic coupling feeding structures.

feeding strips connected to the probes [8, 13]. This feeding method provides additional design flexibility for achieving broadband impedance matching.

Figure 2.5 illustrates three designs featuring EMC feeding structures. In
Figure 2.5(a), a square radiator is electromagnetically coupled to a rectangular strip
fed by a short probe positioned at its bottom [14]. By optimizing the parameters
and location of the strip, an impressive 2:1 VSWR impedance bandwidth of up to
77% can be obtained. Figure 2.5(b and c) showcase triangular and annular planar
monopoles with EMC feeding structures, which exhibit enhanced impedance bandwidths of 2.7:1 and 4.7:1 for a 3:1 VSWR, respectively [4, 7]. When compared to planar monopoles with simple feeding structures, EMC planar monopoles can achieve
broader bandwidths due to the additional design possibilities they offer.

Based on the preceding discussion, it can be inferred that the planar monopole
antenna, with its flat profile and numerous appealing characteristics, including wide
bandwidth, low profile, easy installation, affordability, and simplicity in fabrication,
has garnered significant interest as an internal antenna for wireless communication
in portable electronic devices such as laptop computers, smartphones, and personal
digital assistants (PDAs).

2.3   SINGLE AND MULTIBAND ANTENNAS FOR INTERNAL LAPTOP COMPUTERS

This section discusses the dual-band and multiband antennas for wireless operation in
the laptop computer. Dual-band and multiband antennas are widely used in laptop computers for wireless communication to support multiple frequency bands. These antennas
allow laptops to connect to various wireless networks, such as Wi-Fi and Bluetooth,
operating at different frequency ranges. Dual-band antennas are designed to operate

TABLE 2.1

Frequency Bands Allocated for WLAN and WiMAX Applications

System	Frequency Bands Designation		Range of Frequency Bands Allocation
Wi-Fi, IEEE 802.11	2.4 GHz		2.4–2.5 GHz
	5 GHz	5.2 GHz	5.15–5.5 GHz
		5.5 GHz	5.47–5.725 GHz
		5.8 GHz	5.725–5.875 GHz
Mobile WiMAX IEEE 802.16, 2004	2.3 GHz		2.3–2.4 GHz
	2.5 GHz		2.5–2.69 GHz
	3.3 GHz		3.3–3.4 GHz
	3.5 GHz		3.4–3.6 GHz
	3.7 GHz		3.4–3.8 GHz
Fixed WiMAX IEEE 802.16, 2004	3.7 GHz		3.4–3.8 GHz
	5.8 GHz		5.725–5.850 GHz

in two specific frequency bands, typically the 2.4 GHz band and the 5 GHz band. The 2.4 GHz band is commonly used for Wi-Fi, while the 5 GHz band offers faster data rates and less interference. Dual-band antennas enable laptops to switch between these two bands based on the available networks and optimize the wireless connection.

Multiband antennas, on the other hand, support operation in multiple frequency bands beyond just two bands. These antennas are designed to cover a broader range of frequencies, allowing laptops to connect to various wireless standards and technologies. For example, a multiband antenna may support Wi-Fi, Bluetooth, and cellular frequencies, enabling laptops to communicate across different wireless networks simultaneously.

By incorporating dual-band and multiband antennas in laptop computers, users can enjoy improved connectivity and compatibility with a wide range of wireless networks. These antennas help ensure reliable wireless communication and enhance the overall performance of laptops in various wireless environments.

The demand for high-speed data transfer and seamless data downloads has been experiencing rapid growth. To meet this demand, effective utilization of available bandwidth is essential, and one approach to achieve this is through the implementation of multiple input multiple output (MIMO) systems. Consequently, this section explores MIMO techniques that greatly enhance channel capacity, optimize spectrum utilization, and improve data transfer rates. For reference, Table 2.1 presents the frequency bands designated for WLAN and worldwide interoperability for microwave access (WiMAX) standards, which are commonly employed in wireless applications for laptop computers.

2.3.1 BENT MONOPOLE ANTENNA

In [15], a straightforward implementation of a bent monopole antenna designed for the 2.4 GHz range of the 802.11b WLAN standard is introduced. The antenna

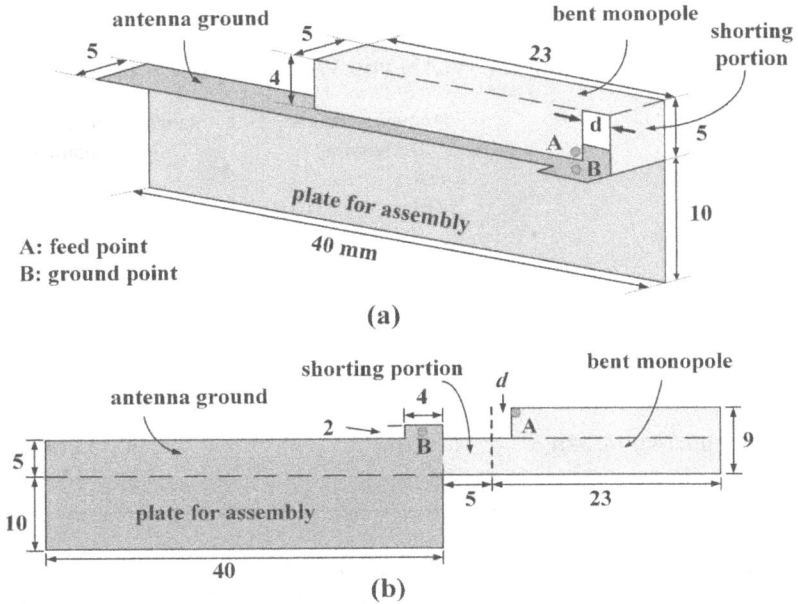

FIGURE 2.6 (a) Geometry of the proposed, bent, shorted, planar monopole antenna for 2.4 GHz band and (b) detailed dimensions of the antenna unbent into a flat plate structure.

comprises a rectangular radiating plate that is connected to a small antenna ground and an assembly plate. The assembly plate serves a dual purpose as a supporting structure for antenna installation and as an antenna ground. The monopole is fed through a coaxial-line feed line. Figure 2.6 illustrates the configuration of the bent monopole antenna specifically designed to operate within the 2.4 GHz frequency band.

Inferences

- Impedance bandwidth measured at -6 dB.
- Complex antenna structure.
- Fabrication difficult.
- Is not compatible with the 802.11a WLAN standard.

2.3.2 AN INVERTED-F ANTENNA WITH COUPLED ELEMENTS

The concept of an inverted-F antenna (IFA) with coupled elements is proposed in [16]. This antenna design combines the advantages of both the IFA and coupled elements to achieve enhanced performance. The antenna structure consists of a radiating element, a parasitic element, and a feed line. By carefully adjusting the length and spacing between these elements, the antenna can achieve improved impedance matching and radiation characteristics. Figure 2.7 showcases the configuration of the IFA with coupled elements, demonstrating its suitability for wireless applications in laptop computers.

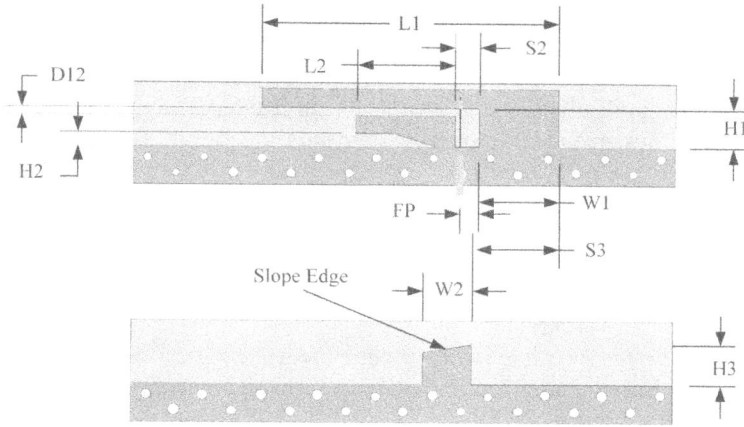

FIGURE 2.7 Inverted-F antenna with coupled elements implemented on the printed circuit board.

The antenna under consideration is a bent-type antenna with closely coupled tri-band functionality. In the low band (2.4 GHz), the antenna exhibits characteristics similar to an IFA antenna, with the majority of the current flowing through the IFA radiating strip. The performance of the low-frequency band remains largely unaffected since the current in the L-shaped strip and tab sections are minimal.

However, in the middle and high bands, the distribution of current changes, with a significant portion flowing either through the L-shaped strip or the tab section. This dominant effect influences the resonance and radiation pattern of the antenna in these frequency ranges. It should be noted that the direct feeding of the IFA section has a notable impact on the middle and high bands.

The behavior of the antenna becomes more complex in the middle and high bands and, depending on the specific application and available space for antenna implementation, the middle and high bands can be interchanged. Figure 2.8 illustrates this

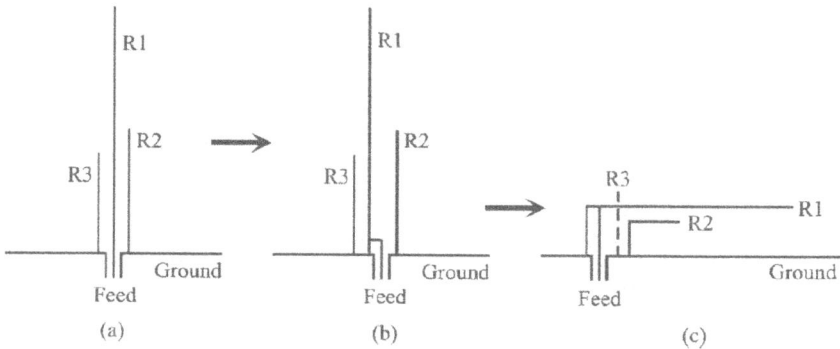

FIGURE 2.8 Evolution of tri-band Inverted-F antenna with coupled elements implemented on the printed circuit board.

concept, where R2 contributes to the middle band while R3 contributes to the high band. The figure also demonstrates a size reduction of the original antenna, resulting in a low-profile tri-band antenna.

For WLAN applications, the middle and high bands are combined to cover the 5 GHz band. Consequently, the tri-band antenna functions as a dual-band antenna in this scenario.

Inferences

- The previously proposed antenna provides inadequate bandwidth to cover entire 5 GHz band, hence, it is unable to be appropriated for the IEEE 802.11a/b/g/n WLAN standard.
- It is not easy to integrate in laptop computers due to bending.
- The bending technique leads to manufacturing complexity and takes more time to reach in the market.

2.3.3 A Dual-Band PCB Antenna with Coupled Floating Elements

The antenna proposed in [17] is based on printed half-wavelength dipole antennas with closely coupled and floating elements for dual-band applications [18, 19]. In this antenna design, the low band is covered by the feeding dipole, while the high band is covered by the coupled elements. Although the antenna size tends to be large due to its half-wavelength dipole configuration at the lower band, it exhibits excellent performance and a wide bandwidth in that frequency range. To cover the high band with a narrower gain bandwidth, multiple coupled elements are employed, ensuring the required bandwidth is achieved.

However, the antenna presented in this study, depicted in Figure 2.9, operates in a different manner. It consists of an IFA antenna and two coupled dipole elements located on the backside of the PCB to cover the high-frequency band. Additionally, a coupled-loop structure is utilized on the front side to cover the low-frequency band. To enhance the gain bandwidth at the high-frequency band, a thin (0.3 mm

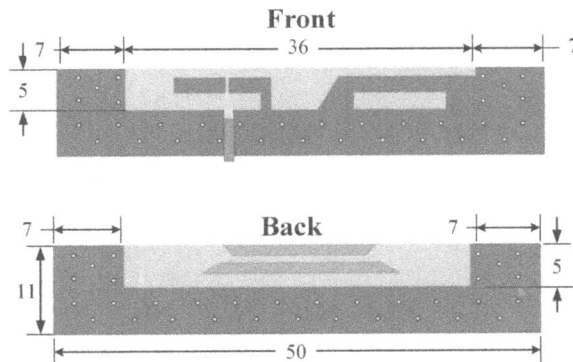

FIGURE 2.9 Dual-band printed circuit board antenna with major dimensions in millimeters.

thickness), low-loss, and low-k FR-4 PCB material with a dielectric constant of 3.5 at 1 GHz and a loss tangent of 0.004 is utilized.

Overall, this antenna design employs a combination of floating and coupled elements, leveraging the characteristics of each component to cover both the low and high frequency bands effectively. The choice of specific materials and structures enables improved performance and wider bandwidth in the desired frequency ranges.

Inferences

- Even though the antenna structure is a half-wavelength long at high frequencies, its dimensions of $50 \times 11 \times 0.3$ mm^3 make it unsuitable for laptop computers due to its size.
- The antenna structure employs a two-layer PCB design, which adds complexity to the manufacturing process.

2.3.4 A LOOP-RELATED DUAL-BAND ANTENNA

Hitachi Cable in Japan developed a simple dual-band antenna design using its proprietary thin film technology called multiframe joiner (MTF) [20, 21]. The antenna structure, which is only 0.1 mm thick, is embedded with a polyamide film (with a 3.0 dielectric constant) along with the ground plane. This provides stability and insulation from other metal devices. Unlike many flexible PCBs, antennas made using Hitachi Cable's thin film technology retain their bent position without damage. Additionally, due to its thin structure, this antenna has its own reliable and spacious ground plane. In laptop applications, the ground plane is positioned between the back side of the LCD panel and the display cover, eliminating the need for extra space.

Figure 2.10 illustrates the dual band antenna and its operating principles. For the 2.4 GHz band, the antenna can be regarded as a variation of the IFA antenna, exhibiting similar performance characteristics. A half-wavelength loop is employed to cover the 5 GHz band, with half of the loop formed by the ground plane. This design enables coverage of both the 2.4 GHz and (5.15–5.825) GHz frequency bands.

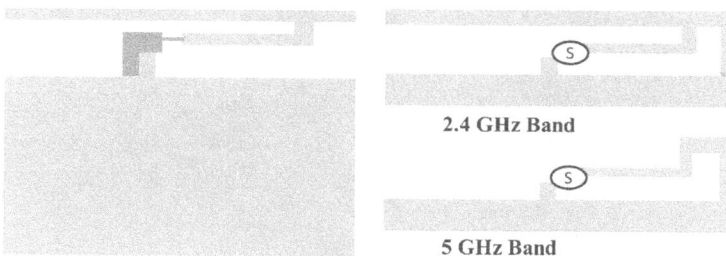

2.4 GHz Band

5 GHz Band

FIGURE 2.10 Hitachi Cable antenna and its equivalent structure.

Inferences

- The antenna's bending requirement introduces potential manufacturing complexities.
- The antenna structure, although very thin, may have limitations in terms of production time and availability, as it relies on Hitachi Cable's proprietary technology and laboratory.
- Reproducing the antenna easily and readily may pose challenges, as it is not easily reproducible anywhere, anytime.

2.3.5 DUAL-BAND METAL-STRIP ANTENNAS

The antenna presented in [22] incorporates two distinct radiating strips: one serving as the driven strip and the other as the parasitic shorted strip, as depicted in Figure 2.11. The driven strip primarily facilitates the excitation of the higher 5 GHz WLAN band, while the parasitic strip contributes to the excitation of the lower 2.4 GHz WLAN band. The antenna's compact size measures $3 \times 5 \times 20$ mm^3, making it suitable for installation inside a laptop computer. However, the antenna's bending requirement of 3 mm introduces manufacturing complexity.

Inferences

- The bending requirement of 3 mm adds complexity to the antenna manufacturing process.
- The antenna necessitates a significant amount of space when being mounted inside a laptop computer.
- With a height of 10 mm above the system ground, the antenna may not be suitable for upcoming next-generation laptop computers.

2.3.6 FLAT-PLATE ANTENNA WITH SHORTED PARASITIC ELEMENT

The proposed dual-band flat metal plate antenna, introduced in [23], features a planar structure consisting of a longer radiating strip, a shorter radiating strip, a shorting

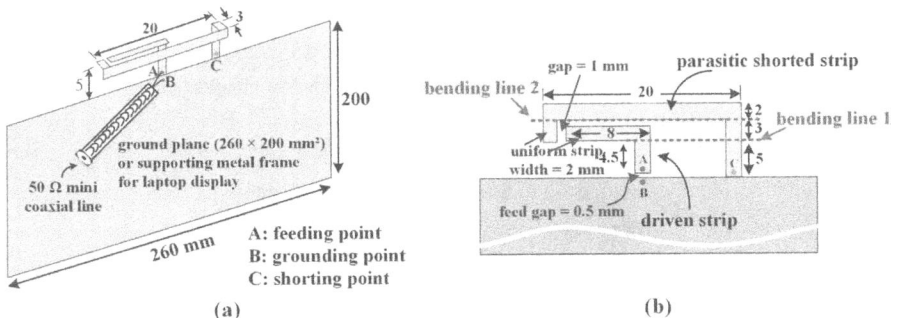

FIGURE 2.11 The dual-band metal-strip antenna structure. (a) Perspective view; (b) front view.

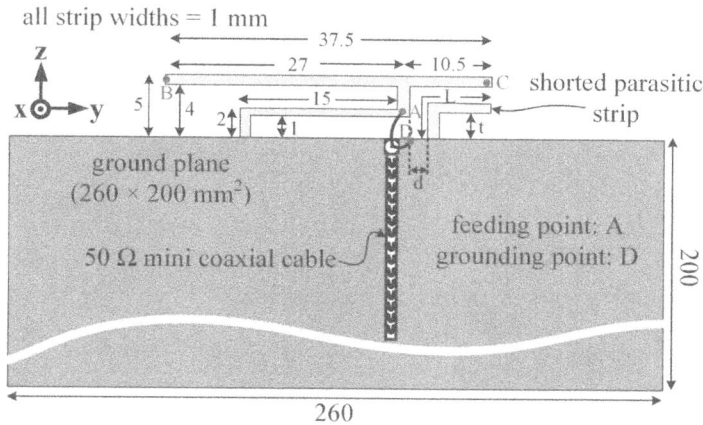

FIGURE 2.12 Geometry of the proposed dual-band flat-plate antenna.

strip, and a parasitic radiating strip, as depicted in Figure 2.12. The longer strip is responsible for controlling the lower 2.4 GHz band, while the shorter strip and the parasitic strip work in tandem to cover a wide operating band of 5 GHz.

Inferences

- The fabrication of the antenna requires line-cutting machines, which may not be readily available everywhere.
- The integration of the antenna inside a laptop computer is a complex process.
- There is a risk of breakage or damage to the isolated parasitic strip, as it is separate from the longer and shorter radiating strips.

2.3.7 COMPACT IFA WITH MEANDERING SHORTING STRIP

The antenna presented in [24] is an internal antenna featuring a C-shaped radiator and two meandering shorting strips. When integrated into a laptop computer, it occupies a size of 35 mm (L) × 3 mm (W), as depicted in Figure 2.13.

Inferences

- The inclusion of an additional meandered short strip and three vias adds complexity to the manufacturing process.
- The utilization of a double-sided PCB and multiple vias significantly increases the manufacturing cost.
- The large dimensions of the antenna make it unsuitable for next-generation laptop computers.

2.3.8 PRINTED MONOPOLE ANTENNA WITH THE EMBEDDED CHIP INDUCTOR

The antenna design proposed in [25], illustrated in Figure 2.14, incorporates a printed monopole antenna with an embedded chip inductor. This design achieves a

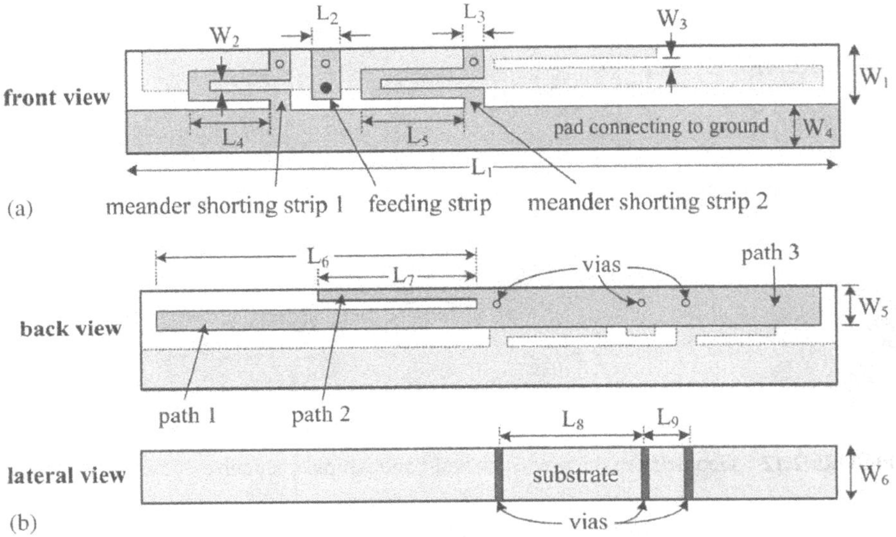

FIGURE 2.13 Geometry of the proposed dual-band flat-plate antenna front view (a) and back view (b).

wide resonant mode through the longer strip, allowing for WLAN operation in the 2.4 GHz band. By combining the shorter strip and the chip-inductor-embedded longer strip, two wide bands are generated to cover WLAN operation in the 2.4 GHz and 5.2/5.8 GHz bands. However, the introduction of an additional reactive component to achieve the desired frequency bands may result in increased power consumption and hardware complexity.

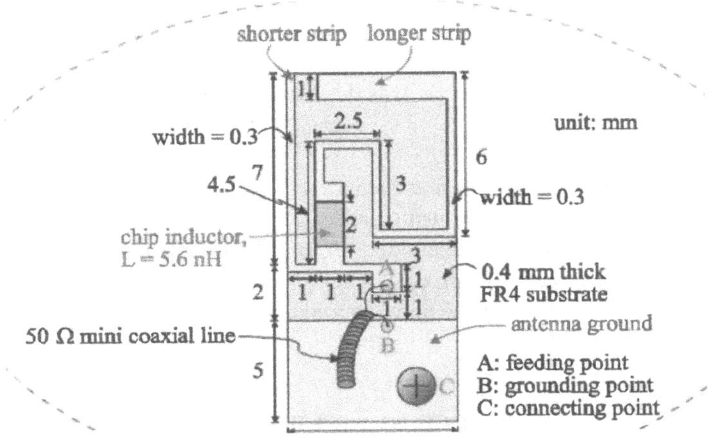

FIGURE 2.14 Geometry of a printed monopole antenna embedded with chip inductor (5.6 nH) for 2.4/5.2/5.8 GHz bands.

Inferences

- The inclusion of an extra reactive component to generate the desired frequency bands can result in increased power consumption, losses due to reactance, and added hardware complexity.
- The antenna dimensions are relatively large, making it unsuitable for next-generation laptop computers.

2.3.9 DUAL-BAND WLAN ANTENNA

The dual-band WLAN antenna, proposed in [26] and depicted in Figure 2.15, is a modified IFA that is printed on a single-sided PCB. While it is based on a conventional IFA, it utilizes a capacitive feed instead of the traditional conducting probe feed. The capacitive feed consists of two parallel strips adjacent to the L-shaped element. The upper strip measures 0.6 mm in width and 4.5 mm in length, while the corresponding dimensions for the lower strip are 1 mm and 4.5 mm, respectively. The gap between the two strips is 0.2 mm and the gap between the upper strip and the L-shaped element is also 0.2 mm. The L-shaped element itself has a height of 5 mm and a width of 1 mm. The substrate on which the antenna is printed has a thickness of 1.6 mm and a dielectric constant of 4.4.
  Inferences

- The fabrication process for the antenna is challenging.
- The antenna has large dimensions.

2.3.10 DUAL-BAND TWO-LAYERED PRINTED ANTENNA

The dual-band antenna, proposed in [27] and depicted in Figure 2.16, utilizes a two-layered printed design. It is constructed on a combined FR-4 substrate and foam, forming a two-layered structure. This antenna is designed to operate in the 2.4/5.2/5.8 GHz frequency range, conforming to the 802.11b/a WLAN standard.
  Inferences

- The inclusion of a two-layered structure adds complexity to the manufacturing process of the antenna.
- The utilization of two layers increases the overall thickness of the antenna, which may not be suitable for next-generation ultra-thin laptops.

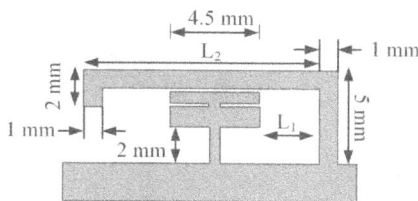

FIGURE 2.15 Geometry of a dual-band WLAN antenna.

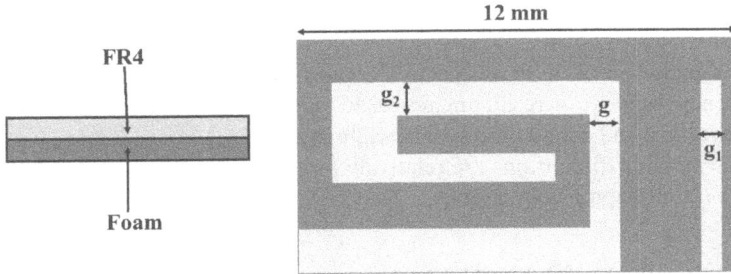

FIGURE 2.16 Geometry of a two-layer printed antenna for 2.4/5 GHz WLAN operation in the laptop computer.

2.3.11 MONOPOLE ANTENNA FOR LARGE SCREEN-TO-BODY RATIO NOTEBOOK COMPUTERS

The proposed monopole antenna in [28] incorporates two inductors for impedance matching. Additionally, the antenna's dielectric substrate, measuring 4 mm × 30 mm, reserves a 1 mm × 6 mm section for grounding purposes. This section facilitates the connection of a grounding copper tape, measuring 6 mm × 16 mm, to the display metal frame (measuring 315 × 182 mm^2), as depicted in Figure 2.17.

Inferences

- The antenna structure requires an additional ground connection to establish a connection with the system ground, leading to increased manufacturing costs and complexity.
- The inclusion of reactive components in the antenna design results in higher power consumption within the system.

2.3.12 PROTRUDING MONOPOLE ANTENNA WITH A SLOT CUT ON THE GROUND PLANE

A protruding monopole antenna [29] with a ground plane slot is proposed to enhance the bandwidth in the higher frequency 5 GHz band. It operates in the 2.4 GHz, 5.2 GHz, and 5.8 GHz bands of the 802.11b/a WLAN standards. The antenna protrudes from the system ground with a height of 2 mm, as illustrated in Figure 2.18. It is fabricated using a 0.2 mm thick brass material.

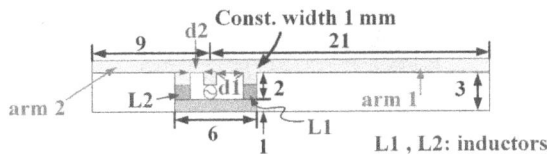

FIGURE 2.17 Photo of the constructed prototype made of a 0.8 mm thick FR-4 substrate and fed by 30 mm long coaxial cable.

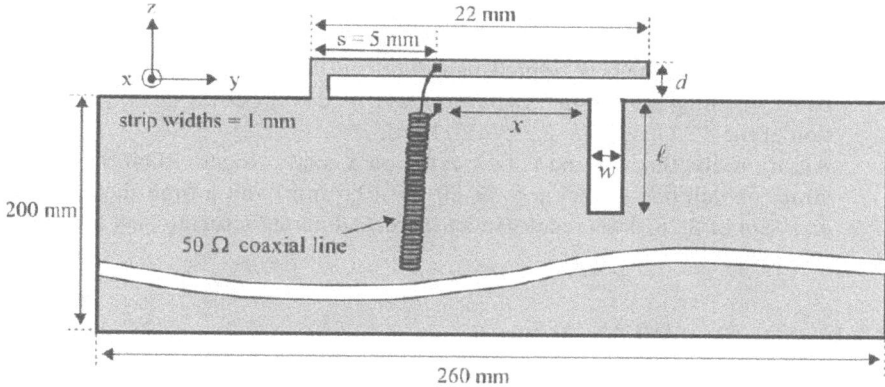

FIGURE 2.18 Geometry of a protruded monopole antenna with a slot on the system ground.

Inferences

- Replacing the system ground in case of non-operation of the antenna becomes challenging due to the presence of the slot on it.
- The manufacturing complexity is increased by creating a slot in the system ground.
- The presence of a slot in the ground plane may potentially impact the performance of other antennas within the laptop computer.

2.3.13 COPLANAR WAVEGUIDE-FED SLOT ANTENNA

The slot antenna with coplanar waveguide (CPW) feeding, proposed in [30] and shown in Figure 2.19, exhibits several promising features such as a compact size, uniplanar structure, favorable radiation characteristics, and minimal mutual coupling, making it suitable for MIMO applications in laptop computers. However, its bandwidth is insufficient to cover the entire 5 GHz band.

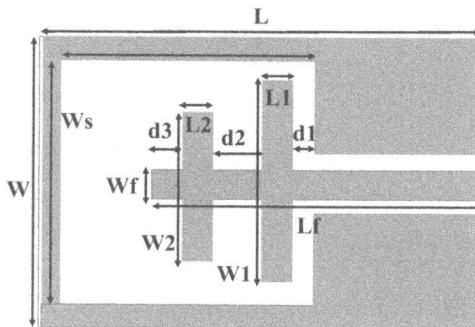

FIGURE 2.19 A -fed slot antenna printed on expensive Rogers RO4003 substrate.

Inferences

- The antenna exhibits a limited bandwidth in the high frequency 5 GHz band, spanning only from 5.1 GHz to 5.4 GHz, thereby covering only a portion of the 5.2 GHz (5.15–5.35 GHz) band.
- Additionally, the antenna is fabricated on a costly Rogers RO4003 substrate, which has a thickness of 20 mil (0.5 mm) and a high dielectric constant of 3.38. This results in an increased manufacturing cost for the antenna.

2.3.14   TRI-BAND WLAN ANTENNA

The proposed tri-band WLAN antenna, developed in [31], was initially fabricated as a flat plate measuring 18.7 mm × 32.5 mm. Subsequently, the flat plate was bent into a compact structure with dimensions of 3.7 mm × 7 mm × 32.5 mm. Figure 2.20 illustrates the design of the antenna, which operates in the 2.4/5.2/5.8 GHz frequency bands according to the 802.11b/a WLAN standards.

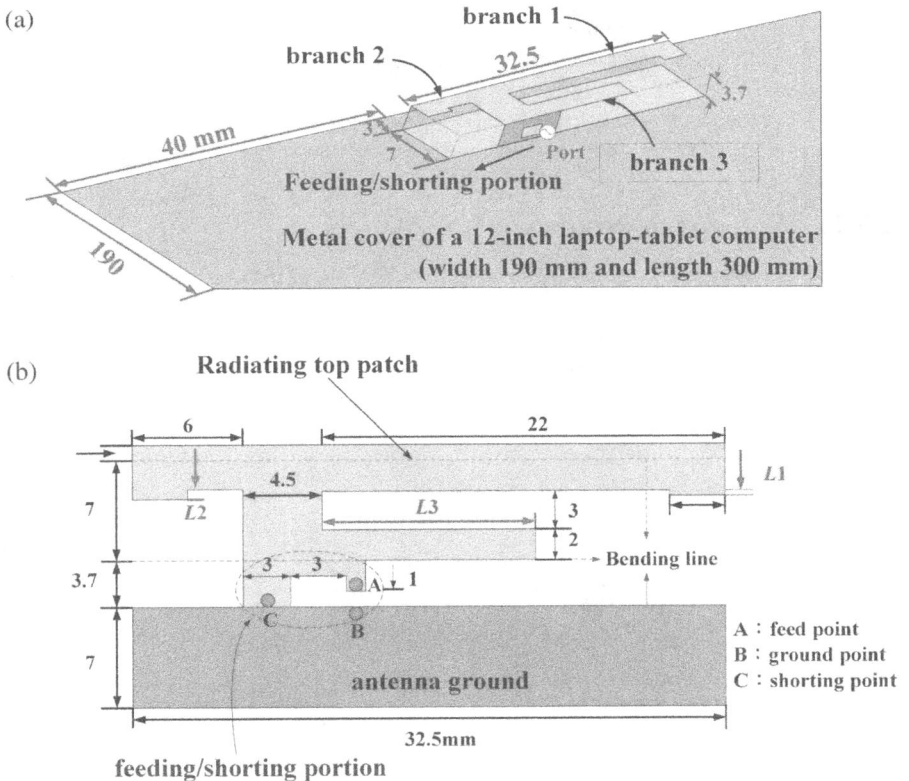

FIGURE 2.20   (a) Geometry of the antenna and (b) detail dimensions of the antenna.

Inferences

- The process of bending the flat plate to achieve antenna miniaturization introduces manufacturing complexity and potentially increases time to market.
- The antenna's thickness exceeds 3.5 mm, which may render it unsuitable for next-generation laptop computers.

2.3.15 OVERVIEW OF THE MIMO SYSTEM

MIMO is a technology used in wireless communication systems to enhance data throughput, improve link reliability, and increase spectral efficiency. It involves the use of multiple antennas at both the transmitter and receiver to transmit and receive multiple data streams simultaneously over the same frequency band.

In a MIMO system, multiple antennas are used to create independent communication paths, known as spatial streams. These spatial streams can be transmitted concurrently, enabling the system to transmit and receive more data in the same amount of time compared to traditional single input single output (SISO) systems.

The key benefits of MIMO technology include:

- **Increased data throughput:** By utilizing multiple antennas, MIMO systems can achieve higher data rates compared to SISO systems. The spatial streams can be combined at the receiver to improve the overall data transmission rate.
- **Improved link reliability:** MIMO systems can mitigate the effects of fading and interference by exploiting the spatial diversity provided by multiple antennas. This improves the link reliability and enhances the system's resistance to signal degradation.
- **Enhanced range and coverage:** MIMO systems can improve the coverage area and range of wireless networks by using spatial multiplexing and beamforming techniques. These techniques allow for better signal propagation and penetration through obstacles.
- **Spectral efficiency:** MIMO systems achieve higher spectral efficiency by transmitting multiple data streams simultaneously within the same frequency band. This results in more efficient use of the available frequency spectrum.

MIMO technology is widely used in various wireless communication standards, including Wi-Fi (IEEE 802.11n, 802.11ac, 802.11ax), long-term evolution (LTE), and 5G. It plays a crucial role in meeting the increasing demand for higher data rates, improved network performance, and seamless connectivity in modern wireless communication systems.

MIMO technology offers a high data-transfer rate and channel capacity that is proportional to the number of antennas, as theoretically explained [32]. This concept, known as space-time technology, utilizes the spatial diversity of multiple antenna elements. However, due to the complex physical structure and diversity requirements, these approaches often involve a large volume of antennas. As a result, size reduction has become an increasingly critical factor in communication systems. In densely populated urban environments with numerous scatterers, the installation of multiple antennas in limited spaces necessitates high isolation to suppress interference. However, when antennas are placed close to each other, there can be coupling effects between them. This coupling can cause unwanted signal leakage and crosstalk, which negatively impacts the system's performance. Isolation techniques, such as antenna decoupling and isolation structures, have already been documented in the literature. The following provides an overview of some commonly employed isolation techniques.

2.3.15.1 Overview of Isolation Techniques Implemented in MIMO Antenna Array for Wireless Operations in the Laptop Computer

2.3.15.1.1 Protruded Ground Plane and a Spiral Open Slot

The isolation technique, as proposed in [33], involves a protruded ground plane and a spiral open slot positioned between two antennas, as depicted in Figure 2.21. Each antenna consists of a driven strip and a shorted strip. Antenna 2 is a mirror image of Antenna 1. These MIMO antennas operate within the 2.4/5.2/5.8 GHz frequency bands, compliant with the IEEE 802.11b/a WLAN standard.

Inferences

- The drawback of this system is that the insertion of isolation between two antennas results in an increased length of the antenna array.
- The dimensions of the MIMO antenna are 55×9 mm^2, which is considered large for next-generation laptop computers.

2.3.15.1.2 "T-shaped" Monopole Slot Acts as an Isolator between Two Antennas

The antenna proposed in [34] consists of two folded monopole slot antenna modules and an isolated element, designed specifically for MIMO applications. Each

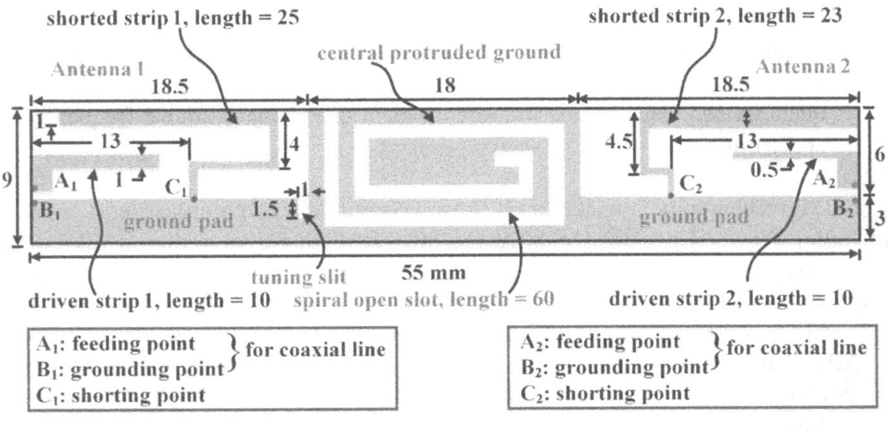

FIGURE 2.21 WLAN MIMO laptop computer antenna array.

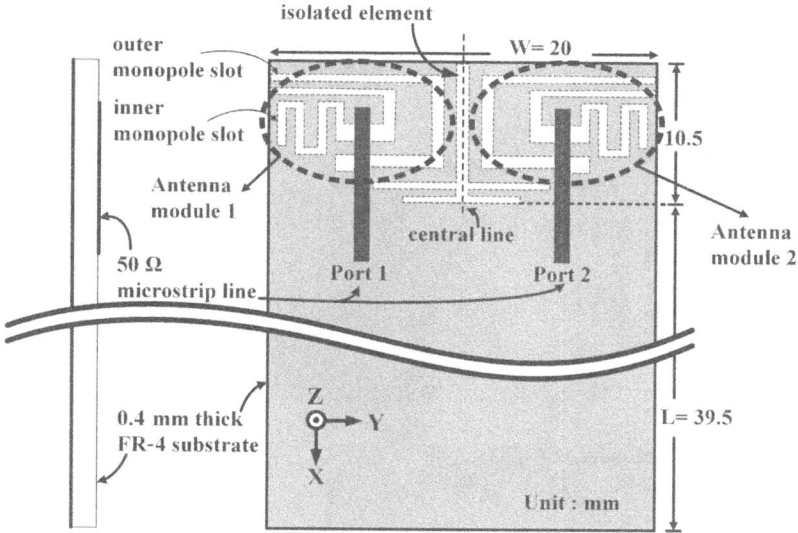

FIGURE 2.22 Geometry of the proposed monopole slot MIMO antenna design for WLAN application.

monopole slot antenna module is comprised of two monopole slots with varying lengths, allowing for two resonant modes, as illustrated in Figure 2.22. The inner monopole slot antenna is encompassed by the outer monopole slot antenna, establishing an isolation mode. Additionally, the isolated element placed between the antenna modules generates another isolation mode. These two isolation modes contribute to an isolation level exceeding 17.8 dB across the operating band.

Inferences

- The level of isolation is relatively low, and to ensure improved efficiency and optimal performance of the MIMO array, it is always recommended to have isolation between two antennas exceeding -20 dB.
- The overall height of the FR-4 substrate is approximately 50 mm, making it unsuitable for next-generation laptop computers.

2.3.15.1.3 Complementary Split-Ring Resonator

The antenna design presented in [35] is specifically intended for MIMO operations in laptop computers, conforming to the 2.4/5 GHz frequency bands of the 802.11b/a WLAN standard. The design incorporates a complementary split-ring resonator (CSRR) as an isolating element between the antennas. The antenna configuration is illustrated in Figure 2.23.

The proposed antenna is printed on a premium substrate called Rogers 4003, which possesses a dielectric constant of 3.55. The dimensions of the antenna are 70 mm × 90 mm^2, and it features a non-metallic bottom layer. However, utilizing the Rogers 4003 substrate increases the overall cost of the antenna design.

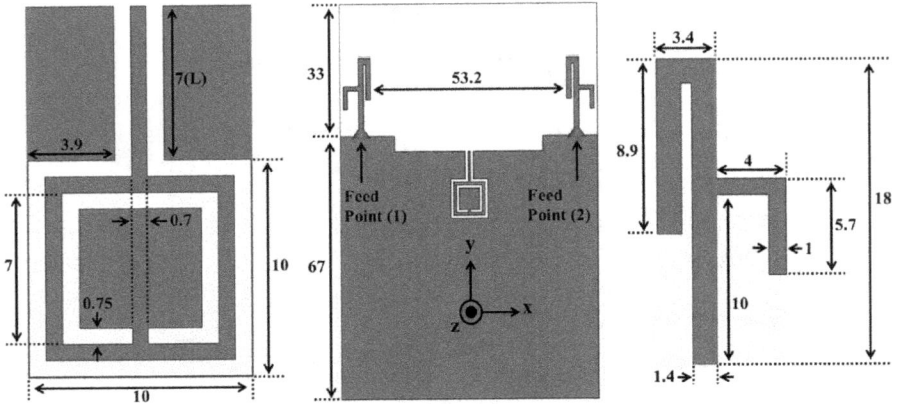

FIGURE 2.23 Dual-band MIMO antenna.

Inferences

- The antenna is printed on a high-cost substrate.
- The dimensions of the MIMO antenna are 70×90 mm^2, which make it unsuitable for ultra-thin laptop computers.

2.3.15.1.4 Distance between Two Antennas Acting as an Isolating Element

The [36] paper introduces a concept for integrating a MIMO antenna array into a laptop computer, utilizing the display ground plane, as depicted in Figure 2.24. The proposed MIMO antenna array comprises two identical WLAN antennas. These antennas have compact dimensions, measuring 80 mm × 2 mm, and are etched onto the display ground plane. They are positioned with a distance of 5 mm from the top edge. To enhance the isolation between the two antennas, there is an 80 mm spacing maintained between them.

Inferences

- The antennas are etched onto the system ground, positioned 5 mm from the top edge, which presents challenges in terms of antenna replacement in case of any issues.

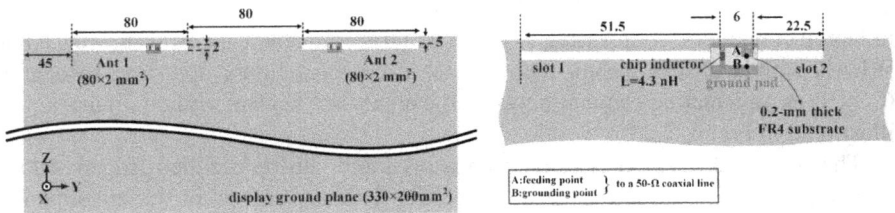

FIGURE 2.24 Geometry of the WLAN MIMO.

- Since the antennas are etched directly onto the system ground plane, there is a potential risk of affecting the performance of other antennas mounted within the laptop computer.

2.4   POSSIBLE OUTCOMES FROM SECTION 2.3

The objectives of this book are established in accordance with the findings presented in Section 2.3 and its Subsection 2.3.1. When designing an effective multiband antenna for wireless operations in laptop computers, it is crucial to consider the following significant design factors:

- **To design a miniaturized multiband antenna for laptop computers:**
  - Integrate the IEEE802.11b/a/g/n/ac/ax WLAN standards into a dedicated single antenna.
  - Decrease the size of the multiband antenna while maintaining key design considerations, such as resonant frequency, impedance bandwidth, radiation patterns, gain, and size.
  - Improve the operating bandwidth without the need for vias, reactive elements, additional matching circuits, or ground plane modifications.
  - Utilize cost-effective substrates or PCBs for antenna design, as the substrate often represents the most expensive component.
  - Ensure the antenna design meets requirements for small or microminiaturized size, low- or ultra-low-profile characteristics, wide operating bandwidth, excellent RF performance, pure polarization, low cost, ease of manufacturing through simple geometry, multiple functionalities, and increased design flexibility.
- **To design an ultra-thin triple-band antenna for next-generation laptop computers:**
  - The antenna should possess an extremely low profile and be ultra-thin to minimize interaction with other electronic components.
  - It is important for the antenna to encompass all emerging and upcoming WLAN and 5G standards, ensuring interference-free operation and supporting high data rates.
  - The routing of the coaxial cable within the laptop computer should be optimized to minimize cable loss and enhance antenna performance.
  - The aim is to cover the maximum number of bands in a single design, thereby reducing the number of antennas required inside the laptop computers.
- **To design the performance-enhanced antenna for MIMO application in the laptop computer:**
  - Enhance the isolation between two antennas without the need for additional isolating elements.
  - Improve data rates and speed to meet the requirements of future communication services.
  - Strategically place multiple antennas to form an array for MIMO applications, optimizing performance.
  - Provide sufficient impedance bandwidth to cover the operating bands.

- Maintain consistent directional or omnidirectional radiation patterns.
- Ensure stable and consistent gain, as well as excellent efficiency across the operating bands.
- Keep design complexity low.
- Achieve low material and manufacturing costs while expediting time to market.

2.5 DESIGN CONSIDERATIONS OF INTEGRATED LAPTOP ANTENNAS

Based on the findings presented in Section 2.3, Chapters 3, 4 and 5 focus on the design of multiband antennas and antenna arrays for MIMO applications in wireless operations for laptop computers. The following features highlight the performance enhancements of these antennas for laptop applications:

- The micro-miniaturized antenna supports all WLAN standards and Bluetooth without the need for lumped elements, matching circuits, or auxiliary grounds. It covers both WLAN bands with wider bandwidth and can be easily installed inside the laptop computer.
- An ultra-thin antenna integrates WLAN, Bluetooth, and 5.8 GHz WiMAX bands, providing wider bandwidth for wireless operations in the laptop computer. It is designed to be ultra-thin, easily integrated without additional substrate or foam, and enhances performance while allowing flexibility for laptop manufacturers.
- An ultra-thin, very low-profile antenna is designed to support the sub-6 GHz 5G band and the 802.11a/ac band in the laptop computer. The antenna design is simple, offers faster time to market, and is cost-effective. It is a suitable solution for emerging 2-in-1s or convertible laptops.
- A novel triple-band antenna integrates WLAN standards, Bluetooth, LTE 2300, and upcoming WiMAX standards (802.16 and 802.16e). This is achieved using only radiating strips without the need for expensive substrates. The operating bands can be individually tuned without the use of diodes.
- An ultra-thin, very low-profile antenna integrates 802.11a and 802.11ac/HiperLAN 2 services into a single antenna for wireless operations in laptop computers, eliminating the need for additional hardware or reactive components.
- The antenna array designed for MIMO operations enables high-speed data transfer without the use of isolating elements. It is compact and efficient, utilizing the available space on the system ground to accommodate multiple antennas.

REFERENCES

1. Chen, Z. N. & Chia, M. Y. W. *Broadband Planar Antennas-Design and Applications*, John Wiley and Sons, England, 2006.
2. Honda, S., Ito, M., Seki, H. & Jinbo, Y. "A disk monopole antenna with 1:8 impedance bandwidth and omnidirectional radiation pattern." *Proceedings of the International Symposium of Antennas and Propagation*, Sapporo, Japan, pp. 1145–1148, 1992.

3. Hammoud, M., Poey, P. & Colombel, F. "Matching the input impedance of a broadband disc monopole." *Electronics Letters*, vol. 29, no. 4, pp. 406–407, 1993.

4. Chen, Z. N. & Chia, M. Y. W. "Impedance characteristics of EMC triangular planar monopoles." *Electronics Letters*, vol. 37, no. 21, pp. 1271–1272, 2001.

5. Dubost, G. & Zisler, S. *Antennas A Large Bande: Theories et Applications*, Paris, Mason, 1976.

6. Agrawal, N. P., Kumar, G. & Ray, K. P. "Wide-band planar monopole antenna." *IEEE Transactions on Antennas and Propagation*, vol. 46, no. 2, pp. 294–295, 1998.

7. Ammann, M. J. & Chen, Z. N. "A wideband shorted planar monopole with bevel." *IEEE Transactions on Antennas and Propagation*, vol. 51, no. 4, pp. 901–903, 2003.

8. Chen, Z. N., Ammann, M. Z., Chia, M. Y. W. & See, S. P. "Circular annular planar monopoles with EM coupling." *IEE Proceedings: Microwave, Antennas and Propagation*, vol. 150, no. 4, pp. 269–273, 2003.

9. Chen, Z. N., Ammann, M. J., Chia, M. Y. W. & See, S. P. "Broadband monopole antenna with parasitic planar element." *Microwave and Optical Technology Letters*, vol. 27, no. 3, pp. 209–210, 2000.

10. King, R. W. P. "Asymmetric driven antenna and the sleeve dipole." *Proceedings of the IRE*, vol. 38, no. 12, pp. 1154–1164, 1950.

11. Chen, Z. N., Hirasawa, K. & Wu, K. "A novel top-sleeve monopole in two parallel plates." *IEEE Transactions on Antennas and Propagation*, vol. 49, no. 3, pp. 438–443, 2001.

12. Chen, Z. N. "Broadband roll monopole." *IEEE Transactions on Antennas and Propagation*, vol. 51, no. 11, pp. 3175–3177, 2003.

13. Chen, Z. N. "Broadband planar monopole antenna." *IEEE Proceedings: Microwave, Antennas and Propagation*, vol. 147, no. 6, pp. 526–528, 2000.

14. Chen, Z. N. & Chia, M. Y. W. "Impedance characteristics of trapezoidal planar monopole antenna." *Microwave and Optical Technology Letters*, vol. 27, no. 2, pp. 120–122, 2000.

15. Su, S.-W. & Chang, F.-S. "A bent, shorted, planar monopole antenna for 2.4 GHz WLAN applications." *Microwave and Optical Technology Letters*, vol. 51, no. 2, pp. 455–457, 2009.

16. Liu, D., Gaucher, B. P., Flint, E. B., Studwell, T. W., Usui, H. & Beukema, T. J. "Developing integrated antenna subsystem for laptop computers." *IBM Journal of Research and Development*, vol. 47, no. 2.3, pp. 355–367, 2003.

17. Fujio, S. & Asano, T. "Dual band coupled floating element PCB antenna." *Proceedings of the IEEE Antennas and Propagation Society International Symposium*, Monterey, CA, vol. 3, pp. 2599–2602, 2004.

18. Collins, B. S., Nahar, V., Kingsley, S. P. & Zhang, S. Q. "A dual-band hybrid dielectric antenna for laptop computers." *Proceedings of the IEEE Antennas and Propagation Society International Symposium*, Monterey, CA, vol. 3, pp. 2619–2622, 2004.

19. Karikomi, M. "Parasitic element excitation of a dual-frequency printed dipole antenna." *Proceedings of the Electronic Information Communication Society National Conference*, B-73, 1989.

20. Ikegaya, M., Sugiyama, T. & Tate, H. "Dual band film type antenna for mobile devices." *Proceedings of the North American Radio Science Meeting*, AP/URSI B Session 133.8, Columbus, OH, 2003.

21. Ikegaya, M., Sugiyama, T., Takaba, S., Suzuki, S., Komagine, R. & Tate, H. "Development of film type antenna for mobile devices." *Hitachi Cable Review*, no. 21, 2002.

22. Chou, L.-C., Wong, K.-L. & Kuo, C.-H. "A small size internal dual-band metal-strip antenna for 2.4/5GHz WLAN operation in the laptop computer." *2008 IEEE Antennas and Propagation Society International Symposium*, San Diego, CA, pp. 1–4, 2008.

23. Wong, K.-L., Chou, L.-C. & Su, C.-M. "Dual-band flat-plate antenna with a shorted parasitic element for laptop application." *IEEE Transactions on Antennas and Propagation*, vol. 53, no. 1, pp. 539–544, 2005.

24. Liu, H.-W., Lin, S.-Y. & Yang, C.-F. "Compact inverted-F antenna with meander shorting strip for laptop computer WLAN applications." *IEEE Antennas and Wireless Propagation Letters*, vol. 10, pp. 540–543, 2011.

25. Kang, T.-W. & Wong, K.-L. "Very small size printed monopole with embedded chip inductor for 2.4/5.2/5.8 GHz WLAN laptop computer', antenna." *Microwave and Optical Technology Letters*, vol. 52, no. 1, pp. 171–177, 2010.

26. Yeo, J., Lee, Y. J. & Mittra, R. "A novel dual-band WLAN antennas for notebook platforms." *IEEE Antennas and Propagation Society Symposium*, vol. 2, pp. 1439–1442, 2004.

27. Abioghli, M. "Dual-band two layered printed antenna for 2.4/5GHz WLAN operation in the laptop computer." *IEICE Electronics Express*, vol. 8, no. 18, pp. 1519–1526, 2011.

28. Su, S.-W. "Very low profile, 2.4/5 GHz WLAN monopole antenna for large screen-to-body-ratio notebook computers." *Microwave and Optical Technology Letters*, vol. 60, no. 5, pp. 1313–1318, 2018.

29. Huang, C.-Y., Tsai, C.-C. & Yang, C.-F. "Low protruding monopole antenna with a slot cut in the ground plane for laptop applications." *Microwave and Optical Technology Letters*, vol. 52, no. 11, pp. 2610–2613, 2010.

30. Naser-Moghadasi, M., Sadeghi Fakhr, R. & Danideh, A. "CPW-FED compact slot antenna for WLAN operation in a laptop computer." *Microwave and Optical Technology Letters*, vol. 52, no. 6, pp. 1280–1282, 2010.

31. Lee, C.-T. & Su, S.-W. "Very-low-profile, stand-alone, tri-band WLAN antenna design for laptop-tablet computer with complete metal-backed cover." *IEEE APCAP*, pp. 189–190, 2016.

32. Nezhad, S. M. A. & Hassani, H. R. "A novel triband E-shaped printed monopole antenna for MIMO application." *IEEE Antennas and Wireless Propagation Letters*, vol. 9, pp. 576–579, 2010.

33. Jiang, H.-J., Kao, Y.-C. & Wong, K.-L. "High-Isolation WLAN MIMO laptop computer antenna array." *Proceedings of APMC*, Kaohsiung, Taiwan, 2012.

34. Chen, W.-S., Lin, C.-H., Lee, B.-Y., Hsu, W.-H. & Chang, F.-S. "Monopole slot antenna design for WLAN MIMO application." *Microwave and Optical Technology Letters*, vol. 54, no. 4, pp. 1103–1107, 2012.

35. Yang, D.-G., Kim, D. O. & Kim, C.-Y. "Design of dual-band MIMO monopole antenna with high isolation using slotted CSRR for WLAN." *Microwave and Optical Technology Letters*, vol. 56, no. 10, pp. 2252–2257, 2014.

36. Chen, S.-C. & Fu, C.-S. "MIMO antenna array for wireless local area network in a laptop computer." *2016 IEEE 5th Asia-Pacific Conference on Antennas and Propagation (APCAP)*, pp. 245–246, 2016.

3 Miniaturized Antenna Designs for Laptop Computers

3.1 IMPORTANCE OF MINIATURIZATION OF ANTENNA

In the era of increasingly compact, sleek, high-speed, and aesthetically appealing laptops, manufacturers are incorporating advanced technologies into these devices. Consequently, laptops have become an integral part of our daily lives, akin to mobile handsets. Modern laptops, equipped with Bluetooth, wireless local area network (WLAN), and Worldwide Interoperability for Microwave Access (WiMAX) standards, seamlessly integrate various services such as communication, video telephony, security, healthcare, entertainment, and medical treatment, among others.

To meet the demands of today's users, laptop computers should possess small form factors, lightweight designs, and cost-effectiveness. Consequently, significant efforts have been dedicated to miniaturizing the antennas employed in these devices, with the objective of reducing overall size and weight. Furthermore, it is essential for laptop antennas to be seamlessly integrated into the devices, rendering them practically invisible to users and viewers alike.

One of the major challenges in antenna design for laptops lies in achieving minimal thickness and height above the system ground. To address this challenge, our proposed monopole antennas have been meticulously engineered to fit seamlessly within the thin bezels of next-generation laptops, ensuring an aesthetically pleasing visual experience.

By addressing these design considerations, we aim to revolutionize the concept of laptop computers, offering users an immersive and seamless computing experience without compromising on performance, style, or functionality.

3.2 EXISTING STATE-OF-THE-ART TECHNOLOGY

Extensive global research has been conducted on antenna miniaturization, exploring novel structures and substrate variations.

One such study proposed a flat plate Inverted-F (IFA) antenna [1-2] to cover Wi-Fi and WiMAX bands in laptops, with dimensions of 46×9.3 mm^2. However, the height of 9.3 mm does not align with the sleek and portable design of modern laptops. Another study [3] introduced a meandered IFA with three reduced-sized vias (35×3 mm^2), but the antenna structure proved to be more complex and expensive. In [4], an IFA was proposed with dimensions of $48 \times 10.2 \times 1.6$ mm^3, utilizing a Taconic substrate, which added to its cost.

DOI: 10.1201/9781003331018-3

The antenna in [5] proposed a stamped brass sheet IFA integrated into the lap-top display screen's metal support frame. While [6] introduced a branched IFA printed on a single-sized printed circuit board (PCB), operating in the 2.4 GHz band. However, both of these antennas did not cover the entire 5GHz WLAN bands.

Planar printed antennas, particularly coupled-fed PIFA designs, have shown promise for laptop applications. Recent reports [7, 8] showcased antennas with dimensions of 13×9 mm^2 and 7×9 mm^2, respectively.

In [9], a branched monopole antenna with dimensions of 27×6 mm^2 was proposed, but placing the antenna alongside hinges could potentially impact its performance when placed on a person's lap. [10] presented a dual-band monopole antenna measuring 20×6 mm^2, positioned at the center of the system ground's top edge. Similarly, [11] and [12] introduced dual-band two-layered printed anten-nas, with dimensions of 5×12 mm^2, but there is a debate regarding mounting the antenna at the center position of the system ground due to the reserved space for the camera.

The antenna in [13] introduced a flat plate IFA with dimensions of 7×45 mm^2, but it required an additional ground plane, increasing its overall dimensions to 32×45 mm^2. Furthermore, the use of expensive Roger RT5800 substrate limited its practicality. [14] adopted a similar concept with an additional ground plane, but used adhesive material for attachment, which had limitations in terms of detachment and lifespan.

Antennas utilizing meander techniques [15, 16] had sizes of 10×19.5 mm^2 and 10×6 mm^2, respectively. The broadband uniplanar antenna proposed by [17] operated in the 2.4/5.2/5.8 GHz WLAN high-frequency bands. A new dual-band antenna printed on both sides of a PCB was introduced by [18], operating in 802.11b, Bluetooth, and 802.11a WLAN standards, with an antenna area of 13×10.2 mm^2. Additionally, [19] presented a dual-band printed microstrip-fed monopole antenna with a broader bandwidth of 120 MHz and coverage across the 2.4 GHz and 4.5–7.5 GHz bands, with dimensions of 16×29 mm^2. A meandered monopole antenna measuring 15×12 mm^2 is proposed in [20]. However, all these antennas had a minimum height of 10 mm above the system ground, posing chal-lenges for mounting in modern laptops.

Uniplanar antennas [21–23] achieved reflection coefficients of -6 dB, which are not suitable for wireless operations in laptop computers. Furthermore, these anten-nas did not cover the 5 GHz high-frequency band of 802.11a. A very small uniplanar printed monopole antenna ($8.7 \times 5.4 \times 0.8$ mm^3) [21] is proposed for wireless opera-tions in laptops, incorporating an open-ring resonator and hook-shaped radiating element. However, the effective height of the antenna above the system ground was 8.7 mm.

A folded antenna design proposed in [24] measured 7×32.5 mm^2, with a mini-mum thickness of 3 mm. This bending technique added manufacturing complexity [25–27] and presented antennas with larger dimensions, making them unsuitable for modern laptops. The cloverleaf-shaped monopole antenna proposed in [28] had a large dimension of 110×60 mm^2, occupying considerable space when mounted inside laptops.

In [29], an internal composite monopole antenna, combining a ceramic chip radiating element with a printed radiating element is proposed. However, the ceramic chip increased fabrication costs and complexity. Monopole antennas in [30, 31], measuring 9×6 mm^2 and 4×30 mm^2 respectively, utilized additional reactive components for impedance matching purposes, leading to increased hardware complexity and power consumption.

All the aforementioned antennas had limitations, such as the use of additional hardware, expensive substrates, lumped elements, or larger dimensions.

In this chapter, we overcome these limitations by designing miniaturized and ultra-thin antennas without the need for expensive substrates, additional hardware, holes or vias, and lumped elements.

3.3 DESIGN OF A LOW-PROFILE, ULTRA-THIN INVERTED H-SHAPED MONOPOLE ANTENNA FOR SEAMLESS WIRELESS OPERATION IN LAPTOP COMPUTERS

This section introduces a cutting-edge H-shaped antenna, featuring an ultra-thin and extremely low-profile design for seamless wireless applications in laptop computers. This innovative antenna has been meticulously crafted without the reliance on expensive substrates, reactive components, or the need for additional hardware.

3.3.1 OVERVIEW OF PROPOSED ANTENNA

A novel and innovative design of an ultra-thin, very low-profile, inverted H-shaped monopole antenna specifically tailored for wireless applications in laptop computers compliant with IEEE 802.11a and IEEE 802.11ac/HiperLAN 2 standards is developed. This antenna boasts a remarkable thickness of only 0.2 mm, utilizing a pure copper strip, measuring 17.5 mm (length) $\times$ 4 mm (width). Its ultra-thin and low-profile structure renders it an exceptional choice for wireless operations in ultra-thin laptop computers.

The proposed antenna features a simple structure, comprising a single radiating strip, a rectangular stub, and two resonating slots referred to as "X" and "Y" with lengths of 7.5 mm and 7 mm, respectively. This design achieves resonance around 5.5 GHz, effectively covering the 5.15–5.35/5.725–5.825 GHz bands of IEEE 802.11a and the 5.15–5.35/5.470–5.725/5.725–5.925 GHz bands of IEEE 802.11ac/HiperLAN 2, making it ideal for wireless operations in laptop computers.

Prototype testing of the fabricated antenna confirms its excellent performance, demonstrating a measured −10 dB impedance bandwidth of 15% 5.10–5.92 GHz, with a voltage standing wave ratio (VSWR) of less than 2 within the operating bands. The radiation patterns exhibit nearly omnidirectional characteristics, accompanied by a stable gain of 5 dBi. Furthermore, the antenna showcases exceptional radiation

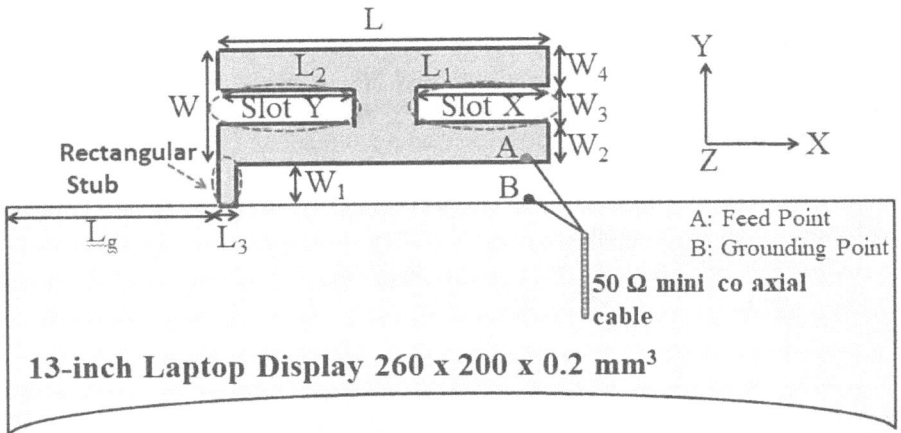

FIGURE 3.1 Schematic geometry of the proposed antenna.

efficiency of approximately 90% across the operating bands. A comparison between simulated and measured results shows good agreement, affirming the suitability of the antenna structure for IEEE 802.11a and IEEE 802.11ac/HiperLAN 2 applications in ultra-thin laptop computers.

3.3.2 PROPOSED ANTENNA GEOMETRY

The proposed antenna's schematic geometry is illustrated in Figure 3.1, showcasing a unique and innovative design. It comprises a single radiating strip, a rectangular stub, and two resonating slots denoted as "X" and "Y." To construct the antenna, we utilize a pure copper material with a thickness of 0.2 mm, which is precisely mounted on the system ground using a rectangular stub with a width denoted as "L_3" and positioned at a distance represented by "L_g." The system ground itself is composed of 91% brass material and supports a 13-inch laptop display measuring $260 \times 200 \times 0.2$ mm³. The selection of a 13-inch laptop display size is based on achieving the optimal balance between device portability and usability.

To facilitate the efficient transmission of signals, the antenna employs a small and low-loss 50 Ω miniature coaxial feed line. The central conductor of the feed line is connected to point "A" on the lower right edge of the H-shaped radiating strip, while the grounding sheath is soldered at point "B" on the system ground. This feeding configuration ensures that the effective dielectric constant of all radiating strips remains equal to 1, enhancing the overall antenna performance.

In the subsequent sections, we will delve into the detailed design of the antenna, covering various crucial aspects to optimize its functionality and effectiveness.

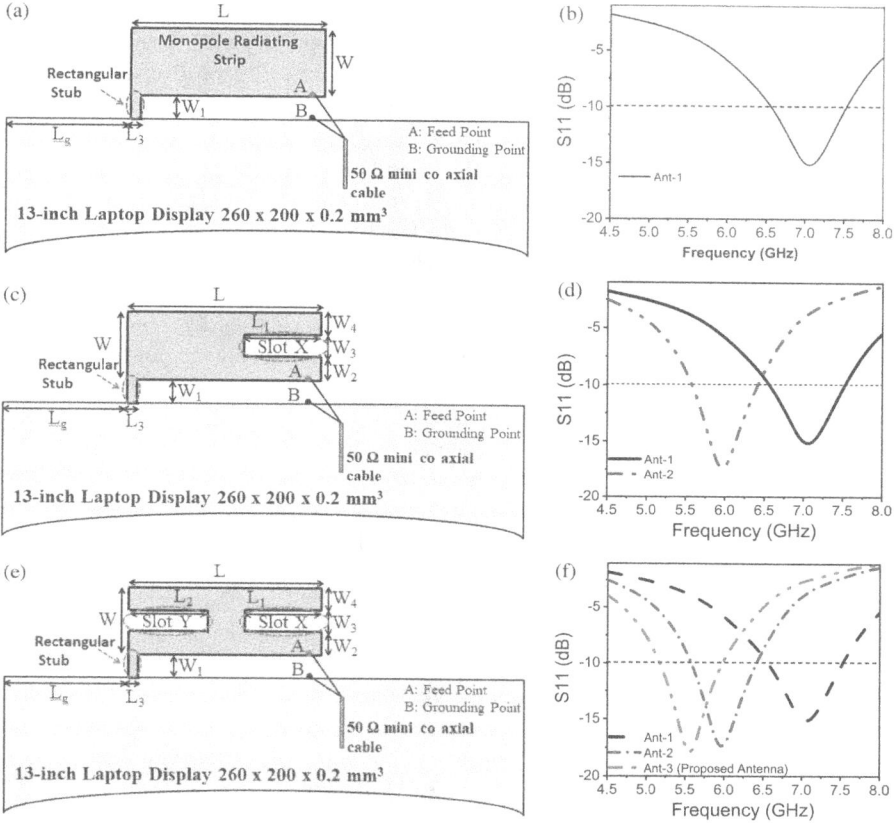

FIGURE 3.2 The step-by-step design process and the corresponding S11 of proposed antenna (dB) results at different stages. (a) Structure of Ant. 1, (b) S_{11} (dB) of Ant. 1, (c) structure of Ant. 2, (d) S_{11} (dB) of Ant. 2 compared with S_{11} (dB) of Ant. 1, (e) structure of Ant. 3, and (f) S_{11} (dB) of Ant. 3 (proposed antenna) compared with S11 (dB) of Ant. 1 and Ant. 2.

3.3.3 STEP-BY-STEP EVOLUTION OF ANTENNA DESIGN PROCESS

To provide a clear understanding of the design process for the H-shaped monopole antenna, Figure 3.2 illustrates the step-by-step evolution of the antenna design and the corresponding simulated reflection coefficient S_{11} (dB) results using Computer Simulation Technology (CST) Microwave Studio (MWS) at different stages. Initially, the antenna, denoted as Ant. 1, was formed by a simple rectangular radiation strip with length "L" and width "W," as depicted in Figure 3.2(a). The length "L" was chosen to be approximately one-quarter wavelength, expected to generate resonance at a frequency of 5.5 GHz, as calculated using Equation (3.1)

$$L = \frac{\lambda}{4} = \frac{C}{4f}$$ (3.1)

In Equation (3.1), "c" represents the speed of light (m/s), and "f" is the resonating frequency. Ant. 1 also included a rectangular stub of size $L_3 \times W_1$ mm², serving as the connection point between the radiating strip and the system ground. However, despite the length (L) of the radiating strip being nearly a quarter wavelength, it did not produce the desired operating resonance. Instead, a single band with a resonance frequency of 7 GHz was generated, as shown in Figure 3.2(b). This impedance mismatch at the resonance frequency of 5.5 GHz resulted from the significant inductive reactance produced by the rectangular stub. The impedance bandwidth of the generated band spanned the range of (6.5–7.5) GHz.

To address this issue, it was recognized that increasing the effective length of the slot resonator would shift the resonance to a lower frequency. The slot length acts as a capacitor, canceling out the additional inductive reactance produced by the antenna. Based on this concept, a resonating slot "X" of size $L_1 \times W_3$ mm² was added from the extreme right of the radiating strip at height W_2, forming Ant. 2, as shown in Figure 3.2(c). Ant. 2 shifted the resonance from 7 GHz to a lower frequency, up to 5.95 GHz, with the impedance bandwidth spanning the range of 5.55–6.45 GHz, as shown in Figure 3.2(d). Although the introduction of slot "X" did not generate the desired resonance, it successfully shifted the band closer to the desired frequency.

Finally, another slot "Y" of size $L_2 \times W_3$ mm² was added to Ant. 2, as depicted in Figure 3.2(e), forming Ant. 3. The inclusion of slot Y further improved impedance balancing by increasing the electrical length of the resonator. As a result, the desired band was obtained, with the resonance shifting from 5.95 GHz to 5.5 GHz, and the impedance bandwidth spanning the range of 5.12–5.92 GHz, as shown in Figure 3.2(f). Therefore, as Ant. 3 achieved the desired band, it was selected for further testing and referred to as the proposed antenna in subsequent sections. The proposed antenna structure operates from 5.12 GHz to 5.92 GHz, covering the IEEE 802.11a and 802.11ac/HiperLAN 2 standards for wireless operations.

3.3.3.1 Impedance Matching of the Proposed Antenna

To provide a more detailed explanation of the antenna design, Figure 3.3 illustrates the input impedance Z_{in} versus the frequency curve. The graph clearly

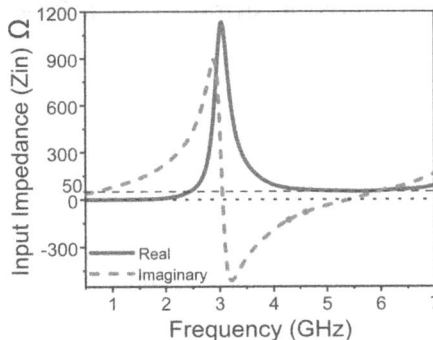

FIGURE 3.3   Input impedance Z_{in} versus the frequency of the proposed antenna.

(a)

L₁ = 0.15 nH
L$_1$ = 0.15 nH

R$_{A1}$ = 40.52 Ω
L$_{A1}$ = 0.58 nH
C$_{A1}$ = 1.5 pF

L$_{s1}$ = 0.178 nH C$_{s1}$ = 64.99 pF

L$_{s2}$ = 0.178 nH
C$_{s2}$ = 64.99 pF

Port 1

C₁ = 0.025 pF

R = 50 Ω

(b)

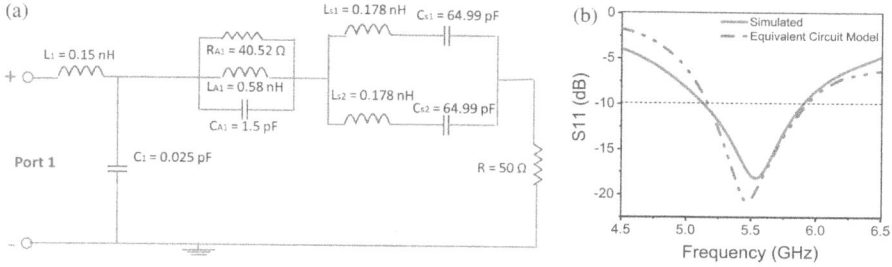

FIGURE 3.4 Schematic of equivalent circuit model and corresponding reflection coefficient S$_{11}$ (dB) of the proposed antenna, (a) e EC model and (b) simulated S$_{11}$ (dB) using.

demonstrates that within the operating bands, the input impedance remains stable at around 50 Ω. The reactance varies between −20 Ω and 38 Ω, reaching precisely 0 Ω at the resonance frequency of 5.5 GHz. This indicates that the proposed antenna achieves outstanding impedance matching at 5.5 GHz, enabling maximum power transmission.

3.3.3.2 Equivalent Circuit Model of the Proposed Antenna

To further validate the design, the input impedance of the proposed antenna was modeled as an (EC) using lumped elements. The circuit consists of a parallel RLC resonator connected in series with a parallel combination of series LC circuits. The monopole radiating strip with length "L" and feeding gap "W₁" forms a parallel electrical resonator circuit. The slot "X" with length "L₁" and another monopole radiating strip of length "L" are represented by a series LC circuit. The values of the circuit components were determined through manual adjustment and optimization techniques in Advanced Design System (ADS) software. Similarly, the slot "Y" of length "L₂" and the radiating strip of length "L" form another series LC circuit with values obtained through optimization techniques. The complete circuit, shown in Figure 3.4(a), is connected to the radiation resistance of the antenna as a 50 Ω load.

This EC model exhibits resonance around 5.45 GHz and operates within the frequency range of 5.13–5.94 GHz, as depicted in Figure 3.4(b). The values of the parallel RLC resonator can be computed using the following expressions [32]:

$$Q_0 = \frac{f}{BW} \tag{3.2}$$

$$Q_0 = 2\pi RC \tag{3.3}$$

$$f = \frac{1}{2\pi\sqrt{LC}} \tag{3.4}$$

In the above equations, the quality factor of the parallel RLC resonator is denoted by Q_0, while f represents the resonating frequency. R, L, and C correspond to the resistor, inductor, and capacitor of the parallel RLC circuit, respectively. The value of the resistor R is determined from Figure 3.3 and is measured to be 52.42 Ω within the operating bands of the proposed antenna.

Next, Q_0 is calculated, and the lumped element values at 5.5 GHz are computed using Equations (3.2), (3.3), and (3.4). These values are derived to construct the EC model, which is then simulated using Agilent ADS software. The corresponding reflection coefficient is depicted in Figure 3.4(b), showcasing the performance of the model.

3.3.3.3 Simulated Surface Current Distribution

The surface current distribution at 5.5 GHz is analyzed through simulation, and the results are depicted in Figure 3.5 for both 0° and 90° phases. It can be observed that the presence of slots X and Y effectively elongates the path of the surface current along the radiating strip, allowing for maximum current distribution across the H-shaped structure. Additionally, from Figure 3.5, it is evident that the proposed antenna radiates in both the 0° and 90° phases, indicating that it exhibits omnidirectional characteristics, as the same magnitude of surface current (measured in A/m) is observed for both phases.

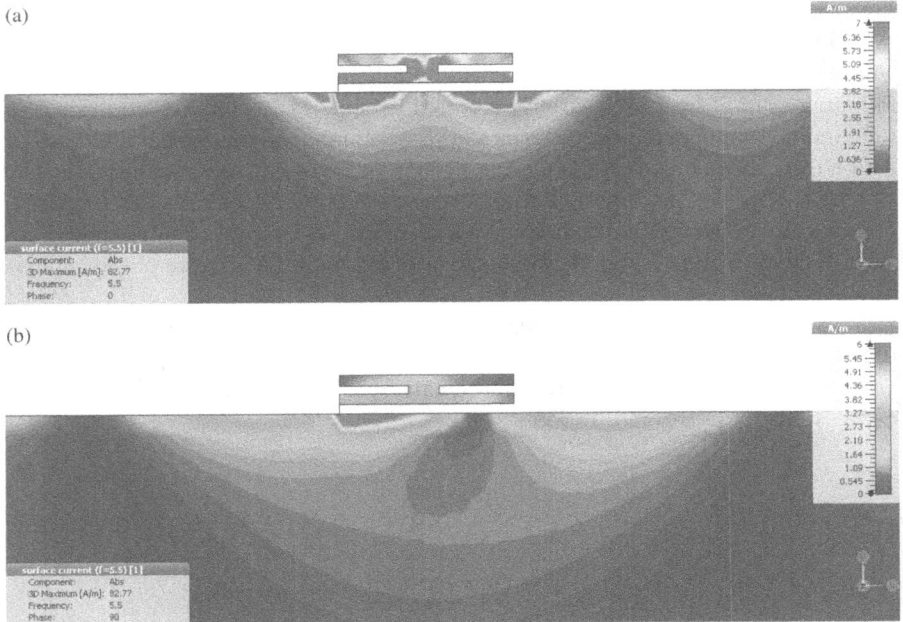

FIGURE 3.5 Simulated surface current (A/m) distribution of the proposed antenna (a) 0° phase and (b) 90° phase.

3.3.4 PARAMETRIC ANALYSIS OF THE PROPOSED ANTENNA

The mirror performance of the proposed antenna is highly sensitive to changes in its geometrical parameters, owing to its very low profile, ultra-thin structure, and compact size. To demonstrate the impact of these essential parameters on the antenna's operating bands, a comprehensive parametric analysis is conducted in this section. Specifically, the analysis focuses on the effective lengths, L_1 and L_2, of the resonating slots "X" and "Y," respectively, as well as the variation in the distance, L_g, for mounting the antenna on the system ground. Additionally, the analysis includes an assessment of the "Effect of varying the sizes of the system ground" on the reflection coefficient and impedance bandwidth across the antenna's operating bands, thereby evaluating its suitability for other portable devices.

3.3.4.1 Effect of Varying the Parameter L_1

To investigate the influence of parameter L_1 on the resonating frequency, a detailed parametric study is conducted on the performance of Ant. 2 (as depicted in Figure 4.15(c)), specifically within the frequency range of 5.55–6.45 GHz. The results, presented in Figure 3.6, clearly demonstrate the crucial role of the precise L_1 value in shifting the resonance from a higher frequency toward a lower frequency. This shift is achieved by utilizing the slot as a capacitance element to effectively balance the impedances. As L_1 increases from 0 mm to 7.5 mm, the reflection coefficient curves exhibit a smooth transition toward the lower frequency range. However, a further increase in L_1 negatively impacts the reflection coefficient, leading to a degradation in impedance matching. This effect becomes particularly evident when L_1 reaches 10 mm. Therefore, to achieve a broader impedance matching at higher frequencies, the optimum value of L_1 is determined to be 7.5 mm.

3.3.4.2 Effect of Varying the Parameter L_2

Figure 3.7 presents the simulated reflection coefficient curves of the proposed antenna as a function of L_2. It is observed that as L_2 increases from 0 mm to 7 mm, all the reflection coefficient curves undergo a downward shift toward lower frequencies. The desired resonance at 5.5 GHz, with a VSWR <2, is achieved when L_2 is set to 7 mm. Therefore, based on these findings, the optimum value of L_2 is determined to be 7 mm.

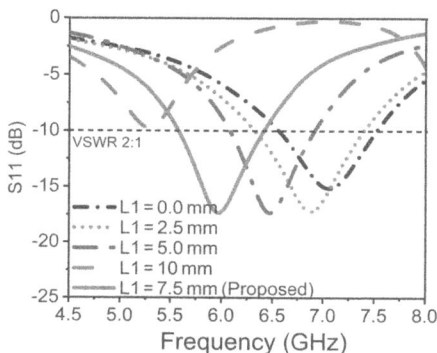

FIGURE 3.6 Simulated reflection coefficient as function of L_1.

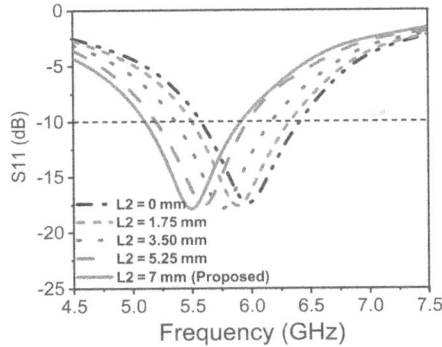

FIGURE 3.7 Simulated reflection coefficient as function of L_2.

3.3.4.3 Effect of Varying L_g for Mounting the Antenna on System Ground

The impact of varying L_g, which represents the mounting position of the proposed antenna on the system ground, is examined in Figure 3.8. The results indicate that the antenna's performance remains unaffected across the operating bands regardless of the mounting position on the system ground. In our specific design, L_g is selected as 40 mm from the left side, taking into consideration the typical placement of laptop components. The top left corner is usually allocated for integrating the laptop assembly, while the center space is commonly reserved for accommodating the camera. Hence, the optimal value of L_g for mounting the antenna is determined to be 40 mm.

3.3.4.4 Effect of Varying Sizes of the System Ground on the Performance of the Proposed Antenna

To investigate the impact of varying sizes of the system ground (display screen) on the proposed antenna's performance, several standard sizes, namely, "5-inch" (mobile), "10-inch" (tablet/notebook), and "14-inch" (laptop), are compared with the existing system ground. The results, as depicted in Figure 3.9, demonstrate that the

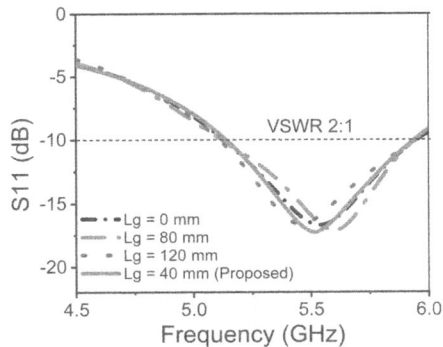

FIGURE 3.8 Simulated reflection coefficient as a function of L_g for mounting the antenna on the system ground.

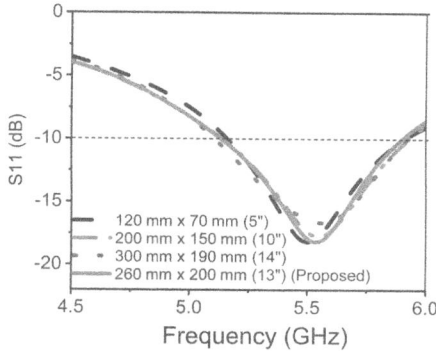

FIGURE 3.9 Simulated reflection coefficient for various sizes of system ground on the performance of the antenna.

proposed antenna maintains consistent performance across different system ground sizes. There is negligible effect on the impedance bandwidth and reflection coefficient, indicating that the antenna performs equally well regardless of the display screen size. Based on these findings, it can be concluded that the proposed antenna is a suitable choice for wireless operations in various portable devices beyond laptops.

Table 3.1 contains the parameters obtained from the above parametric study and other optimized parameters using CST MWS of the proposed antenna.

3.3.5 PERFORMANCE CHARACTERIZATION OF PROPOSED ANTENNA

The proposed antenna, designed with optimal dimensions listed in Table31, was fabricated, and subjected to rigorous testing, as illustrated in Figure 3.10. The measurement of reflection coefficient (S_{11}) in dB and VSWR was performed using a Rohde & Schwarz network analyzer with a frequency range of 9 KHz to 16 GHz. The radiation performance of the antenna, encompassing radiation patterns, gain, and radiation efficiency, was evaluated in an anechoic chamber that had been calibrated. The dimensions of the anechoic chamber were $8 \times 4 \times 4$ m³.

TABLE 3.1
Optimized Parameters Obtained from Parametric Study and Remaining Optimized Parameters of the Proposed Antenna

Optimized Parameters using Parametric Study		Optimized Parameters using CST MWS			
Parameters	Value (mm)	Parameters	Value (mm)	Parameters	Value (mm)
L_1	7.5	L	17.5	$W_2 = W_1$	1
L_2	7	L_3	0.2	W_3	0.8
L_g	40	W	4	W_4	1.2

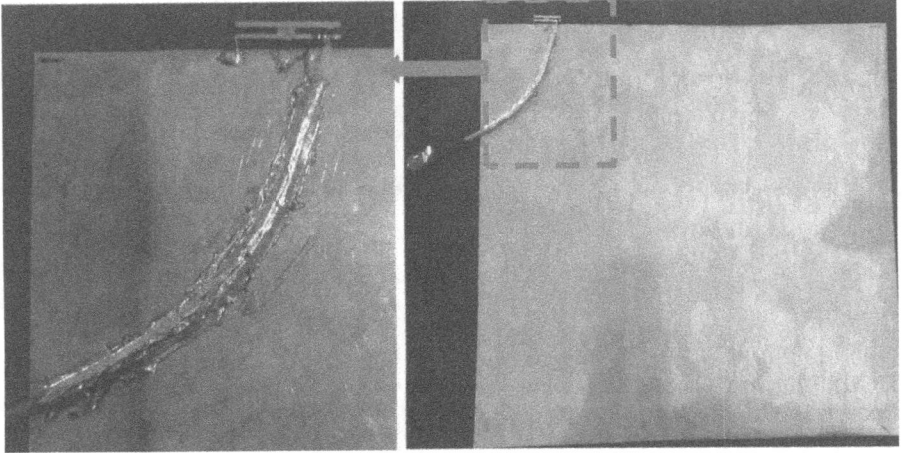

FIGURE 3.10   Fabricated photo of the proposed antenna.

3.3.5.1   Reflection Coefficient (S_{11}) dB

Figure 3.11 presents the S_{11} (dB) values for the simulated, measured, and EC model of the proposed antenna. The comparison between these values, as well as the operating bands, is summarized in Table 3.2. It is worth noting that there is a remarkable alignment among the simulated, measured, and EC model, all of which cover the necessary bandwidth across the entire 5 GHz spectrum with minimal deviation. Any slight deviation observed may be attributed to factors such as soldering or fabrication tolerances. Consequently, the proposed antenna effectively operates within the entire 5 GHz band, meeting the bandwidth requirements specified by IEEE 802.11a and 802.11ac/HiperLAN 2 standards for wireless operations in laptop computers.

3.3.5.2   VSWR

Figure 3.12 showcases the simulated and measured VSWR of the proposed antenna. The results indicate that both the simulated and measured VSWR values remain

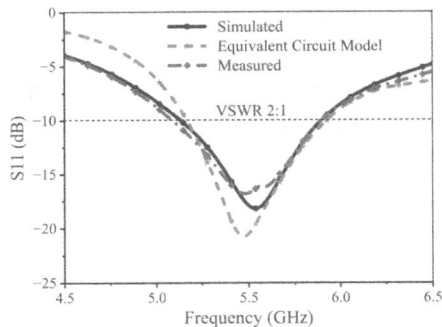

FIGURE 3.11   Simulated versus measured S_{11} (dB) of the proposed antenna.

TABLE 3.2

Comparison of Reflection Coefficient and Operating Bands of the Proposed Antenna

	Magnitude of S_{11} (dB)	Operating Bands (GHz)	Fractional Impedance Bandwidth (%)
Simulated	−18	5.12–5.92	14.49
Measured	−16.8	5.10–5.92	14.88
Equivalent circuit model	−21	5.13–5.94	14.63

below two (VSWR <2) across the operating bands of the antenna. This observation signifies that the impedance of the proposed antenna is accurately matched, allowing for efficient power transmission and reception.

3.3.5.3 Radiation Patterns of the Proposed Antenna

To verify the optimal radiation characteristics of the proposed antenna across the operating bands, both simulated and measured radiation patterns were analyzed and compared. Figure 3.13 presents the normalized radiation patterns of the proposed antenna in three principal planes, namely the x-y, y-z, and x-z planes, specifically at the resonating frequency of 5.5 GHz. The simulated and measured results exhibit a strong agreement, validating the antenna's performance.

During the measurement process, a double ridge horn antenna with a frequency range of 450 MHz to 8.50 GHz was employed as a reference antenna. The proposed antenna was positioned on a rotating mast, placed face-to-face with the reference antenna. By rotating the proposed antenna through 360 degrees with a step interval of 2 degrees, a larger number of measurement points were acquired, leading to enhanced pattern uniformity in the final plots.

At the resonating frequency of 5.5 GHz, the radiation patterns of the proposed antenna demonstrate near-omnidirectional coverage in all three principal planes.

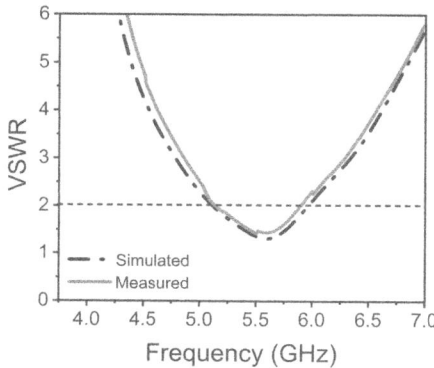

FIGURE 3.12   Simulated and measured VSWR of the proposed antenna.

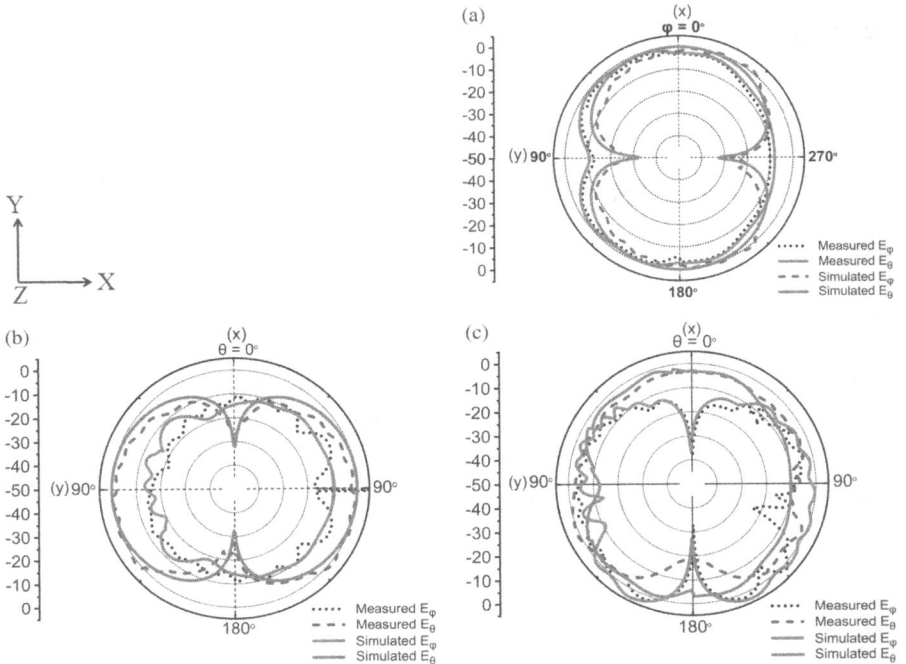

FIGURE 3.13 Simulated and measured radiation patterns of the proposed antenna (a) at 5.5 GHz (x-y plane), (b) at 5.5 GHz (y-z plane), and (c) at 5.5 GHz (x-z plane).

This characteristic ensures broad coverage of the surrounding area, enabling the antenna to deliver maximum energy for wireless operations compliant with IEEE 802.11a and 802.11ac/HiperLAN 2 standards in laptop computers. Consequently, the proposed antenna proves to be highly suitable and applicable for wireless operations in laptop computers.

3.3.5.4 Simulated and Measured Gain and Radiation Efficiency

Figure 3.14 presents the simulated and measured values of gain and efficiency for the proposed antenna, while Table 3.3 provides a comprehensive comparison between the two. It is worth noting that there is a slight deviation between the simulated and

TABLE 3.3

Comparison of Simulated and Measured Gain and Efficiency of the Proposed Antenna

	Gain (dBi)		Efficiency (%)	
Operating band	**Simulated**	**Measured**	**Simulated**	**Measured**
5.15–5.92 GHz	4.75–5.40	4.45–5.22	88.50–90.00	83.50–89.20

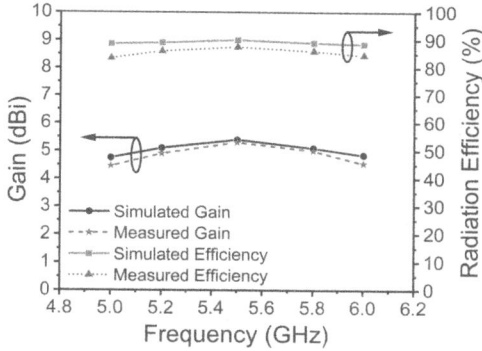

FIGURE 3.14 Simulated and measured gain and radiation efficiency of the proposed antenna.

measured values. However, this discrepancy can be attributed to the fact that the simulations were conducted under ideal conditions, and such minor deviations can be considered negligible.

Analyzing Table 3.3, it can be confidently concluded that the proposed antenna consistently exhibits stable gain and excellent efficiency across the operating bands. This signifies that the antenna possesses a high-quality signal reception capability, which is crucial for reliable wireless operations in laptop computers.

3.3.6 EXPERIMENTAL ANALYSIS OF PROPOSED ANTENNA

To investigate the impact of the laptop's metallic or plastic cover on the performance of the proposed antenna, a practical study was conducted by integrating the antenna onto an actual laptop computer. The specific laptop model utilized for this experiment was the L412 ThinkPad manufactured by Lenovo. As depicted in Figure 3.15(a), the antenna was strategically mounted on the top edge of the laptop's display ground.

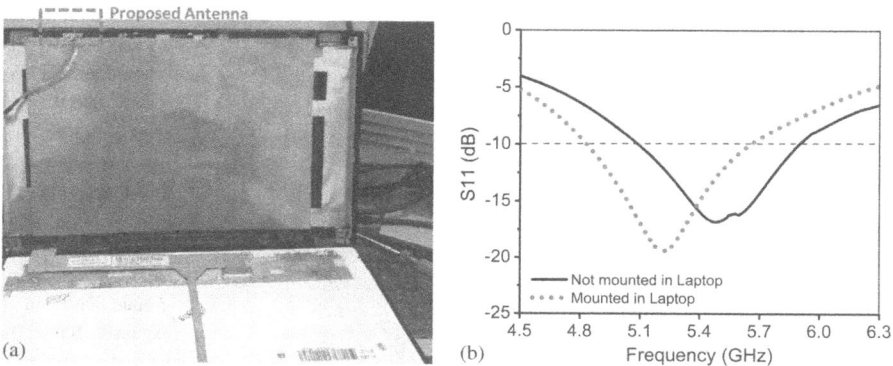

FIGURE 3.15 (a) Proposed antenna installed inside practical laptop and (b) comparison of measured S_{11} results when proposed antenna is installed inside the laptop and when it is not installed in the laptop.

Figure 3.15(b) showcases the measured reflection coefficient (S_{11}) in dB as a function of frequency for the proposed antenna. It is noteworthy that the resonant mode exhibits a slight shift toward lower frequencies due to the material properties of the laptop cover. As a result, the measured operating bandwidth spans from 4.81 GHz to 5.67 GHz. Nonetheless, by making minor adjustments to the lengths of the slots (L_1 and L_2), the desired operating bands of the proposed antenna, when mounted on the actual laptop, can be effectively restored.

3.3.7 PERFORMANCE COMPARISON OF PROPOSED ANTENNA WITH OTHER STATE-OF-THE-ART TECHNOLOGY

To validate the effectiveness and capabilities of the innovative H-Shaped antenna design, a comprehensive performance comparison was conducted, as presented in Table 3.4. This table showcases a comparison between the proposed antenna and

TABLE 3.4
Comparison of the Proposed Antenna with Existing State-of-the-Art Technology

Ref.	Antenna Area (mm²)	BW (%)	Gain (dBi)	Eff. (%)	CSPD	HASG	Remarks
[3]	105	17.7	4.63	55	N	3 mm	Complex antenna structure as it uses vias.
[7]	117	28.19	5.4	92	N	9 mm	Not suitable for ultra-thin laptops as height is 9 mm.
[8]	63	30.68	4.0	67	N	9 mm	Not suitable for ultra-thin laptops as height is 9 mm.
[11]	70	13.91	$	$	N	5 mm	Unapt for ultra-thin as thickness is 1.6 mm. Uses additional foam material.
[12]	100	27	3.4	90	N	5 mm	Unapt for ultra-thin laptop due to 3 mm thickness, complex to manufacture due to bending.
[13]	315	13.38	$	$	N	7 mm	Uses additional ground plane and expensive RT Duroid substrate.
[16]	72	20	3.51	80	N	10 mm	Unapt for ultra-thin as shows the height of 10 mm.

(Continued)

TABLE 3.4 (*Continued*)
Comparison of the Proposed Antenna with Existing State-of-the-Art Technology

Ref.	Antenna Area (mm²)	BW (%)	Gain (dBi)	Eff. (%)	CSPD	HASG	Remarks
[24]	227.5	12.30	6.3	76	N	0 mm	Unapt for ultra-thin laptops due to thickness of 3.7 mm, bending is required, complex structure.
[25]	315	23	2.72	$	N	9	Unapt for ultra-thin laptop due to height of 9 mm.
[30]	54	20	3.6	60	N	9 mm	Unapt for ultra-thin laptop as 1.6 mm thick and height is 9 mm.
[31]	120	12.30	3.5	65	N	4 mm	Uses additional 6 × 16 mm² copper tape, complex structure, and also uses reactive components.
sProposed Antenna	70	15	4.45	90	Y	4 mm	Meets the requirement of ultra-thin laptops, does not require any additional hardware or component.

Abbreviations: $ = represents the information is not mentioned in reference, AGP = Adhesive ground plane, CSPD = Checked suitability for other portable devices, HASG = Height of antenna above system ground.

several recently reported wireless antennas. Key performance parameters such as size, bandwidth, gain, efficiency in the 5.15–5.92 GHz band, and antenna height above the system ground are included in Table 3.4.

Upon analysis, it becomes evident that the proposed antenna possesses several distinct advantages over the other reported antennas. Notably, it exhibits a remarkably compact size, a simple and straightforward structure, and excels in terms of radiation efficiency within the higher frequency bands. Moreover, the proposed antenna offers complete coverage of the IEEE 802.11a and 802.11ac/HiperLAN 2 frequency bands. These notable advantages further validate the viability and potential of the proposed antenna for wireless applications.

3.3.8 FEATURES OF PROPOSED ANTENNA

The proposed antenna boasts several noteworthy features that set it apart from state-of-the-art designs. These salient features are as follows:

- **Unparalleled thinness:** The proposed antenna stands out as the thinnest antenna ever developed, surpassing the capabilities of previously reported state-of-the-art designs.
- **Utilization of pure copper radiating strips:** The antenna's construction incorporates pure copper radiating strips, enabling easy replication, fine-tuning, and cost-effectiveness. This feature also renders it suitable for mass production, ensuring a fast time-to-market advantage not found in other reported antennas.
- **Complete coverage of the 5 GHz band:** Unlike previous designs [9, 29], the proposed antenna offers comprehensive coverage across the entire 5 GHz band, further enhancing its versatility and applicability.
- **Impressive gain and efficiency:** The antenna achieves a gain of more than 4 dBi and an efficiency exceeding 90%, surpassing the performance of antennas presented in Ref. [8, 16, 31].
- **Ultra-thin profile:** With a height of less than 5 mm, the proposed antenna proves to be an excellent choice for next-generation ultra-thin laptop computers, outperforming designs discussed in Ref. [8, 16, 25, 30].
- **Compatibility with various laptop display sizes:** The proposed antenna exhibits suitability for deployment on laptop computers of varying display sizes, distinguishing it from other antennas documented in existing literature.
- **Minimal volume occupied:** The proposed antenna boasts the smallest volume requirements among all antennas reported in the literature, ensuring efficient utilization of limited space.
- **Hardware simplicity:** Unlike alternative designs, the proposed antenna achieves the desired frequency band without the need for additional hardware, lumped elements, or expensive substrates, enhancing its cost-effectiveness and practicality.

These distinctive features collectively establish the superiority and innovation of the proposed antenna, positioning it as a promising solution for wireless applications, particularly in the realm of laptop computer technology.

3.3.9 CONCLUDING REMARKS

A groundbreaking and cutting-edge H-Shaped antenna, distinguished by its extremely low profile and ultra-thin structure, was conceptualized, fabricated, and rigorously tested. The antenna demonstrates remarkable performance within the measured frequency bands of 5.10–5.92 GHz, encompassing a substantial −10 dB bandwidth of 15%. This outstanding capability ensures compliance with the standard bandwidth requirements specified for 802.11a and 802.11ac/HiperLAN 2, making it ideal for wireless applications in laptop computers.

The proposed antenna features an exceptionally low profile, occupying a mere 17.5 × 4 mm² footprint. Moreover, its ultra-thin design boasts an unprecedented thickness of a mere 0.035 mm, outclassing previously reported antennas in the state-of-the-art literature. Despite its compact size, the antenna delivers the necessary radiation performance across the entire 5 GHz wireless band, effectively meeting the demands of laptop applications.

The extraordinary attributes of the proposed antenna extend beyond its diminutive profile. Its reduced form factor contributes to cost-effectiveness, while its superior performance within the frequency domain facilitates a faster time to market. These qualities position the proposed antenna as an optimal solution for wireless operations in modern, ultra-thin laptop computers, ensuring seamless connectivity and enhanced user experiences.

3.4 ADVANCING WLAN/WIMAX CONNECTIVITY IN LAPTOP COMPUTERS WITH MICROMINIATURIZED PRINTED ANTENNA DESIGN

In response to the aforementioned requirements, this section introduces the design and development of a novel microminiaturized, printed, and dual-band monopole antenna in a maze-shaped configuration, tailored specifically for wireless operations in laptop computer devices. Notably, this antenna is engineered without the need for any lumped elements or supplementary hardware.

3.4.1 OVERVIEW OF PROPOSED ANTENNA

In this section, we present a cutting-edge design for a microminiaturized, printed dual-band monopole antenna with a unique maze-shaped configuration. The proposed antenna measures 6 × 4 × 1.6 mm³ and incorporates an air-filled orifice thickness between the antenna and the system ground. The innovation of this design lies in achieving dual-band operation and miniaturization, while maintaining excellent radiation performance within the operating bands.

To achieve dual-band operation and enhanced bandwidth, we employed several key features in the antenna design. Firstly, the air-filled orifice thickness between the antenna and system ground helps optimize the performance across the two frequency bands of interest: the lower frequency band (2.40–2.48 GHz), which includes Bluetooth, and the higher frequency band (5.15–5.85 GHz), which corresponds to WLAN standards.

Furthermore, the maze-shaped radiating structure in the antenna design serves multiple purposes. It increases the surface current path length on the radiating surface, thereby enhancing the overall bandwidth. Additionally, an open-ended protruded tuning stub further contributes to achieving the desired radiation performance.

To validate the proposed design, we derived a simplified EC model of the antenna and conducted extensive simulations. The results demonstrate that the antenna exhibits a gain ranging from 2.72 dBi to 6.62 dBi and a desirable radiation efficiency ranging from 61% to 91% across the operating bands. Moreover, the simulated and measured results are in excellent agreement, confirming the suitability of the proposed antenna structure for WLAN/WiMAX applications in laptop computers.

3.4.2 PROPOSED ANTENNA GEOMETRY

The antenna is printed on a single FR-4 substrate with a thickness of 1.6 mm and a relative dielectric constant (ε_r) of 4.3, along with a loss tangent (tan δ) of 0.025. The substrate size is a mere 6×4 mm^2. The core structure of the antenna consists of a maze-shaped radiation patch composed of two folded monopole radiating strips, referred to as "X" and "Y," supporting the lower and higher frequency bands, respectively.

The proposed antenna is mounted on a system ground measuring 260×200 mm^2 with an air-filled orifice thickness (g), and it is positioned at a height (h) above the system ground. The system ground, 0.2 mm thick, is primarily made of 91% brass material. The antenna is fed through a 50 Ω miniature coaxial cable, with its central conductor connected to point "A" on the antenna, and the grounding sheath connected to point "B" on the system ground.

In the subsequent sections, we will provide a detailed analysis of the proposed antenna design from two perspectives, further elucidating its design principles and performance characteristics (Figure 3.16).

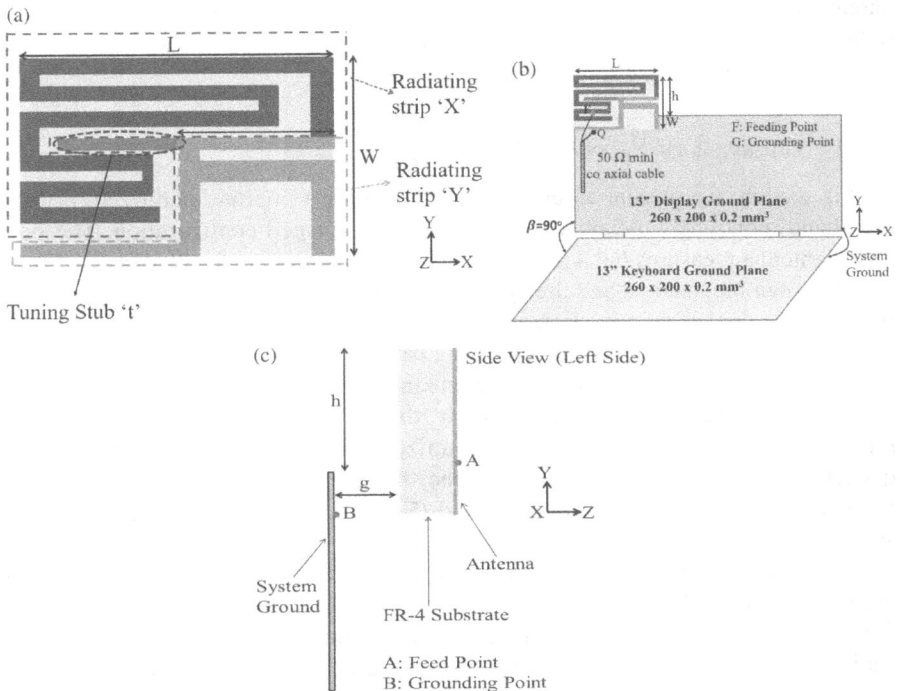

FIGURE 3.16 Proposed schematic geometry of antenna. (a) Front view of proposed antenna without system ground, (b) perspective view when antenna is mounted on the system ground, and (c) left-hand side view of proposed antenna.

3.4.3 ANTENNA DESIGN PRINCIPLE

The utilization of folding techniques for radiating patches represents a highly efficient approach in the design of miniaturized multiband antennas, as it allows for significant size reduction and improved performance without incurring additional costs or expenses [15, 16]. In Figure 3.17, it is evident that the folding of the radiating strip alters the path of surface current distribution on the radiating patch, effectively increasing the length of the current path while maintaining a fixed strip length. This characteristic of the surface current distribution results in a substantial reduction in antenna size at a specific resonating frequency. Moreover, the folding technique leads to a broadening of the antenna's bandwidth due to the decreased value of Q, which is directly influenced by the folding of the radiating strip.

Through precise dimensional adjustments in the folding technique, the proposed antenna offers the capability to flexibly tune the desired bands with the necessary bandwidth. The total lengths, L_{X1} and L_{Y1}, of the folded monopole radiating strips denoted as "X" and "Y" respectively, are designed to be approximately one-quarter wavelength long at resonating frequencies of 2.45 GHz and 5.5 GHz, respectively. These dimensions can be expressed mathematically as per Equations (3.5) and (3.6) below:

$$L_{X1} = L_1 + 2L_2 + 2L_3 + W_2 + W_3 + W_4 + W_5 + W_6 + W_7 + L \qquad (3.5)$$

$$L_{Y1} = L_4 + W_9 + 2L_5 + W_{10} \qquad (3.6)$$

By precisely determining these lengths, the proposed antenna achieves optimal tuning for the desired bands, allowing for efficient wireless communication within the specified bandwidth.

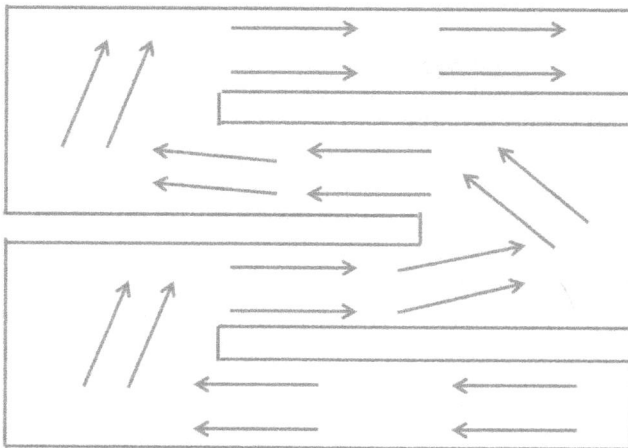

FIGURE 3.17   Schematic diagram of the surface current distribution on a folding radiating patch.

3.4.4 Working Mechanism of Proposed Antenna

To provide a clear understanding of the design process behind the microminiatur-ized, printed dual-band maze-shaped monopole antenna, Figure 3.18 depicts the step-by-step evolution of the antenna design, conducted using CST MVS. The corresponding simulated reflection coefficient (dB) results at different stages are also presented.

Initially, Ant. 1 was created by printing only the folded monopole radiating strip "X" on the FR-4 substrate. Ant. 1 was mounted on the system ground with an air-filled orifice thickness "g" and excited at points "A" and "B" using a 50 Ω miniature coaxial feed line. The air-filled orifice thickness "g" between Ant. 1 and the system ground acts as an air gap capacitance. This configuration successfully achieved reso-nance at 2.45 GHz, as shown in Figure 3.18(a).

To achieve resonance around 5.5 GHz, the folded monopole radiating strip "Y" was introduced from point "D" without disturbing Ant. 1. This resulted in the com-plete structure, denoted as Ant. 2, as shown in Figure 3.18(b). Ant. 2, along with

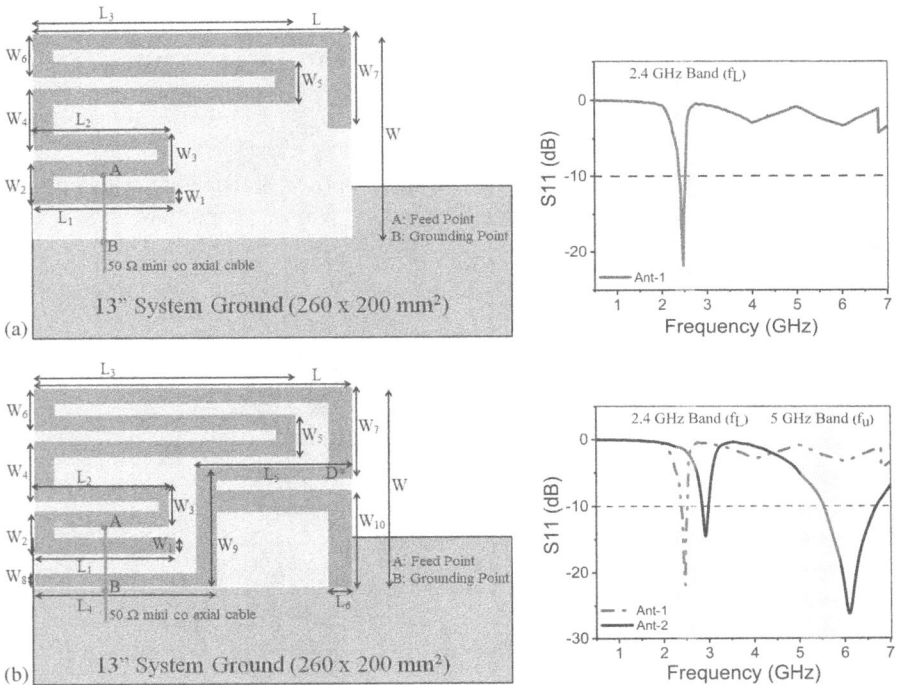

FIGURE 3.18 Step-by-step evolution of the design process of proposed antenna and cor-responding reflection coefficient (dB). The left column gives the geometries of Ant. 1, Ant. 2, and Ant. 3 (proposed antenna) involved in the design evolution process. The right col-umn gives the simulated reflection coefficient (dB) results for the different proposed antenna geometries. (a) Step 1; (b) step 2; and (c) step 3. (*Continued*)

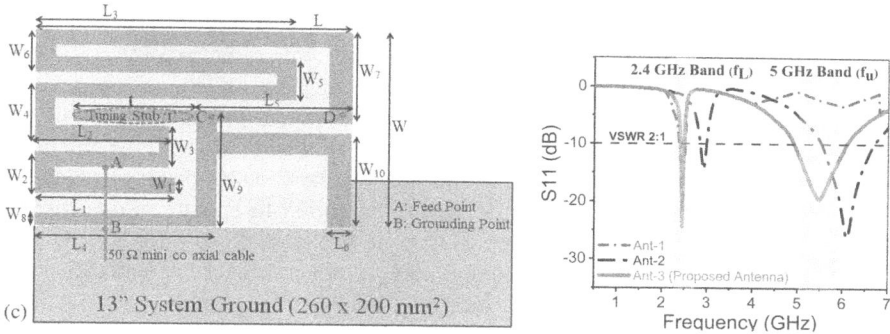

FIGURE 3.18   (*Continued*)

the air gap, resonated at a higher frequency of 6 GHz. However, this shift in resonance also caused the resonance at 2.45 GHz to shift toward a higher frequency of 3 GHz, which did not meet the requirements of the 2.4 GHz WLAN standard and the Bluetooth band due to impedance mismatching.

To improve impedance matching and achieve the desired bands, an open-ended tuning stub of length "t" was protruded from point "C," forming Ant. 3, which represents the proposed antenna design. This "t" length tunes the lower frequency band (f_l) within the range of 2.39–2.51 GHz and the upper frequency band (f_u) within the range of 4.90–6.09 GHz. As a result, the proposed antenna operates in the dual bands of f_l and f_u, covering the 2.4 GHz and 5 GHz operating bands of WLAN, as well as the 5.8 GHz band of WiMAX.

This step-by-step evolution demonstrates the systematic design approach employed to optimize the performance of the antenna, ensuring resonance within the desired frequency bands and achieving efficient wireless communication.

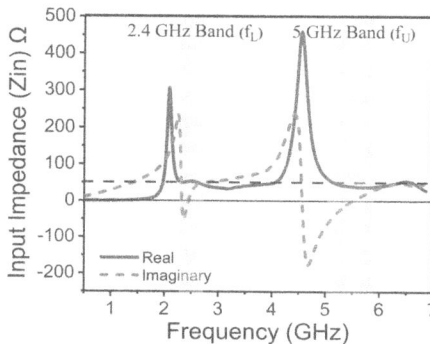

FIGURE 3.19   Input impedance Z_{in} versus the frequency of the proposed antenna.

3.4.4.1 Input Impedance Z_{in} Versus Frequency of Proposed Antenna

To provide a comprehensive understanding of the operating principle of the proposed antenna structure, Figure 3.19 illustrates the curve of input impedance (Z_{in}) versus frequency. Notably, it is evident that when the proposed antenna operates within the f_l and f_u bands, the input impedance remains stable around 50 Ω, with the reactance nearly equal to 0 Ω. This observation signifies the excellent impedance matching achieved by the proposed antenna, along with its ability to operate in dual bands. Consequently, the antenna structure proves to be well-suited for WLAN standards and WiMAX applications in laptop computers, demonstrating its potential for reliable and efficient wireless communication.

3.4.4.2 Surface Current Distribution

To gain a deeper understanding of the design principles underlying the proposed dual-band antenna, we examine the simulated surface current distribution diagrams at the resonating frequencies of 2.45 GHz and 5.5 GHz, corresponding to the f_l and f_u bands, respectively. These diagrams are analyzed and presented in Figure 3.20.

Figure 3.20(a) illustrates the surface current distribution for the folded monopole radiating strip "X" at 2.45 GHz. It is evident that the folding of the radiating strip

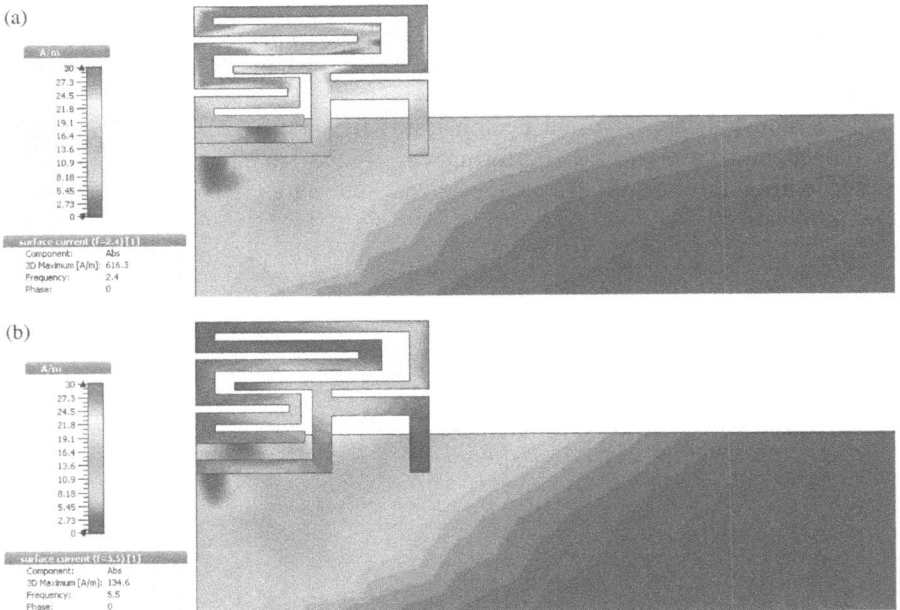

FIGURE 3.20 Surface current distribution (A/m) of the proposed antenna (a) 2.45GHz and (b) 5.5 GHz.

significantly lengthens the path of the current on the radiation patch. The maximum current flows through the strip "X," which indicates efficient energy transfer and resonant behavior at this frequency.

Similarly, Figure 3.20(b) depicts the surface current distribution at 5.5 GHz for the radiating strip "Y." In this case, the maximum current is observed to be distributed around the center and various other locations along the strip "Y." This observation signifies that the proposed antenna achieves a well-matched response and resonates effectively in both the lower and higher frequency bands of interest.

The surface current distribution diagrams further validate the suitability of the proposed antenna for dual-band operation. The distinct patterns and the efficient utilization of the radiating strips "X" and "Y" demonstrate the antenna's capability to operate in the desired frequency bands with optimal current flow, resulting in enhanced performance and reliable wireless communication.

3.4.5   PARAMETRIC STUDY

The mirroring performance of the proposed antenna is highly sensitive to variations in its geometrical parameters, owing to its compact size. To assess the impact of crucial parameters on the antenna's performance within the desired bands of f_l (2.39–2.51 GHz) and f_u (4.90–6.09 GHz), a comprehensive parametric study has been conducted.

In this study, while keeping all other parameters constant, the following parameters have been selected for analysis: the air-filled orifice thickness (g), the length (t) of the protruded tuning stub, and the effective height (h) of the antenna above the system ground. By examining these specific parameters, we can gain valuable insights into their influence on the reflection coefficient (dB) and impedance bandwidth of the proposed antenna. Additionally, the parametric study also explores the impact of various sizes of the system ground on the antenna's performance, aiming to assess its suitability for implementation in other portable devices.

This detailed parametric study contributes to a comprehensive understanding of the proposed antenna's behavior and performance characteristics. By carefully analyzing and manipulating these key parameters, it is possible to optimize the antenna's performance within the desired frequency bands, thus enabling its successful integration into a variety of portable devices.

3.4.5.1   Effects of Air-Filled Orifice Thickness (g)

To thoroughly investigate and optimize the air-filled orifice thickness (g) between the proposed antenna and the system ground, a comprehensive parametric study was conducted in the desired frequency bands of f_l and f_u. The study involved varying (g) from 0 to 1.6 mm in increments of 0.4 mm, utilizing CST MWS. Figure 3.21 showcases the reflection coefficient (dB) versus frequency for different values of (g) for the proposed antenna.

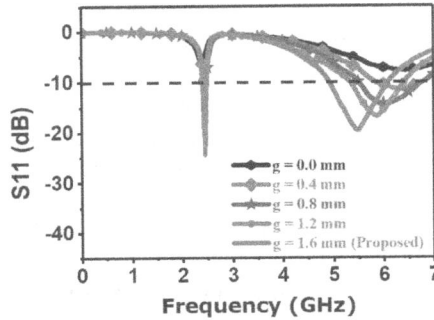

FIGURE 3.21 Effects of air filled-orifice thickness (g) on the simulated reflection coefficient of the proposed antenna.

Observing the results in Figure 3.21, it is evident that when (g) is set to 0 mm, none of the bands resonate due to the absence of air gap capacitance. However, as (g) increases, the desired impedance bandwidth for VSWR <2 is achieved for both f_l and f_u bands. Notably, a (g) value of 1.6 mm provided the desired performance characteristics within the required frequency bands.

3.4.5.2 Effects of Protruded Tuning Stub of Length (t)

In addition, the protruded tuning stub (t) was examined to optimize its length. The study involved varying (t) from 0 to 2.0 mm in increments of 1 mm, as depicted in Figure 3.22. The simulation results indicate that as (t) increases, f_l and f_u bands experience a linear shift from 2.85 to 2.39 GHz and from 5.5 to 4.90 GHz, respectively. This shift is attributed to the capacitive coupling between the tuning stub and width W_{12}, resulting in improved impedance matching. Based on these findings, a tuning stub length of 2 mm was determined as optimal, covering the desired WLAN and WiMAX bands for wireless applications in laptop computers.

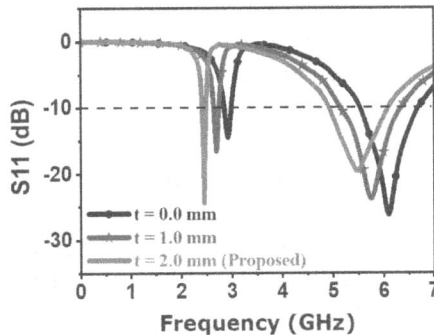

FIGURE 3.22 Effects of tuning decoupling stub length (t) on simulated reflection coefficient of the proposed antenna.

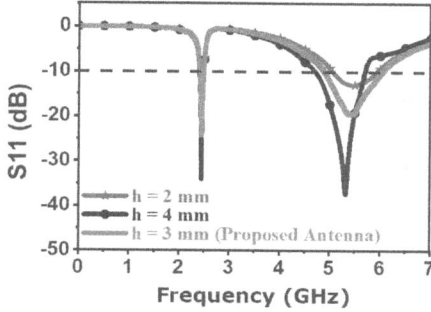

FIGURE 3.23 Simulated reflection coefficient of the proposed antenna as a function of height (h) with respect to system ground.

3.4.5.3 Effects of Effective Height (h) of Proposed Antenna above the System Ground

The effective height (h) of the proposed antenna above the system ground was investigated. The parametric study involved varying (h) from 2 mm to 4 mm in increments of 1 mm, as illustrated in Figure 3.23. The results indicate that for (h) values of 2 mm and 4 mm, f_l remains unaffected, while the bandwidth of f_u is impacted and does not cover the entire 5 GHz high-frequency band. Therefore, an effective (h) of 3 mm was selected to ensure the desired f_l and f_u bands were obtained.

3.4.5.4 Effects of Various Sizes of System Ground Plane

To evaluate the suitability of the proposed antenna for various sizes of system ground planes, a parametric study was conducted while keeping other parameters constant. Figure 3.24 displays the simulated reflection coefficient for system ground sizes ranging from 5 inches to 17 inches. Remarkably, the proposed antenna exhibits consistent performance across all the studied system ground sizes, without compromising the impedance bandwidth and magnitude of the reflection coefficient within the f_l and f_u bands.

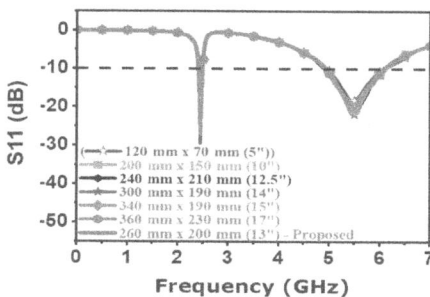

FIGURE 3.24 Simulated reflection coefficient as a function of attaching proposed antenna to various sizes of system ground.

TABLE 3.5

Optimized Dimensions of Essential Parameters of Proposed Antenna Obtained from Parametric Study

Name of the Parameter	Notation	Value (mm)
Air-filled orifice thickness	g	1.6
Open-ended protruded tuning stub length	t	2
Effective height of the antenna above the system ground	h	3

3.4.6 Dimensions of Proposed Antenna

Table 3.5 presents the optimized parameter values derived from the parametric study, specifically for the f_l and f_u bands.

The remaining parameters were optimized using the CST MWS optimization technique and are detailed in Table 3.6. These results provide valuable insights for achieving optimal antenna performance and inform the design process for wireless applications in laptop computers.

3.4.7 EC Model of Proposed Antenna

The proposed antenna's EC model, represented by lumped elements, is illustrated in Figure 3.25. The distributed parameters of the coaxial feed line, namely, series inductance L and capacitance C, are essential components of the model and can be calculated using the following expressions [32]. It is important to note that in our design, we have utilized a very short coaxial cable, only 10 cm in length, resulting in negligible transmission line losses. Therefore, the impact of transmission line losses can be disregarded in our analysis.

$$R = \frac{R_s}{(2\pi)^2}\left\{ \int_{\phi=0}^{2\pi} \frac{1}{a^2} a d\phi + \int_{\phi=0}^{2\pi} \frac{1}{b^2} b d\phi \right\} = \frac{R_s}{2\pi}\left(\frac{1}{a} + \frac{1}{b} \right) = 0 \ .... \ \Omega/m \qquad (3.7)$$

TABLE 3.6

Optimized Dimensions of Proposed Antenna

Parameter	Value (mm)	Parameter	Value (mm)	Parameter	Value (mm)
L_1	2.7	W_2	0.8	W_8	0.35
L_2	4.8	W_3	0.9	W_9	0.5
L_3	6.0	W_4	1.2	W_{10}	2.4
L_4	3.5	W_5	0.8	W_{11}	2.0
L_5	3.0	W_6	0.8	W_{12}	0.2
W_1	0.3	W_7	1.8		

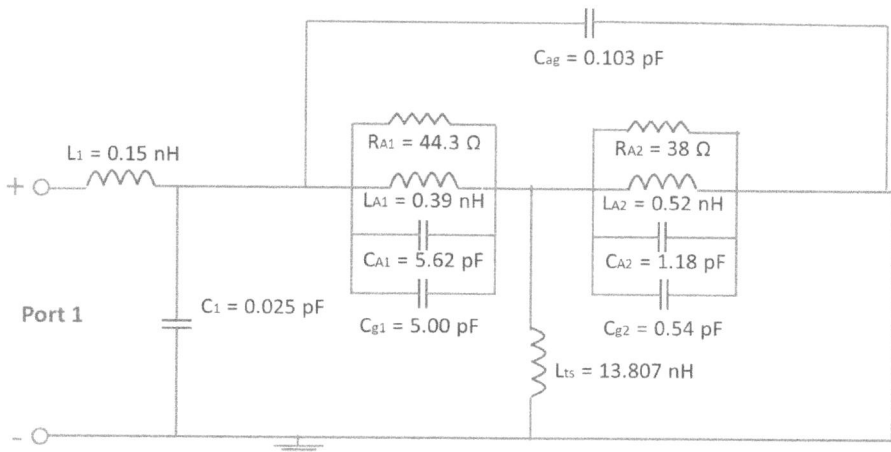

FIGURE 3.25 EC model of proposed antenna.

$$L = \frac{\mu}{(2\pi)^2} \int\limits_{\phi=0}^{2\pi} \int\limits_{\rho=a}^{b} \frac{1}{\rho^2} \rho d\rho d\phi = \frac{\mu}{2\pi} \ln\left(\frac{b}{a}\right) \dots H/m \qquad (3.8)$$

$$C = \frac{\epsilon'}{\left(\ln\left(\frac{b}{a}\right)\right)^2} \int\limits_{\phi=0}^{2\pi} \int\limits_{\rho=a}^{b} \frac{1}{\rho^2} \rho d\rho d\phi = \frac{2\pi \,\epsilon'}{\ln\left(\frac{b}{a}\right)} = \frac{2\pi\epsilon_0\epsilon_r}{\ln\left(\frac{b}{a}\right)} \dots F/m \qquad (3.9)$$

$$G = \frac{\omega\epsilon''}{\left(\ln\left(\frac{b}{a}\right)\right)^2} \int\limits_{\phi=0}^{2\pi} \int\limits_{\rho=a}^{b} \frac{1}{\rho^2} \rho d\rho d\phi = \frac{2\pi\omega\epsilon''}{\ln\left(\frac{b}{a}\right)} = 0 \dots S/m \qquad (3.10)$$

where "R" is the series resistance per unit length (Ω/m), "L" is the series inductance per unit length (H/m), "C" is the shunt capacitance per unit length in F/m, and "G" is the shunt conductance per unit length (S/m). Also, "a" and "b" are inner and outer radii of the coaxial cable, $\epsilon' = \epsilon_0\epsilon_r$ and $\epsilon'' = 0$ are the real and imaginary part of complex permittivity, respectively.

The "C_{ag}" realized as air gap capacitance is produced by air-filled orifice thickness "g" and computed using Equation 3.11 as below:

$$C_{ag} = 0.5H.Q.exp\left(-1.86\frac{g}{h}\right) \times \left[1 + 4.09\left\{1 - exp\left(0.75\sqrt{\frac{H}{W}}\right)\right\}\right] \qquad (3.11)$$

The folded monopole radiating strips, namely, X and Y are modeled by parallel RLC circuit. RA1, LA1, CA1 and RA2, LA2, CA2 are the lumped elements for

"X" and "Y," respectively, whose values are calculated according to the following expressions [33]:

$$Q = \frac{f}{BW} \tag{3.12}$$

$$Q = R\sqrt{\frac{C_p}{L_p}} = \omega RC \tag{3.13}$$

$$\omega = \frac{1}{\sqrt{CL}} \tag{3.14}$$

where, Q is quality factor, f is resonating frequency, R is real part of impedance at resonance frequency obtained from Figure 3.19 and BW is the bandwidth of the respective bands f_l and f_u.

The gap between radiating strips of X and Y due to folding is modeled by the total parallel capacitance C_{g1} and C_{g2}, respectively, and computed by using Equation 3.7. The length (t) of the protruded tuning stub is modeled by inductor L_{ts}, which plays a very important role in tuning the f_l and f_u bands. The values of L_{ts} are obtained by manually adjusting and using an optimization technique in Agilent ADS software.

3.4.8 Performance Characterization of Proposed Antenna

The proposed antenna's RF characteristics and radiation performance were evaluated through simulations conducted in CST MWS, while the EC model was analyzed using Agilent ADS software. To verify the antenna's performance with the optimized dimensions specified in Table 3.5 and 3.6, a physical prototype was fabricated and tested, as depicted in Figure 3.26. The antenna's radiation performance,

FIGURE 3.26 Fabricated prototype of the proposed antenna.

TABLE 3.7
Comparison of S₁₁ (dB) and Operating Bands (GHz)

	S$_{11}$ (dB)		Operating Bands (GHz)	
	f$_l$	f$_u$	f$_l$	f$_u$
Simulated	−25	−20	2.39–2.51	4.90–6.09
Measured	−20	−18	2.37–2.53	5.05–5.90
Equivalent circuit	−25	−16	2.40–2.60	4.87–6.08

encompassing radiation pattern, gain, and radiation efficiency, was measured in an anechoic chamber measuring $8 \times 4 \times 4$ m³.

3.4.8.1 Reflection Coefficient Characteristics

The reflection coefficient of the proposed antenna was analyzed through simulations, measurements, and the EC model, as illustrated in Figure 3.12. A comparison of the reflection coefficient along with the covered bands is presented in Table 3.7. It is noteworthy that there is a high level of agreement among the simulated, measured, and EC model results. All three approaches demonstrate excellent performance, with the antenna covering the required bandwidth for VSWR <2 in both the f$_l$ and f$_u$ bands, exhibiting only minimal deviation. The slight deviation observed could be attributed to factors such as soldering or fabrication tolerances, as well as potential inaccuracies in dimensions during the fabrication process, considering the extremely small size of the antenna, and losses in the substrate.

3.4.8.2 VSWR

The VSWR of the proposed antenna was evaluated through simulations and measurements, as shown in Figure 3.28. Both the simulated and measured values were found to be below two across the two bands of interest. This indicates that the impedance of the proposed antenna is accurately matched, ensuring maximum power transmission.

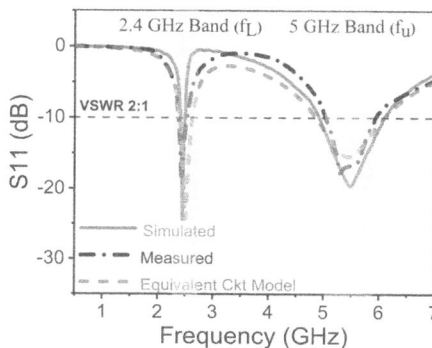

FIGURE 3.27 Simulated and measured reflection coefficient of the proposed antenna.

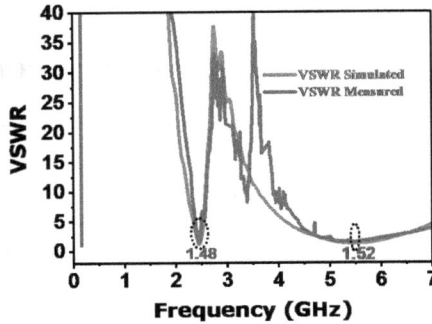

FIGURE 3.28 Simulated and measured VSWR values as a function of frequency of proposed antenna.

3.4.8.3 Radiation Pattern

To verify the radiation capabilities of the fabricated prototype antenna within the desired frequency bands, the measured radiation patterns were analyzed in three principal planes: the x-y, x-z, and y-z planes. The radiation patterns at the center frequencies of 2.45 GHz and 5.5 GHz for the f_l and f_u bands, respectively, were normalized and plotted in Figure 3.29.

3.4.8.4 Simulated and Measured Gain and Radiation Efficiency

Figure 3.30((a and b)) presents the simulated and measured variations of gain and radiation efficiency for the f_l and f_u bands of the proposed antenna. A comparison of these values is provided in Table 3.8. A slight deviation between the measured and simulated results is observed, which can be attributed to the simulation being conducted under ideal conditions. Nonetheless, this deviation is considered acceptable. It is worth noting that the gain and efficiency in the f_l band are relatively lower due

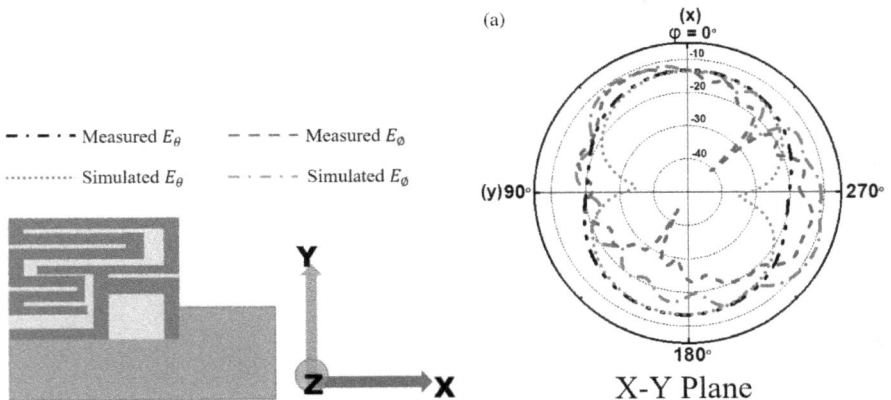

FIGURE 3.29 Measured radiation patterns of proposed antenna (a) at 2.45 GHz and (b) at 5.5 GHz. (*Continued*)

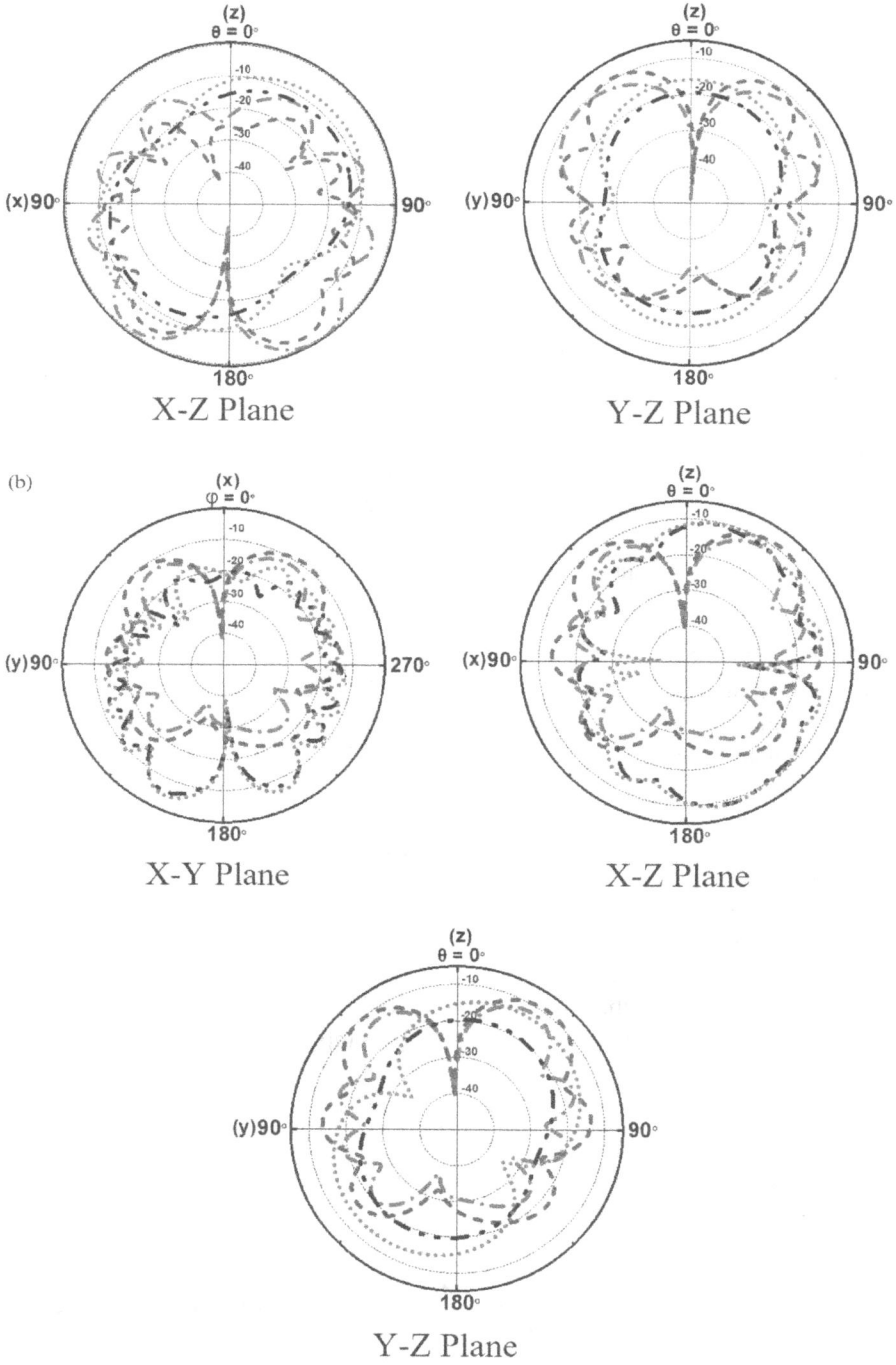

X-Z Plane

Y-Z Plane

(b)

X-Y Plane

X-Z Plane

Y-Z Plane

FIGURE 3.29 (*Continued*)

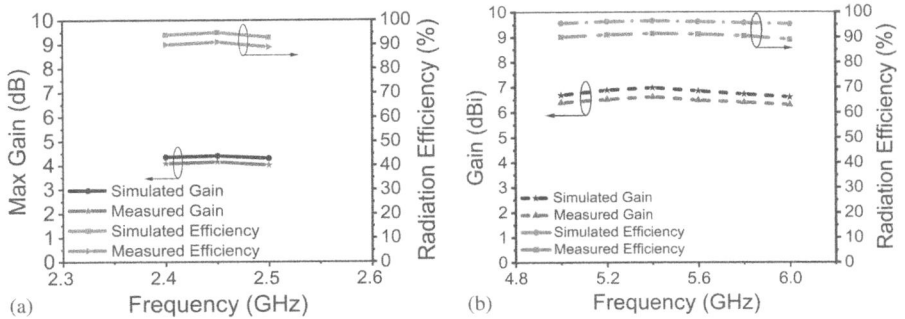

FIGURE 3.30 Simulated and measured gain and radiation efficiency of the proposed antenna.

to the antenna's minuscule size, which inherently reduces the gain and efficiency at the lower resonating frequency. However, the proposed antenna exhibits satisfactory gain and excellent radiation efficiency in the higher frequency f_u band. Overall, it is evident that the proposed antenna maintains signal reception quality suitable for wireless operations in laptop computers.

3.4.9 Experimental Analysis of Proposed Antenna

To assess the impact of the laptop's plastic/metallic cover on the performance of the proposed antenna, a practical study was conducted by integrating the antenna onto an actual laptop computer. The chosen laptop for this implementation was the L412 ThinkPad by Lenovo. As depicted in Figure 3.31(a), the antenna was positioned on the top left edge of the display ground.

Figure 3.31(b) presents the measured results of the reflection coefficient S_{11} (dB) as a function of frequency (GHz) for the proposed antenna. It was observed that the laptop cover material had a minor effect on the f_L band. However, in the f_U band, there was a slight shift toward lower frequencies. Despite this shift, the band still encompassed the desired 5 GHz range. With a slight adjustment to the length of the tuning stub "t," it is possible to restore the desired lower band of interest for the proposed antenna integrated onto the actual laptop computer.

TABLE 3.8

Comparison of Simulated and Measured Value of Gain and Radiation Efficiency of the Proposed Antenna

Operating Bands (GHz)	Gain (dBi)		Radiation Efficiency (%)	
	Simulated	Measured	Simulated	Measured
f_l	2.95–3.21	2.72–3.05	65.00–68.50	61.20–66.37
f_u	6.60–6.98	6.30–6.62	95.20–96.55	89.00–91.50

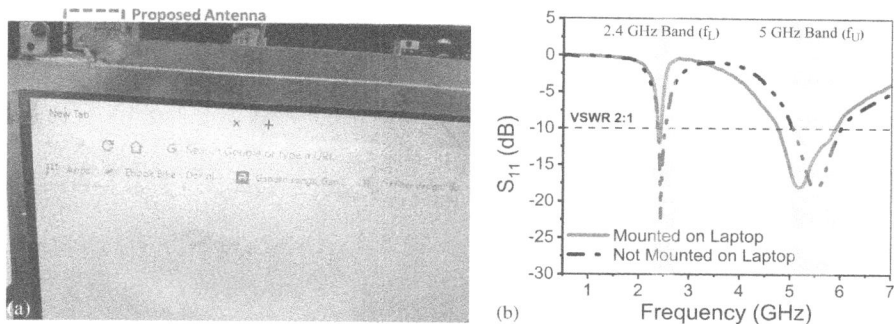

FIGURE 3.31 (a) Proposed antenna installed inside practical laptop and (b) comparision of measured S_{11} (dB) results when proposed antenna is installed inside the laptop and outside the laptop.

3.4.10 PERFORMANCE COMPARISON OF PROPOSED ANTENNA WITH OTHER STATE-OF-THE-ART TECHNOLOGY

To assess the effectiveness of the proposed microminiaturized printed dual-band monopole antenna, a performance comparison was conducted with several recently reported WLAN/WiMAX antennas, as presented in Table 3.9. Table 3.9 provides details such as size, bandwidth, gain, efficiency, and antenna height above the system ground for the 2.4 GHz (2.40–2.48 GHz) and 5 GHz (5.15–5.85 GHz) bands. Compared to the other antennas, the proposed antenna exhibits distinct advantages, including its extremely compact size, excellent radiation efficiency in the higher

TABLE 3.9
Performance Comparison of Proposed Antenna with Existing State-of-the-Art Technology

Ref.	BW 2.4/5 GHz (%)	Max Gain 2.4/5 GHz (dBi)	Eff 2.4/5 GHz (%)	Antenna Size (mm³)	System Ground (mm²)	Height of Antenna with Reference to System Ground
[2]	16.3/13.43	2.56/3.43	99/99	$46.5 \times 9.3 \times 1.6$	-	-
[3]	6.5/17.7	2.17/4.63	54/55	$35 \times 3 \times 0.4$	260×200	3 mm
[14]	8.8/16.8	1.47/3.15	79/83	$23 \times 6 \times 0.4$	23×35 (AGP)	-
[16]	12/20	2.08/3.51	80/80	$12 \times 6x \times 0.4$	300×190	12 mm
[21]	6.4/19.5	3.8/5.6	62/87	$8.7 \times 5.4 \times 0.8$	200×150	8.7 mm
[24]		6.1/7.0	70/76	$3.7 \times 7 \times 32.5$	300×190	-
[31]	3.43/12.30	5.4/3.8	68/65	$4 \times 30 \times 0.8$	$315 \times 182 \times 2.2$	4 mm
Proposed	6.5/16	3.05/6.4	61/85	$6 \times 4 \times 1.6$	260×200	3 mm

frequency band (f_u), and complete coverage of WLAN frequency bands. It is worth noting that the proposed antenna, comprised of simple radiating patches on a FR-4 substrate, offers benefits such as ease of replacement, fine-tunability, cost-effectiveness, mass production feasibility, and a rapid time-to-market.

3.4.11 FEATURES OF PROPOSED ANTENNA

The proposed antenna showcases a range of distinctive features, highlighting its innovative design and functionality:

- **Compact and simplified design:** With dimensions measuring only 6 mm × 4 mm, the antenna offers a compact form factor, ensuring efficient integration within laptop computers while minimizing space requirements.
- **Seamless band operation:** The antenna operates within the desired frequency bands without the need for additional matching circuits, lumped elements, or extra vias. This streamlined design enhances overall performance and simplifies implementation.
- **Versatility for portable devices:** The antenna's versatility extends beyond laptops, making it suitable for various sizes of portable devices. Its compatibility with different form factors ensures reliable wireless connectivity across a range of portable electronic devices.
- **Broad percentage bandwidth:** The proposed antenna achieves an impressive percentage bandwidth, with coverage spanning 6.5% to 16% across the operating bands. This wide bandwidth enables robust and reliable wireless communication in diverse environments.
- **Enhanced gain performance:** In all the operating bands, the proposed antenna achieves a gain above 3 dBi. This noteworthy gain performance enhances signal reception and transmission, optimizing the overall wireless performance of the antenna.

Through these distinct features, the proposed antenna offers a compelling solution that combines compactness, simplified design, versatility, broad bandwidth, and enhanced gain performance, making it well-suited for integration in laptop computers and other portable devices.

3.4.12 CONCLUDING REMARKS

The proposed minuscule maze-shaped printed dual-band monopole antenna, measuring $6 \times 4 \times 1.6$ mm^3, along with a protruded tuning stub of 2×0.2 mm^2 and an air-filled orifice thickness of 1.6 mm between the antenna and the system ground, has been successfully proposed and validated for wireless operations in laptop computers. The influence of various parameters on the antenna's performance in the f_l and f_u bands has been thoroughly studied. The antenna exhibits a measured impedance bandwidth of -10 dB of 6.5% (2.37–2.53) GHz for f_l and 16% (5.05–5.90) GHz for f_u, making it suitable for WLAN/WiMAX applications in laptop computers. Notably, varying sizes of the system ground have no impact on the impedance bandwidth and reflected coefficient (S_{11} dB) of the proposed antenna prototype, making it an excellent choice for small-sized laptop computers with limited space for mounting multiple antennas for wireless applications. Additionally, an EC model of the proposed antenna was developed to analyze its RF performance, demonstrating good agreement between the simulated, EC model, and measured reflection coefficients. The simulated and measured RF performance of the proposed antenna confirms its suitability for WLAN/WiMAX applications in laptop computers.

3.5 DESIGN OF DUAL-BAND ULTRA-THIN INVERTED E-SHAPED MONOPOLE ANTENNA FOR WLAN/ WIMAX IN LAPTOP COMPUTERS

In this section, we introduce a cutting-edge and highly compact dual-band antenna designed specifically for WLAN/WiMAX applications in laptop computers.

3.5.1 Overview of the Proposed Antenna

Our proposed design features a novel inverted E-shaped monopole antenna with an exceptionally low profile and ultra-thin structure. The antenna measures a mere 0.2 mm in thickness and is constructed using a pure copper strip measuring 20×6 mm^2.

The key innovation of our design lies in its miniaturized size and enhanced impedance bandwidth in dual-band operation. This achievement is made possible by incorporating two monopole radiating strips, namely RS (inverted J) and PQ (inverted Z), along with an open-ended vertical tuning stub measuring 4.5×1 mm^2. Through extensive measurements, we have confirmed that the proposed antenna exhibits an impressive impedance bandwidth of 11.24% (2.35–2.63) GHz in the lower band (F_l) and 18.78% (4.92–5.94) GHz in the upper band (F_u) while maintaining a VSWR <2. These bandwidths effectively cover the 2.4/5 GHz WLAN standards as well as the 5.8 GHz WiMAX bands.

Furthermore, our proposed antenna demonstrates exceptional radiation performance. It generates nearly omnidirectional radiation patterns, ensuring reliable signal reception from various angles. The antenna also maintains a stable gain of approximately 4 dBi across the operating bands, contributing to robust signal transmission. Additionally, we have achieved an outstanding efficiency of around 90% in both F_l and F_u bands, indicating highly effective power utilization.

The remarkable radiation characteristics and compact form factor of our antenna make it an ideal choice for WLAN/WiMAX applications in modern ultra-thin laptop computers.

3.5.2 Antenna Geometry of Proposed Antenna

Figure 3.32 showcases the schematic geometry of our innovative antenna design tailored for WLAN and WiMAX operations in laptop computers.

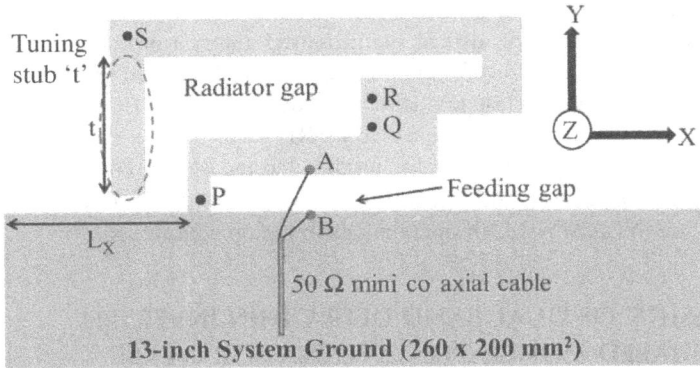

FIGURE 3.32 Schematic geometry of the proposed antenna.

The antenna configuration includes two distinctive radiating strips, namely RS (Inverted J) and PQ (Inverted Z), accompanied by an open-ended vertical tuning stub with a length denoted as "t." For mounting, the antenna is positioned on the system ground at point P, located at a specific distance denoted as "L_x."

The system ground itself is composed of 91% brass material, meticulously crafted to support a 13-inch laptop display measuring $260 \times 200 \times 0.2$ mm^3. To ensure efficient signal propagation, we employ a small and low loss 50 Ω miniature coaxial feed line, where the central conductor is connected to point "A" along the lower edge of strip PQ, while the grounding sheath is soldered to point "B" on the system ground. This particular feeding arrangement introduces a feeding gap and maintains an effective dielectric constant of 1 for all the radiating strips [34]. This critical characteristic plays a crucial role in achieving the desired operating bands for our antenna design.

3.5.3 Design Principle of Proposed Antenna

The design methodology for the proposed antenna, implemented in CST MWS, is outlined, and the corresponding reflection coefficient (S_{11}) dB versus frequency results are depicted in Figure 3.33.

To form the structure of Ant. 1, as illustrated in Figure 3.33A, a straightforward approach was employed. A monopole radiating strip PQ (inverted-Z) with a length approximately equal to a quarter wavelength, resonating at 5.5 GHz, was computed using the expression [35]:

$$L = \frac{\lambda}{4} = \frac{C}{4f} \qquad (3.15)$$

Here, "C" represents the speed of light (m/s), and "f" denotes the resonating frequency. Through optimization techniques in CST, the length of strip PQ was fine-tuned to 16 mm. The input impedance of Ant. 1 is dominated by the inductive reactance generated by strip PQ. This dominant inductive reactance is

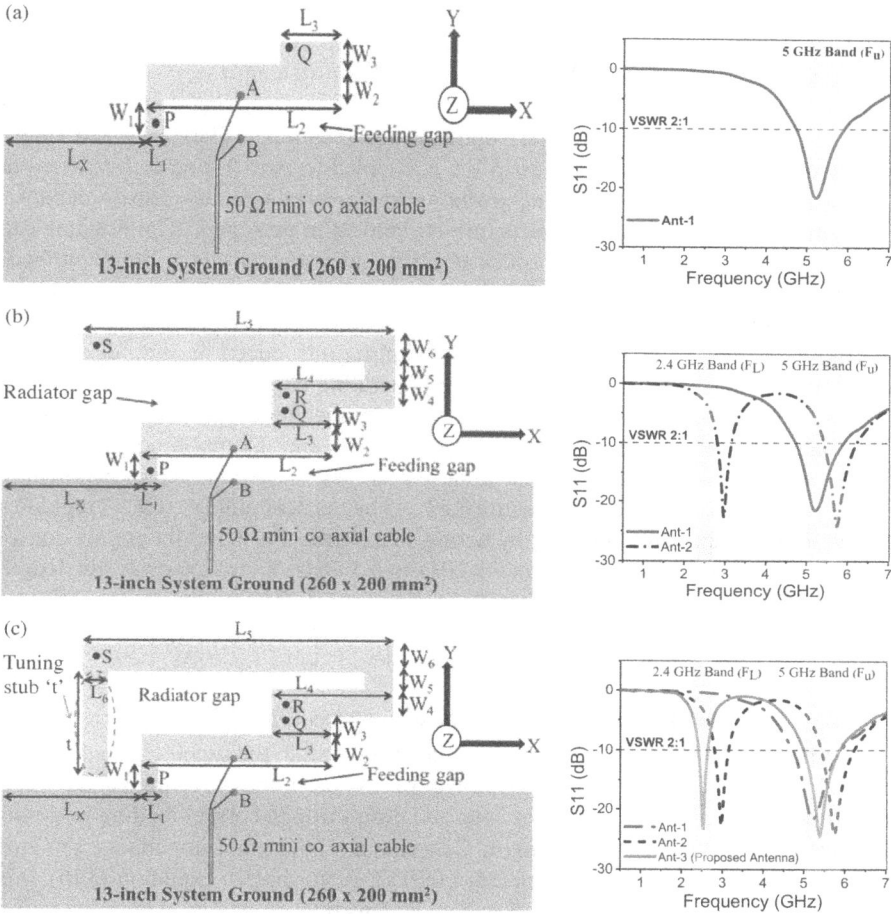

FIGURE 3.33 Complete design process of the proposed antenna (a) Ant. 1, (b) Ant. 2, and (c) Ant. 3 (proposed antenna).

counterbalanced by an opposite capacitive reactance produced by the feeding gap (W_1) at 5.5 GHz. Consequently, proper impedance matching is achieved through accurate dimensions of the feeding gap (W_1) and feeding positions, facilitating maximum power transfer. As a result, Ant. 1 successfully resonates at 5.5 GHz with a wider bandwidth, as shown in Figure 3.18A. The wider bandwidth is attained due to the small quality factor Q_0 of the antenna. Q_0 is defined as:

$$Q_0 = \frac{X}{R} = \frac{f}{BW} \tag{3.16}$$

Where X represents reactance, and R is the radiation resistance of the antenna. The small value of reactance, approximately 0 Ω, and the radiation resistance equal to 50 Ω lead to a small Q_0 for Ant. 1, thereby broadening the bandwidth. This verifies that Ant. 1 achieves resonance at 5.5 GHz with a wider impedance bandwidth spanning from 4.92–5.90 GHz, as shown in Figure 3.33A.

Next, a monopole radiating strip RS (inverted J) with a length approximately equal to a quarter wavelength and resonating at 2.45 GHz was computed using Equation (3.15) and optimized using the CST optimization technique. The optimized length was determined as 28 mm. The strip RS is coupled to Ant. 1, connecting at point "Q," without disturbing the feeding positions, and creates a radiator gap, as depicted in Figure 3.33B. This complete structure is referred to as Ant. 2. The radiator gap and monopole radiating strip RS act as an electrical resonator due to the shunt feeding, and in conjunction with Ant. 1, exhibit dual resonances at 2.9 GHz and 5.8 GHz, as shown in Figure 3.18B. However, impedance matching is disrupted at 2.45 GHz and 5.5 GHz due to additional capacitive coupling introduced by the radiator gap, preventing Ant. 2 from covering the desired operating bands.

Finally, Ant. 3 was designed to achieve proper impedance matching at the desired resonances by incorporating a vertical open-ended tuning stub denoted as "t" at point "S," without compromising the dual resonances, wider impedance bandwidth, and compactness, as depicted in Figure 3.33C. The vertical tuning stub "t" regulates the additional capacitive coupling by acting as a barrier for the radiator gap, consequently shifting the resonance from 5.8 GHz to 5.5 GHz. Simultaneously, the length of the stub "t" increases the path for electrical current, shifting the resonance from 2.9 GHz to 2.45 GHz. Ant. 3 covers two frequency bands, namely F_l (2.39–2.65 GHz) and F_u (4.96–5.92 GHz), with wide impedance bandwidths for VSWR <2.

3.5.3.1 Achievement of Wider Bandwidth across Operating Bands

To further validate the wider bandwidth feature of Ant. 3, the graph of input impedance Z_{in} versus frequency is depicted in Figure 3.34(a). In the operating bands F_l and F_u, proper impedance matching has been achieved through correct feeding positions and accurate dimensions of W_1 and t. Consequently, the input impedance remains stable around 50 Ω, with reactance close to 0 Ω. According to Equation (3.16), this results in a very small Q_0 for Ant. 3, which widens the bandwidth in both operating bands. Figure 3.34(b) also demonstrates that due to excellent impedance matching, there are numerous densely populated current paths in Ant. 3, indicating varying electrical lengths. The merging of these current paths leads to a decrease in Q_0, further widening the bandwidth.

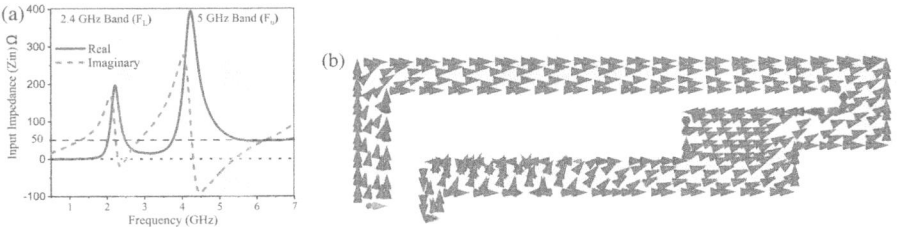

FIGURE 3.34 Wider bandwidth achievements in operating bands of proposed antenna (a) input impedance Z_{in} versus frequency and (b) vector current distribution.

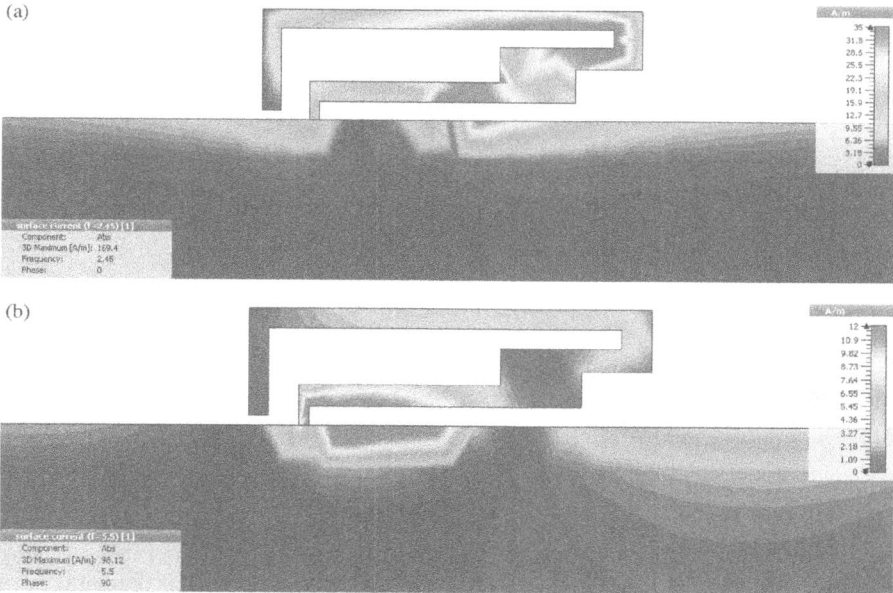

FIGURE 3.35 Simulated surface current (A/m) distribution (a) at 2.45 GHz and (b) 5.55 GHz of the proposed antenna.

3.5.3.2 Surface Current Distribution of the Proposed Antenna

To gain a better understanding of the desired dual-band characteristic of the proposed antenna, the simulated surface current distribution at 2.45 GHz and 5.5 GHz is analyzed in Figure 3.35. It is observed that at 2.45 GHz, the maximum current flows through the strip RS, with relatively small current flowing through the vertical tuning stub "t." This confirms that strip RS and tuning stub "t" contribute to the antenna's operation in the F_l band. On the other hand, at 5.5 GHz, the maximum current flows through the strip PQ, while almost zero current flows through the vertical tuning stub. This indicates that the tuning stub acts as a barrier, enabling the antenna to operate in the F_u band.

As the antenna operates in a dual-band configuration with a wider bandwidth, meeting the requirements for WLAN and WiMAX operations in laptop computers, it was selected for further testing and will be referred to as the proposed antenna in the subsequent sections.

3.5.4 PARAMETRIC ANALYSIS OF PROPOSED ANTENNA

Parametric analysis was conducted to investigate the effect of varying the open-ended vertical tuning stub "t" and the distance "L_x" for mounting the antenna on the system ground. Additionally, the effect of varying the sizes of the system ground (display screen) on the antenna's performance was analyzed.

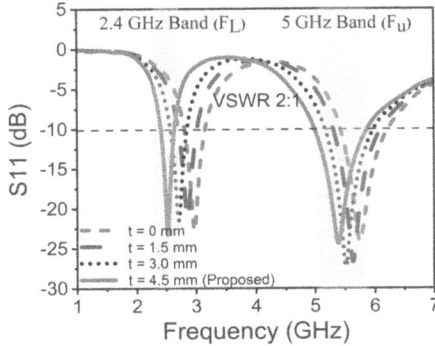

FIGURE 3.36 Effect of varying "t" on performance of the proposed antenna.

3.5.4.1 The Effect of Varying Length "t" of Tuning Stub on the Performance of the Proposed Antenna

The length "t" primarily affects the resonances of the operating bands. Figure 3.36 demonstrates that as the value of "t" increases, the resonances at 2.9 GHz and 5.8 GHz shift toward lower frequencies. At a length of 4.5 mm for "t," the desired resonances of 2.45 GHz and 5.5 GHz are achieved with the desired bandwidth.

3.5.4.2 The Effect of Varying Distance "Lx" for Mounting the Proposed Antenna on System Ground

The effect of varying "L_x" for mounting the proposed antenna on the system ground over the F_l and F_u bands is analyzed in Figure 3.37. It is confirmed that the mounting position of the antenna on the top edge of the system ground does not significantly affect its performance. In this design, "L_x" is chosen as 35 mm

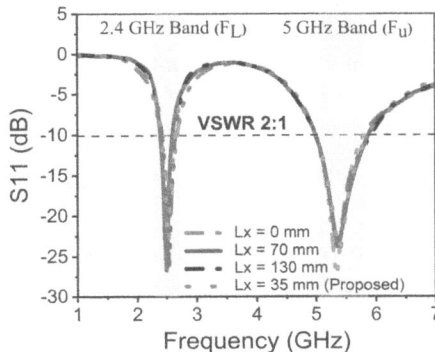

FIGURE 3.37 Effect of varying "L_x" on performance of the proposed antenna.

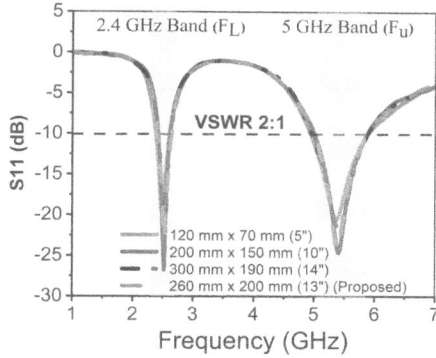

FIGURE 3.38 Effect of varying size of system ground (display screen) on the performance of the proposed antenna.

from the left, considering that the top left edge is typically reserved for installing the laptop assembly, while the center space is reserved for embedding the camera.

3.5.4.3 The Effect of Varying the Sizes of the System Ground (Display Screen) on the Performance of Proposed Antenna

To assess the impact of different sizes of the system ground (display screen) on the proposed antenna, standard sizes such as "5-inch" (mobile), "10-inch" (tablet/notebook), and "14-inch" (laptop) display screens are compared with the existing system ground. Figure 3.38 confirms that the proposed antenna performs equally well for different sizes of the system ground, with negligible effects on impedance bandwidth.

3.5.5 Dimensions of the Proposed Antenna

Based on the parametric study and using optimization techniques in CST, the optimized parameters for the proposed antenna are specified in Table 3.10.

TABLE 3.10
Optimized Parameters of the Proposed Antenna

Parameter	Value (mm)	Parameter	Value (mm)	Parameter	Value (mm)
L_1	0.5	L_6	1.0	W_4	1.2
L_2	14	L_x	35	W_5	1
L_3	4	W_1	1	W_6	1
L_4	7.5	W_2	1.1	T	4.5
L_5	17.5	W_3	0.6		

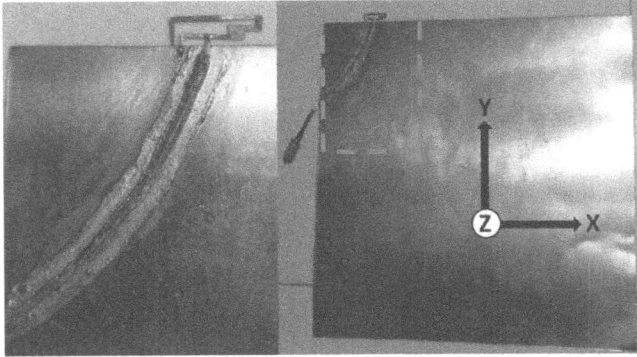

FIGURE 3.39 Fabricated photo of the proposed antenna.

3.5.6 Performance Characterization of Proposed Antenna

The proposed antenna prototype, constructed according to the optimized dimensions outlined in Table 3.10, underwent fabrication and testing, as illustrated in Figure 3.39. As previously discussed, coaxial cable was employed for feeding the antenna due to its robust nature, flexibility in choosing the feed point, and reliable performance characteristics. To minimize losses and enhance performance, a short length of coaxial cable measuring 10 cm was utilized. The coaxial cable had a diameter of 1.8 mm and supported a maximum operating frequency of up to 8 GHz.

When working with coaxial cable, it is important to consider the possibility of unbalanced current flowing through the outer conductor, resulting in the formation of standing waves on the cable and causing secondary radiation. This phenomenon can lead to increased losses and attenuation, subsequently impacting overall performance. To mitigate these effects, the coaxial cable was carefully routed over the system ground. The bending radius of the cable is a crucial parameter to consider in this regard. While every coaxial cable has a defined bending radius, it is advantageous to minimize bending as much as possible to avoid further losses and attenuation, even within the specified bending radius. Consequently, the coaxial cable was routed with the least amount of bending to overcome the impact of standing waves and achieve optimal performance.

VSWR and S_{11} (dB) measurements were conducted using a network analyzer (Rohde & Schwarz) capable of operating up to 16 GHz. The radiation performance of the antenna, including radiation patterns, gain, and radiation efficiency, was evaluated in a calibrated anechoic chamber measuring $8 \times 4 \times 4$ m^3, as depicted in Figure 3.40.

3.5.6.1 Reflection Coefficient (S_{11}) dB Characteristics of the Proposed Antenna

Figure 3.41 showcases the simulated and measured S_{11} (dB) values of the proposed antenna, with a comparison provided in Table 3.11. A minimal deviation between the

FIGURE 3.40 Radiation measurement set-up in the anechoic chamber.

simulated and measured values is observed, which can be attributed to soldering and fabrication tolerances. Nevertheless, both the simulated and measured performance in the F_l and F_u bands align with the 2.4/5 GHz WLAN standards and the 5.8 GHz WiMAX band, demonstrating a wider bandwidth.

3.5.6.2 Radiation Pattern of the Proposed Antenna

Figure 3.42 demonstrates the excellent agreement between the simulated and measured radiation patterns of the proposed antenna in the x-y plane at 2.45 GHz and 5.5 GHz. The measured patterns exhibit smooth and clean characteristics, attributed to the proper routing of the coaxial cable. The E_θ patterns exhibit bi-directional radiation with a higher energy distribution in the horizontal direction, ensuring maximum

TABLE 3.11
Comparison of S_{11} (dB) and Operating Bands (GHz)

	Amplitude (dB)		Operating Bands (GHz)	
	F_l	F_u	F_l	F_u
Simulated	−23	−25	2.39–2.65	4.96–5.92
Measured	−22	−23	2.35–2.63	4.92–5.94

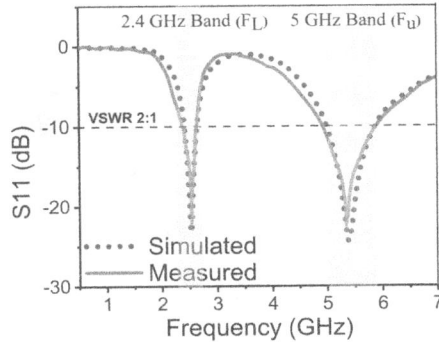

FIGURE 3.41 Comparison of simulated and measured reflection coefficient versus frequency.

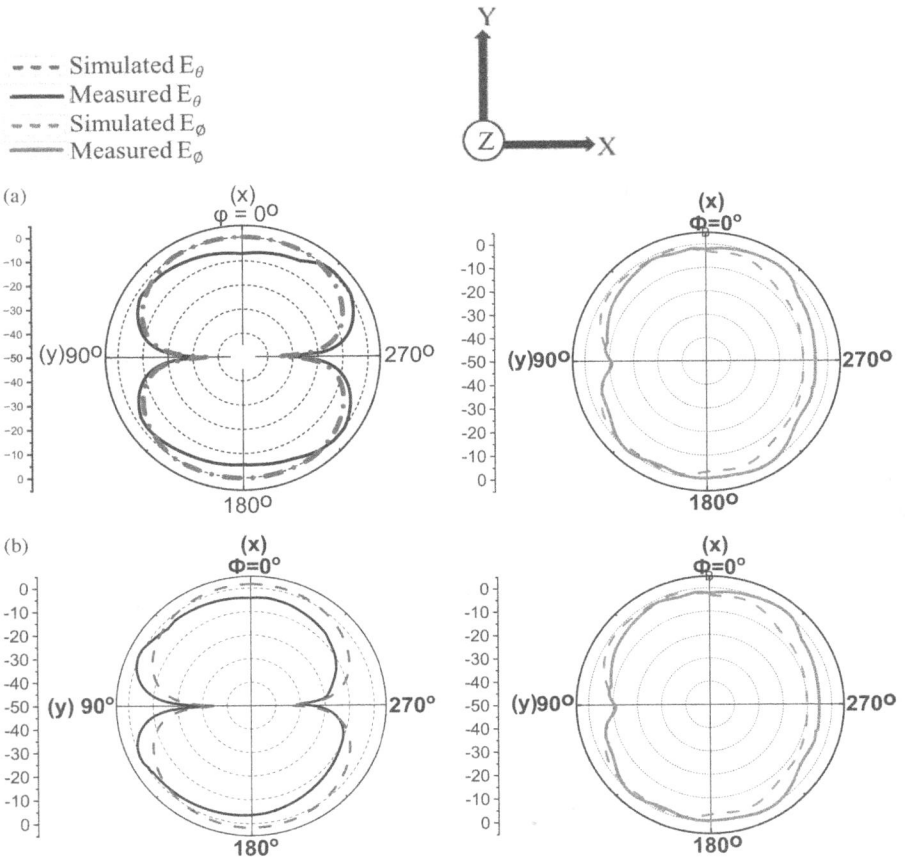

FIGURE 3.42 The simulated and measured radiation patterns of the proposed antenna (a) at 2.45 GHz (x-y plane) and (b) at 5.5 GHz (x-y plane).

energy for WLAN/WiMAX applications in laptop computers. On the other hand, the $E_\varnothing$ radiation patterns are nearly omnidirectional, enabling broad coverage of the area.

3.5.6.3 Simulated and Measured Gain and Efficiency

The measurement of gain requires the same facility as radiation pattern measurement and is based on the Friis transmission equation, as expressed in Equation (3.17).

$$G_{2dB} = 20log_{10}\left(\frac{4\pi r}{\lambda}\right) + 10log_{10}\left(\frac{p_2}{p_1}\right) - \left(G_{1dB}\right) \qquad (3.17)$$

where, G_{2dB} and G_{1dB} are gains of the proposed and the reference antenna, respectively, r is the distance between the reference and proposed antenna, and p_2 and p_1 are the received and transmitted power of a proposed and reference antenna, respectively.

The radiation efficiency of the proposed antenna is measured using the gain/directivity method employing the anechoic chamber. The radiation efficiency of the antenna is the ratio of the gain (G) to the directivity (D) and is mathematically expressed as given in Equation (3.18).

$$\eta_r\ (\%) = \frac{Gain}{Directivity} \qquad (3.18)$$

Here, the gain is obtained using Equation (3.17), whereas the directivity is obtained using the CST MWS simulation tool. As directivity does not contain any antenna losses, the simulated directivity is almost equal to the measured directivity.

The simulated and measured values of gain and efficiency for the proposed antenna are graphically represented in Figure 3.43, with a comparison provided in

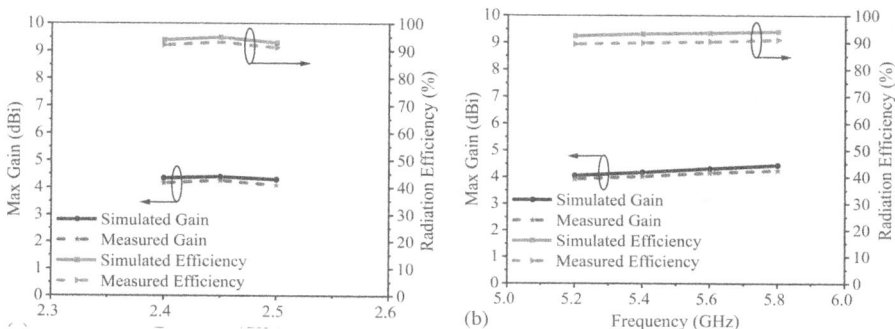

FIGURE 3.43   Simulated and measured gain and efficiency of the proposed antenna (a) in the F_l band and (b) in the F_u band.

TABLE 3.12
Comparison of Simulated and Measured Gain and Efficiency

Operating Bands (GHz)	Gain (dBi)		Efficiency (%)	
	Simulated	Measured	Simulated	Measured
F_l	4.30–4.35	4.09–4.26	93.00–94.00	91.22–93.20
F_u	4.06–4.45	3.93–4.25	92.43–94.00	89.44–91.00

Table 3.12. A small deviation between the simulated and measured values is observed due to the ideal conditions in simulation. However, this deviation remains acceptable for wireless operations in laptop computers. Table 3.12 concludes that the proposed antenna exhibits stable gain and excellent efficiency in both operating bands (F_l and F_u), demonstrating a high-quality signal reception necessary for wireless operations in laptop computers.

3.5.7 EXPERIMENTAL ANALYSIS OF PROPOSED ANTENNA

To evaluate the impact of the plastic/metallic cover of a laptop computer on the proposed antenna, the antenna was mounted on an L412 ThinkPad by Lenovo, as illustrated in Figure 3.44(a). The measured results of the reflection coefficient S_{11} (dB) versus frequency (GHz) are depicted in Figure 3.44(b). It is observed that the resonant mode of both F_l and F_u experiences a slight shift toward the lower frequency range due to the influence of the laptop cover material. In this scenario, the measured operating bands for F_l and F_u span from 2.22–2.41 GHz and 4.53–5.51 GHz, respectively. However, by tuning the length "t" of the tuning stub, the desired bands of interest for the proposed antenna installed in an actual laptop can be restored.

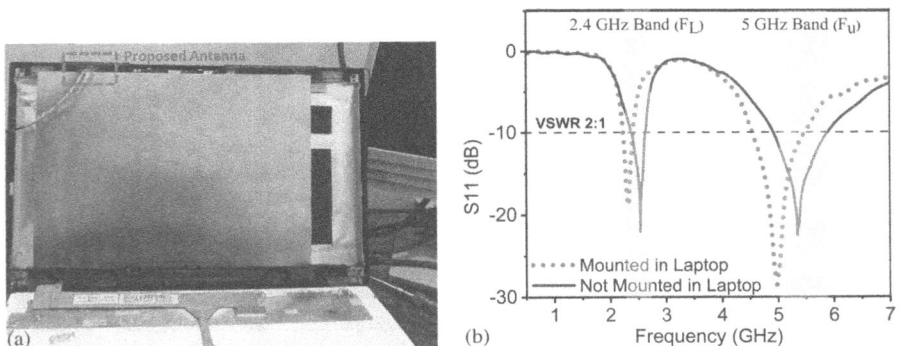

FIGURE 3.44 (a) Proposed antenna installed inside practical laptop and the (b) comparison of measured s_{11} (dB) results when proposed antenna is installed in the laptop and when it is not installed in the laptop.

3.5.8   COMPARISON OF PROPOSED ANTENNA WITH EXISTING STATE-OF-THE-ART TECHNOLOGY

The comparison of the proposed antenna with the existing state-of-the-art technology is presented in Table 3.13

3.5.9   FEATURES OF PROPOSED ANTENNA

The proposed antenna exhibits several remarkable features, which are as follows:

- **Wide band support:** The antenna supports the WLAN/Wi-Fi/WiMAX bands without requiring any additional matching circuits. This simplifies the design and ensures compatibility with these communication standards.
- **Extended operating bands:** The antenna offers wider operating bands with a percentage bandwidth of 11.27% and 18.7% in comparison to previously reported antennas [3, 14]. This expanded bandwidth enhances the versatility and coverage of the antenna.
- **Compact size:** The proposed antenna boasts a compact size of 17.5 mm × 6 mm, setting it apart from antennas documented in previous studies [2, 8, 30]. Its smaller dimensions enable easier integration into various devices and systems.
- **High efficiency:** With efficiency exceeding 90% in both operating bands, the proposed antenna outperforms antennas discussed in [3, 14, 30]. This exceptional efficiency ensures optimal signal transmission and reception, maximizing the antenna's performance.
- **Ultra-thin profile:** The proposed antenna achieves an unprecedented thickness of only 0.2 mm, distinguishing it as the thinnest antenna in the existing literature. This ultra-thin profile facilitates seamless integration into space-constrained devices, where thickness is a critical factor.
- **Overall**: The proposed antenna stands out due to its wide band support, extended operating bands, compact size, high efficiency, and ultra-thin profile, making it a compelling choice for various wireless communication applications.

3.5.10   CONCLUDING REMARKS

A groundbreaking and highly innovative design introduces an inverted E-shaped monopole antenna with exceptional characteristics. This antenna, measuring a mere $20 \times 6 \times 0.2$ mm³, achieves an impressively wide impedance bandwidth. In the lower band (F_1), it covers 11.24% (2.35–2.63) GHz, while in the upper band (F_u), it spans 18.78% (4.92–5.94) GHz, all while maintaining a VSWR below 2. These bandwidths perfectly align with the standard requirements of 2.4/5.2/5.8GHz WLAN and 5.8 GHz WiMAX bands, rendering the proposed antenna ideal for wireless applications in laptop computers.

Moreover, extensive radiation performance testing validates the antenna's capability to deliver high-quality signal reception within the desired F_1 and F_u bands. The

TABLE 3.13

Comparison of Proposed Antenna with Existing State-of-the-Art Technology

Ref.	Type of Antenna	Thickness in (mm)	Operating Bands (GHz)	% BW	Gain (dBi) in Operating Bands	η (%)	Height of Antenna above the System Ground (mm)	Remarks
[2]	Inverted-F	1.6	2.4/3.5/5	-	2.56/3.39/3.73	99/98/88–99	9.3	Printed on hard and expensive Taconic substrate.
[3]	Inverted-F with meander shorting	0.4	2.4/5	6.50/17.71	2.17/4.63	>54/>55	3	Two-sided structure, hence, complex hardware and expensive.
[7]	Coupled-fed PIFA	0.8	2.4/3.5/5	13/14.08/28.19	4.7/5.3/6.6	82–92/90–95/92	9	Fabrication process is difficult.
[8]	Coupled-fed PIFA	1.6	2.4/5	5.48/30.68	4.3/5.5	67/67	9	
[9]	Antenna with metal housing	0.8	2.4/5	10.04/20.25	-/-	>80/>80	6	Antenna excitation technique is not mentioned.
[10]	Monopole	0.8	2.4/5	8.03/19.21	3.91/5.9	>84/>92	6	Antenna is placed at center, which is normally reserved for camera.
[14]	Antenna with adhesive ground	0.4	2.4/5	8.8/16.78	1.47/3.15	79/83	6	Uses additional ground plane, hence, making it difficult to integrate in laptops.
[30]	Printed monopole	0.4	2.4/5	7.43/20	3.9/4.3	80/74	9	Uses additional reactive components for impedance matching.
Proposed Antenna	Inverted-E Monopole	0.2	2.4/5	11.24/18.7	4.26/4.25	93.2/91.0	4	Low profile antenna with simple structure.

proposed antenna not only satisfies the required performance criteria but also offers distinct advantages such as an ultra-thin profile, low cost, superior performance, and accelerated time to market. These attributes position it as a prime candidate for deployment in modern ultra-thin laptops.

In summary, the novel, ultra-thin, and low-profile inverted E-shaped monopole antenna showcases outstanding impedance bandwidth, fulfilling the demands of WLAN and WiMAX bands. Its remarkable radiation performance ensures optimal signal reception, making it an excellent choice for wireless operations in laptop computers. With its unique blend of ultra-thin form factor, cost-effectiveness, enhanced performance, and rapid time to market, the proposed antenna emerges as a compelling solution for the latest generation of ultra-thin laptops.

3.6   ULTRA-THIN DUAL-BAND MONOPOLE ANTENNA FOR 5G (SUB-6 GHZ) AND WLAN IN LAPTOP COMPUTERS

3.6.1   IMPORTANCE OF 5G (SUB-6 GHz) BAND

Currently, laptop computers integrated with long term evolution (LTE) technology have two significant limitations:

- The battery life of these laptops is insufficient for enterprise use, making them less suitable for professional users.
- The data speeds provided by 4G LTE networks are not sufficient for a consistently connected experience, mainly due to latency issues and the power consumption of the current generation of processors.

To address these drawbacks, the implementation of sub-6 GHz-5G technology is highly anticipated. Sub-6 GHz-5G offers numerous advantages, including lower latency, improved battery life, highly efficient processors similar to those found in smartphones, and seamless migration between Wi-Fi 802.11ac and 5G cellular connectivity. The integration of 5G into laptops is particularly appealing to business users who require ruggedized notebooks for fieldwork, video production, and high-definition video conferencing on the go. The expectation is that laptops should always be connected to mobile data networks, setting the stage for future products.

Intel's laptops and two-in-one devices will utilize eighth-generation Core i5 processors along with XMM 8000 series 5G modems, which operate on both sub-6 GHz and millimeter wave cellular networks. This series of devices will enable a wide range of devices to connect to 5G laptops, including PCs, phones, fixed wireless consumer premise equipment, and vehicles. Intel is heavily investing in its wireless portfolio and collaborating with partners to bring 5G-connected mobile PCs to the market. The integration of 5G (sub-6 GHz) technology into laptops offers the following advantages:

- Significantly increased peak bit rate.
- Higher system spectral efficiency, resulting in a larger data volume per unit area.

- Enhanced capacity to accommodate concurrent and instantaneous connectivity for multiple devices.
- Lower battery consumption.
- Improved connectivity regardless of the geographic region.
- Support for a larger number of devices.
- Reduced infrastructural development costs.
- Enhanced reliability of communication.

Furthermore, laptops equipped with 5G (sub-6 GHz) technology offer a wide range of features that benefit various groups of people, including students, professionals (such as doctors, engineers, teachers, and administrative bodies), and the general public. Some of the advantages of these laptops for the general public include:

- **Parallel multiple services:** Users can access multiple services simultaneously, such as checking the weather and location while having a conversation.
- **Enhanced education:** Students located anywhere in the world can attend classes through their laptops.
- **Improved healthcare:** Doctors can treat patients located in remote areas.
- **Efficient monitoring:** Governmental organizations and law enforcement agencies can monitor any part of the world.
- **Missing person location:** The laptops can assist in locating and searching for missing individuals.
- **Faster natural disaster detection:** The laptops can aid in the rapid detection of natural disasters like tsunamis and earthquakes.

Considering the significance of 5G (sub-6 GHz), it is essential to design antennas that meet the bandwidth requirements of sub-6 GHz 5G bands and can be easily installed in laptop computers. It would be advantageous if the antenna could support both sub-6 GHz 5G and WLAN standards in a single antenna. Since 5G technology is not yet fully matured, considerable efforts are being made to make it widely available worldwide. If the antenna also covers WLAN standards, users can remain connected to seamless internet even in areas where 5G technology is not available.

Therefore, the objective of this section is to design an ultra-thin, dual-band 5G (sub-6 GHz) and WLAN antenna for wireless operations in the next generation of laptop computers.

3.6.2 Overview of Proposed Antenna

We have developed a novel, ultra-thin monopole antenna for next-generation laptop computers, operating in the sub-6 GHz 5G and WLAN frequency bands. This antenna has an impressively thin profile, measuring only 0.2 mm in thickness. It is constructed using a pure copper strip with dimensions of 17.5×6 mm². The antenna's design is based on a simple structure comprising two monopole radiating strips labeled AC and CD, along with an open-ended rectangular tuning stub labeled BE, measuring 9 mm in length.

This innovative antenna structure enables the generation of two resonating modes at frequencies of 3.45 GHz and 5.5 GHz. Through careful design, we have achieved an impedance bandwidth of 20% (3.21–3.91) GHz in the lower band (F_l) of the sub-6 GHz 5G spectrum and 15% (5.05–5.875) GHz in the upper band (F_u) conforming to the 802.11a/ac WLAN standards, all while maintaining a VSWR below 2. These frequency bands cover the 3.3–3.6 GHz sub-6 GHz 5G range and the 5 GHz WLAN standards.

The measured radiation performance of our proposed antenna is exceptional, featuring nearly omnidirectional radiation patterns, a stable gain of approximately 5 dBi, and excellent efficiency of around 90% in both operating bands. To validate the practicality and effectiveness of our antenna structure, we have derived a simplified EC model and performed simulations. The simulated results align well with the measured data, demonstrating the suitability of our antenna structure for sub-6 GHz and WLAN operations in ultra-thin laptop computers.

3.6.3 PROPOSED ANTENNA GEOMETRY

Figure 3.45 illustrates the schematic geometry of our proposed antenna, specifically designed for sub-6 GHz 5G and WLAN operations in laptop computers.

The proposed antenna design comprises two monopole radiating strips, namely, strip AC and strip CD, along with an open-ended rectangular tuning stub labeled BE, which has a length denoted as L6. The antenna is positioned on the top edge of the system ground plane at point A, at a distance L_x, and has a height of 6 mm. The system ground plane is constructed from 91% brass material and is formed by the display ground and keyboard ground, both measuring $260 \times 200 \times 0.2$ mm^3. This system ground plane supports a 13-inch laptop display screen, which has been chosen for its optimal balance between device portability and usability. The display ground lies on the x-y plane, while the keyboard ground lies on the x-z plane, forming an angle denoted as "β" as depicted in Figure 3.45. The antenna design incorporates a 50 Ω miniature coaxial feed line, where the central conductor is connected to point "P" on strip AC, and the grounding sheath is soldered at point "Q" on the display ground. This feeding configuration ensures that the effective dielectric constant of all radiating strips is equal to 1 [34].

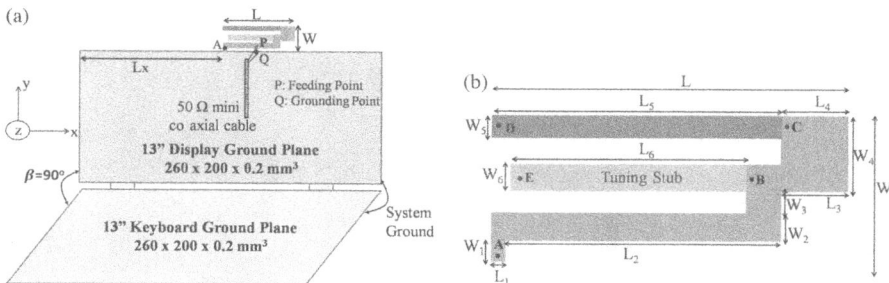

FIGURE 3.45   Schematic geometry of the proposed antenna (a) mounted on system ground and (b) geometry of proposed antenna (B).

3.6.4 EVOLUTION OF PROPOSED ANTENNA

To illustrate the design process of the "3"-shaped monopole antenna, Figure 3.46 provides a step-by-step evolution and the corresponding reflection coefficient (S_{11}) results obtained using CST simulations.

Initially, to excite resonance at 3.45 GHz, a step-shaped radiating strip AC is computed with a length LAC that is approximately a quarter wavelength long. The length LAC is determined using Equation (3.19) and is further optimized through CST MWS simulations as described in Equation (3.20).

$$L_{AC} = \frac{\lambda}{4} = \frac{c}{4f} \qquad (3.19)$$

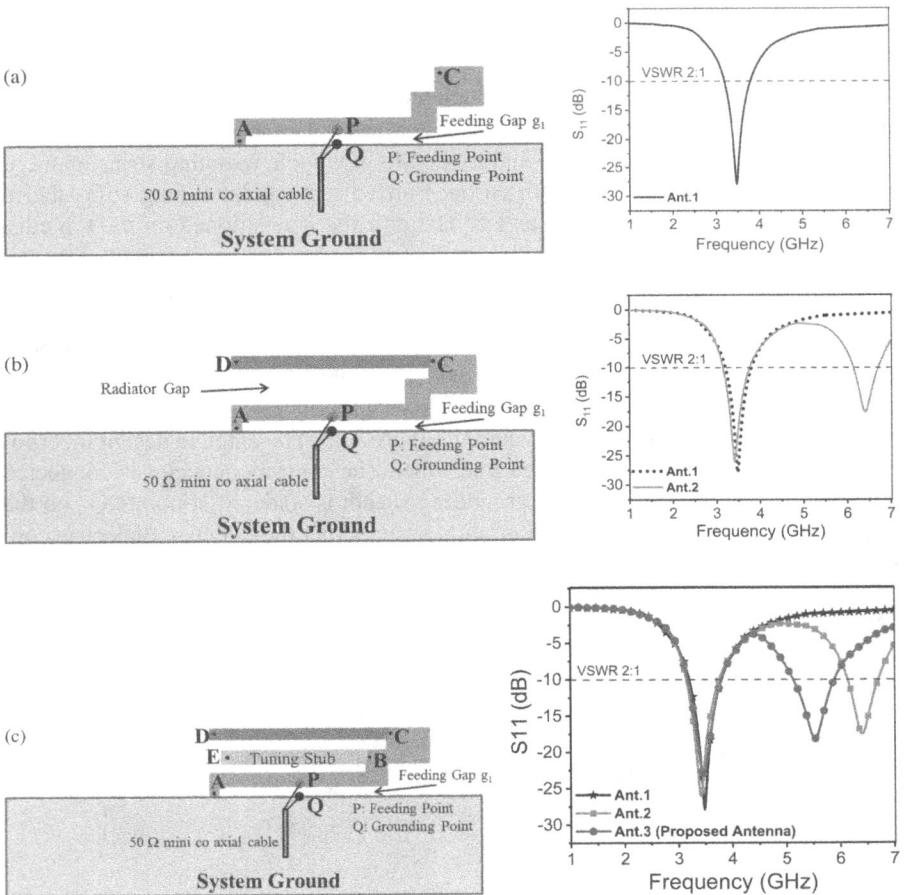

FIGURE 3.46 The step-by-step design process of the proposed antenna (a) Ant. 1, (b) Ant. 2, and (c) Ant. 3 (proposed antenna).

In (3.19), "c" is the speed of light (m/s) and "f" is corresponding resonating frequency (GHz).

$$L_{AC} = \left(W_1 + L_2 + W_2 + L_3 + W_3 + W_4 \right) \tag{3.20}$$

The shorter arm of strip AC is placed on the system ground, creating an open feeding gap labeled g1 between the monopole radiating strip and the system ground. This configuration forms Ant. 1, as shown in Figure 3.46(a). Ant. 1 is excited at points P and Q, and the careful selection of feeding points and feeding gap (g_1) results in excellent impedance matching at 3.45 GHz.

To generate resonance at 5.50 GHz without affecting the behavior at 3.50 GHz, a quarter-wavelength long monopole radiating strip CD at 5.50 GHz is connected at point C of strip AC, while maintaining the feeding position. This introduces a radiator gap between strip AC and strip CD, as depicted in Figure 3.46(b). The length L_{CD} of strip CD is calculated using Equation (3.19) and optimized through CST MWS simulations as described in Equation (3.21).

$$L_{CD} = \left(L_5 \right) \tag{3.21}$$

This structure is referred to as Ant. 2 and exhibits dual resonances due to strip AC and CD, the feeding position, feeding gap (g_1), and radiator gap. The dual resonance is observed at 3.45 GHz and 6.50 GHz, where the resonance at 6.50 GHz is a result of the additional capacitive reactance offered by the radiator gap.

To mitigate the additional capacitive reactance, an open-ended tuning stub labeled BE is added between strip AC and strip CD. This tuning stub reduces the radiator gap without compromising the compactness of the antenna. This modified structure is referred to as Ant. 3, as shown in Figure 3.46(c). The addition of the tuning strip shifts the resonance from 6.50 GHz to 5.50 GHz, while maintaining the resonance at 3.50 GHz. Consequently, Ant. 3 achieves the desired resonances at 3.45 GHz and 5.50 GHz, covering the frequency bands F_1 (3.17–3.84) GHz and F_u (5.12–5.90) GHz, respectively, for wireless operations in the laptop computer. Due to the successful achievement of the desired bands, Ant. 3 is selected for further testing and will be referred to as the proposed antenna in subsequent sections.

3.6.4.1 Input Impedance Versus Frequency of Proposed Antenna

To further evaluate the design of the proposed antenna, Figure 3.47 illustrates the graph of input impedance (Z_{in}) versus frequency. It is evident that within the operating bands F_1 and F_u, the input impedance remains stable around 50 Ω, with reactance close to 0 Ω. This characteristic confirms that the proposed antenna exhibits excellent impedance matching in the operating bands, enabling maximum power transmission.

3.6.4.2 Simulated Surface Current Distribution (A/m) of Proposed Antenna

To validate the design of the proposed antenna, the simulated surface current distribution at 3.45 GHz and 5.50 GHz is depicted and analyzed in Figure 3.48. From

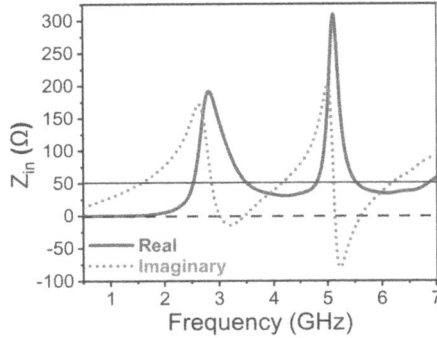

FIGURE 3.47 Input impedance Z_{in} (Ω) versus frequency (GHz).

Figure 3.48(a), it is clearly observed that at 3.45 GHz, the strip AC carries the maximum current, indicating its dominant role in this operating frequency. Conversely, Figure 3.48(b) demonstrates that at 5.50 GHz, the strip CD exhibits the highest current flow, accompanied by a relatively smaller current flowing through specific regions of strip AC and tuning stub BE. These results affirm the proposed antenna's capability to operate effectively in the desired dual bands, rendering it suitable for 5G/WLAN operations in laptop computers.

3.6.4.3　EC Model

Figure 3.49(a) presents the EC model of the proposed antenna, which can be described as two parallel RLC circuits and an open-ended tuning stub.

The series inductance (L) and capacitance (C) are the distributed parameters of the coaxial feed line, computed according to Equations (3.8–3.9). The parallel

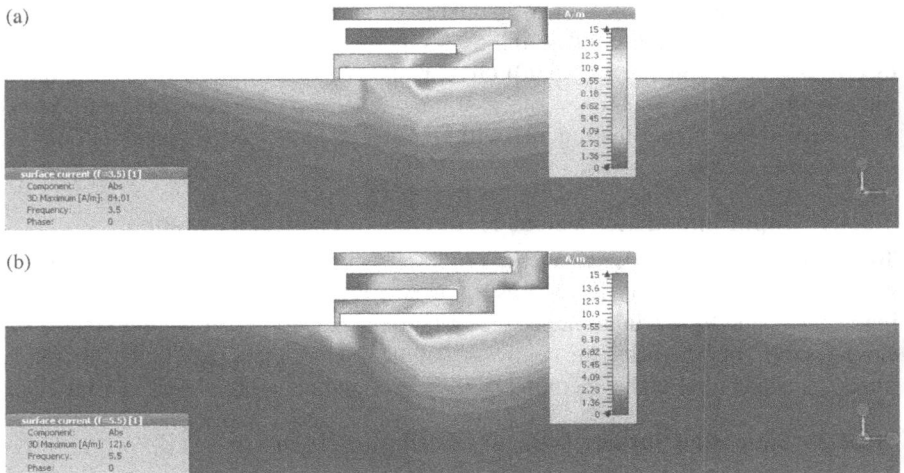

(a)

(b)

FIGURE 3.48 Simulated surface current (A/m) distribution (a) at 3.45 GHz and (b) 5.5 GHz (B).

(a)

(b)

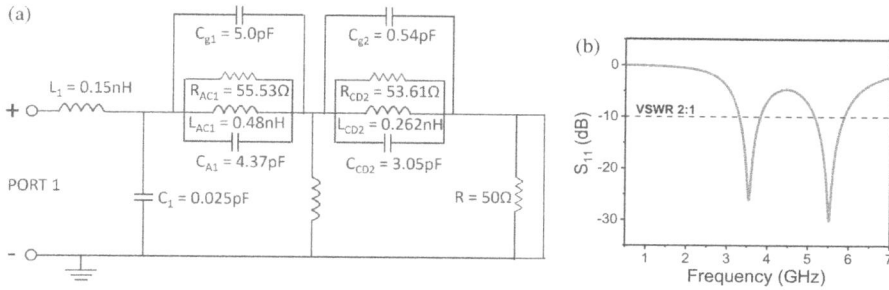

FIGURE 3.49 EC model of the proposed antenna: (a) circuit model and (b) simulated S_{11} using.

resonator circuit formed by radiating strip AC and feeding gap g_1 is represented by R_{AC1}, L_{AC1}, and C_{AC1}. Similarly, the second resonator circuit is formed by strip CD and the radiating gap between strip CD and AC, represented by R_{CD2}, L_{CD2}, and C_{CD2}. The open-ended tuning stub BE is realized using inductor L_{BE}, with its value determined through manual adjustment and optimization techniques in ADS. The gaps created by tuning stub BE are represented by C_{g1} and C_{g2}, and their values are determined using Equation (3.11).

These two parallel RLC circuits are connected through tuning stub L_{BE}, and the complete circuit is connected in series with a 50 Ω radiation resistance, acting as the load for the proposed antenna. The first and second parallel RLC circuits resonate at 3.45 GHz and 5.50 GHz, respectively. The values of the parallel RLC circuits are obtained using the following expressions:

$$Q_0 = \frac{f}{BW} = \frac{R}{2\pi f L} \tag{3.22}$$

$$f = \frac{1}{2\pi\sqrt{LC}} \tag{3.23}$$

In the above equations, "Q_0" represents the quality factor, "f" is the resonating frequency, and "BW" is the simulated bandwidth in the F_l and F_u bands. The value of resistor R is obtained from Figure 3.47 at the resonances of 3.45 GHz and 5.50 GHz. The simulated S_{11} (dB) of the EC model using ADS software is depicted in Figure 3.49B. It is evident that the EC model also covers the desired F_l and F_u bands, with an impedance bandwidth spanning the range of 3.28–3.83 GHz and 5.14–5.88 GHz, respectively.

3.6.5 PARAMETRIC STUDY

The parametric study is conducted to investigate the impact of varying the length "L_6" of the open-ended rectangular tuning stub on the operating bands F_l and F_u. It also examines the effect of changing the angle "β" on the impedance bandwidth of the proposed antenna. Additionally, the study analyzes the "Effect of varying the

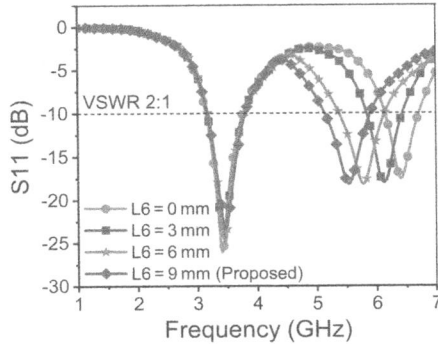

FIGURE 3.50   Effect of varying "L_6" on proposed antenna.

sizes of the system ground" on the impedance bandwidth across the two operating bands to assess the antenna's suitability for other portable devices.

3.6.5.1   Effect of Varying "L_6" on the Performance of Proposed Antenna

The length "L_6" plays a crucial role in shifting the resonance of 6.50 GHz toward the lower frequency, as depicted in Figure 3.50. At a length of 9 mm, the proposed antenna achieves a resonance of 5.50 GHz with an impedance bandwidth of (5.12–5.90) GHz, while maintaining the desired performance in the F_1 band.

3.6.5.2   Effect of Varying an Angle "β" on the Performance of the Antenna

Considering that users may not always maintain an angle "β" of 90 degrees while using the laptop, the effect of varying "β" on the antenna's performance is analyzed. Figure 3.51 demonstrates that changing "β" does not significantly impact the performance of the proposed antenna. This finding validates the antenna's suitability for wireless operations in laptops where the angle "β" may vary according to user convenience, including convertible or two-in-one laptops.

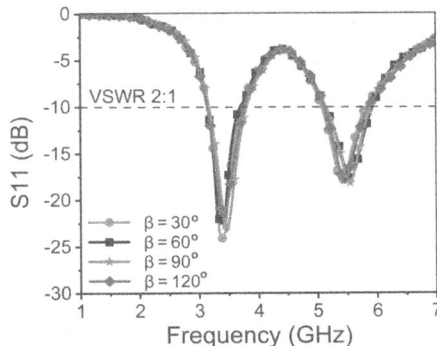

FIGURE 3.51   Effect of varying "β" on the performance of proposed antenna.

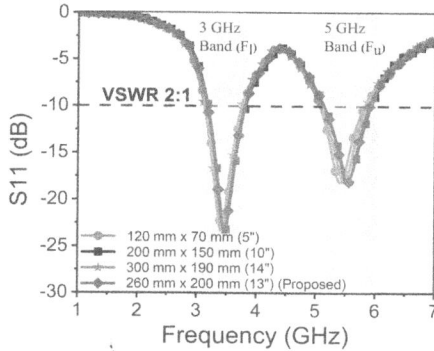

FIGURE 3.52   Effect of varying sizes of the system ground.

3.6.5.3   Effect of Varying the Sizes of the System Ground on the Performance of Proposed Antenna

To evaluate the impact of different sizes of the system ground on the proposed antenna, standard sizes such as "5-inch" (mobile), "10-inch" (tablet/notebook), and "14-inch" (laptop) display screens are compared to the existing system ground. Figure 3.52 verifies that the proposed antenna maintains excellent performance across different sizes of the system ground, with minimal effect on the impedance bandwidth.

3.6.6   Dimensions of Proposed Antenna

Based on the parametric study and utilizing optimization techniques in CST MWS, the optimized parameters for the proposed antenna are specified in Table 3.14.

3.6.7   Performance Characterization of Proposed Antenna

The fabricated prototype of the proposed antenna, utilizing the optimal dimensions specified in Table 3.14, was tested as shown in Figure 3.53. The reflection coefficient (S_{11}) was measured using an Agilent network analyzer (Rohde & Schwarz) up to 16 GHz. The radiation performance, including radiation pattern,

TABLE 3.14
Optimized Parameters of Proposed Antenna

Parameter	Value (mm)	Parameter	Value (mm)	Parameter	Value (mm)
L_1	0.5	L_6	9	W_4	3
L_2	13	L_x	130	W_5	1
L_3	4.5	W_1	1	W_6	1.3
L_4	3	W_2	1.1	L	17.5
L_5	14.5	W_3	0.9	W	6

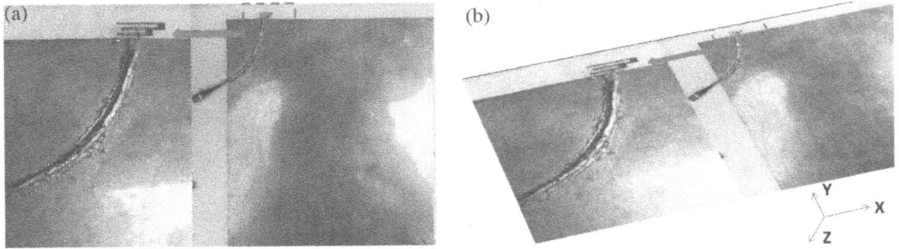

FIGURE 3.53   (a) 2D view and (b) 3D view fabricated photo of the proposed antenna.

gain, and radiation efficiency, was assessed in a calibrated anechoic chamber measuring $8 \times 4 \times 4$ m³.

3.6.7.1   Reflection Coefficient (S_{11}) dB

Figure 3.54 presents the simulated and measured S_{11} (dB) of the proposed antenna, along with their comparison in Table 3.15. A minimal deviation is observed, possibly due to soldering. However, both the simulated and measured F_l and F_u bands cover the sub-6 GHz bands of 5G (3.30–3.60) GHz and the 5 GHz bands of 802.11a/ac WLAN standards, exhibiting a wider impedance bandwidth. Thus, the proposed antenna demonstrates its suitability for dual-band wireless operations in laptop computers.

3.6.7.2   Radiation Patterns

Figure 3.55 shows a slight deviation between the simulated and measured radiation patterns at 3.45 GHz and 5.50 GHz, which could be attributed to the influence of the coaxial cable directly connected at points P and Q during measurement but not considered during simulation, as is typically the case. At these frequencies, the H-plane (x-z plane) radiation pattern is nearly omnidirectional, providing broad coverage, while the E-plane (x-y plane) radiation pattern is bi-directional, radiating more energy horizontally. This ensures maximum energy for 5G (sub-6 GHz) and WLAN

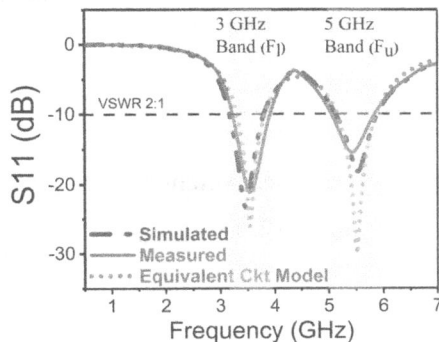

FIGURE 3.54   Simulated and measured S_{11} (dB) versus frequency curves of proposed antenna.

TABLE 3.15

Comparison of S₁₁ (dB) and Operating Bands of Proposed Antenna

	Amplitude (dB)		Operating Bands (GHz)	
	F_l	F_u	F_l	F_u
Simulated	−23.31	−18.20	3.17–3.84	5.12–5.90
Measured	−21.07	−15.5	3.21–3.91	5.05–5.875
Equivalent circuit	−26.10	−30	3.28–3.83	5.14–5.88

standards applications in laptop computers. Additionally, the antenna exhibits pure polarization, with co-polarizations having larger magnitudes than cross-polarization.

3.6.7.3 Simulated and Measured Gain and Efficiency

The simulated and measured values of gain and efficiency for the proposed antenna are illustrated in Figure 3.46, with a comparison in Table 3.16. A small deviation is observed between the simulated and measured values due to the ideal conditions assumed in the simulation. However, this deviation can be accepted. Table 3.16 confirms that the proposed antenna maintains stable gain and excellent efficiency in

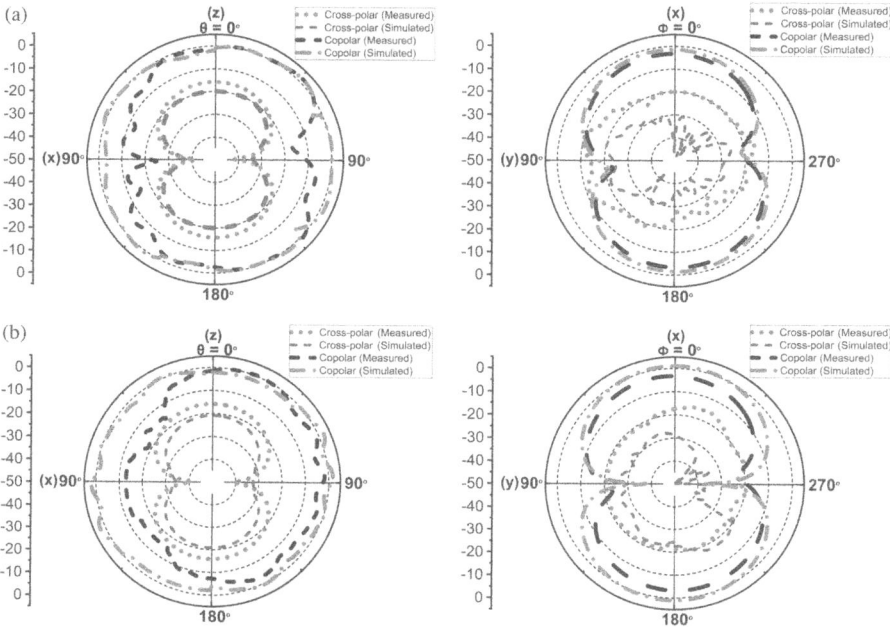

FIGURE 3.55 Simulated and measured radiation patterns (a) at 3.45 GHz and (b) at 5.5 GHz.

TABLE 3.16

Comparison of Simulated and Measured Gain and Efficiency of the Proposed Antenna

Operating Bands (GHz)	Gain (dBi)		Efficiency (%)	
	Simulated	**Measured**	**Simulated**	**Measured**
F_l	5.22–5.50	5.01–5.30	89.50–91.00	87.00–88.25
F_u	5.0–5.42	4.8–5.25	86.43–91.00	83.5–86.80

both operating bands, indicating good signal reception quality, which is essential for wireless operations in laptop computers (Figure 3.56).

3.6.8 Experimental Analysis of Proposed Antenna

To study the effect of the plastic/metallic cover of the laptop on the proposed antenna, it was mounted on the L412 ThinkPad Lenovo, as shown in Figure 3.57(a). Figure 3.57(b) displays the measured result of the reflection coefficient S_{11} (dB) versus frequency (GHz) for the proposed antenna. It is observed that the resonant mode of F_l is slightly shifted toward a lower frequency due to the material effect of the laptop cover. The operating bandwidth spans 2.91–3.45 GHz, while there is negligible effect on the F_u band. However, by slightly tuning the length L_3, the bands of interest for the proposed antenna mounted on the actual laptop can be restored.

3.6.9 Effect of Other Electronic Components on the Performance of the Proposed Antenna

In this section, we investigate the impact of additional electronic components on the performance of the proposed antenna. Specifically, we focus on the battery, speaker, and USB connector, which are mounted on the laptop alongside the proposed

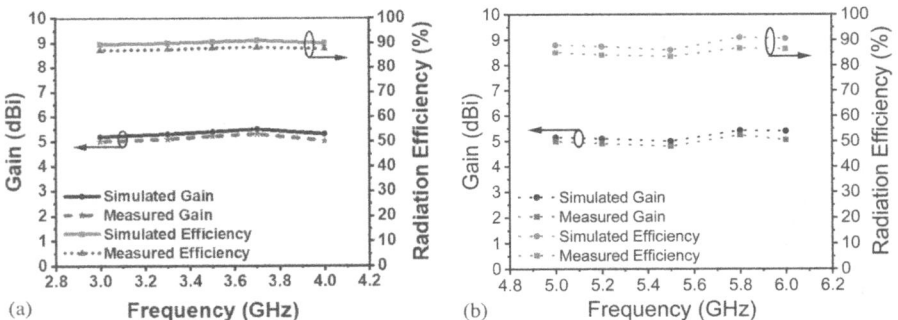

FIGURE 3.56 Simulated and measured gain and efficiency (a) at 3.45 GHz and (b) at 5.5 GHz.

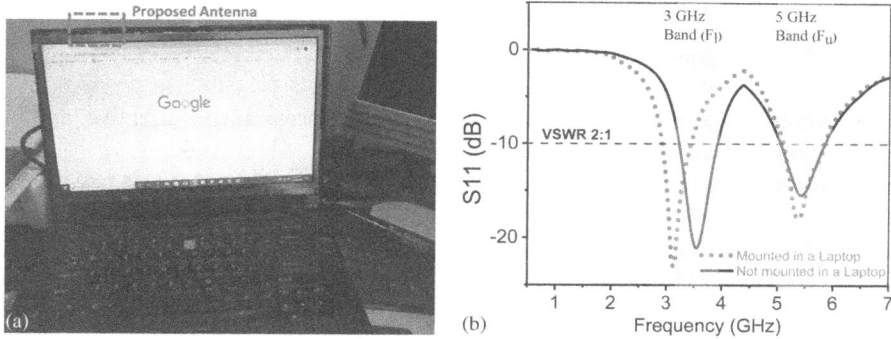

FIGURE 3.57   (a) The proposed antenna installed inside practical laptop and (b) a comparison of measured S_{11} (dB) results when the proposed antenna is installed on the laptop and when it is not installed on laptop.

antenna, as depicted in Figure 3.58(a). To analyze the antenna's performance, we conduct a simulation of the S_{11} (dB) versus frequency (GHz) using CST MWS.

The battery used in the experiment has dimensions of $150 \times 30 \times 4$ mm³ and is placed on the keyboard ground plane, with metal shielding employed for protection. The rectangular speaker, also made of metal, is mounted on the right side of the keyboard ground and has dimensions of $100 \times 20 \times 4$ mm³. Additionally, the USB connector is positioned on the left edge of the keyboard ground.

From the simulation results illustrated in Figure 3.58(b), it is evident that these components have a negligible effect on the antenna's performance. This is attributed to the proposed antenna being mounted on the top edge of the display ground, maintaining a sufficient distance from these electronic components. Thus, their presence does not significantly impact the performance of the antenna.

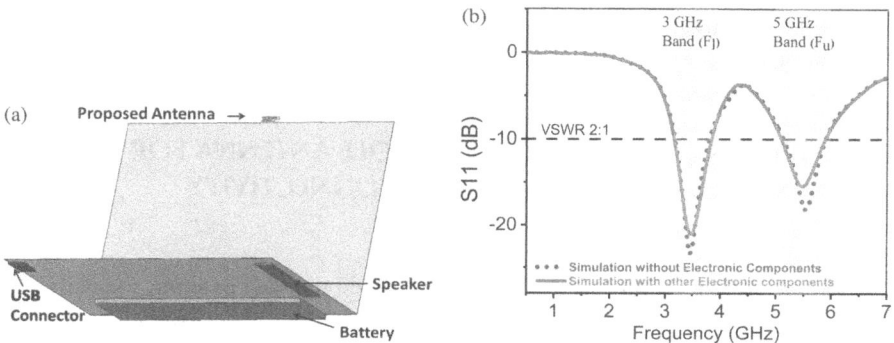

FIGURE 3.58   (a) The placement of electronics components and (b) the simulation results S_{11} (dB) versus frequency effect of other electronic components on the presented antenna.

3.6.10 Features of the Proposed Antenna

The proposed antenna offers several notable features:

- **Simple design:** The antenna design is straightforward, cost-effective, and requires minimal time to bring to market.
- **Compact size:** With dimensions of 17.5 mm × 6 mm, the antenna is remarkably compact, while still achieving a gain and efficiency above 5 dBi and 83%, respectively.
- **Integration of 5G and WLAN:** The antenna seamlessly combines both 5G (sub-6 GHz) and WLAN functionalities, enhancing the performance and coverage of laptop computers in terms of capacity and coverage.
- **Component-free design:** The antenna is engineered without the need for reactive components, additional circuits, or matching circuits.
- **Multiservice compatibility:** The antenna supports various services such as WLAN, WiMAX, and 5G (sub-6 GHz), making it versatile and adaptable to diverse communication requirements.
- **Compatibility with portable devices:** The antenna demonstrates consistent performance across different sizes of system ground, making it suitable for integration with various portable devices.

3.6.11 Concluding Remarks

In conclusion, a novel "3"-shaped monopole antenna with a very low profile and ultra-thin design was successfully developed, fabricated, and tested. The antenna achieved an impedance bandwidth of 20% (3.21–3.91) GHz in F_1 and 15% (5.05–5.875) GHz in F_u, ensuring compliance with the required standard bandwidth of the sub-6 GHz 5G and 5 GHz 802.11a/ac WLAN bands in laptop computers (VSWR <2). The antenna's compact size of 17.5×6 mm^2 and ultra-thin profile, with a thickness of only 0.2 mm, make it exceptionally suitable for modern ultra-thin laptops. Moreover, the antenna exhibits excellent radiation performance for dual-band operations, guaranteeing high-quality signal reception in wireless laptop operations. Overall, due to its low profile, ultra-thin structure, simplicity, cost-effectiveness, superior performance, and quick time to market, the proposed antenna is a promising choice for contemporary ultra-thin laptops.

3.7 EFFICIENT MULTIBAND MONOPOLE ANTENNA FOR SEAMLESS GSM/WLAN/WIMAX CONNECTIVITY

3.7.1 Antenna Overview

Presenting a novel multifrequency printed monopole antenna, specifically designed for GSM, WLAN, and WiMAX standards in laptop devices, the proposed antenna stands out for its simplistic monopole design, achieving multiband operation without the need for reactive components, expensive substrates, or additional hardware. With compact dimensions of only 17.5×9 mm^2, this antenna features

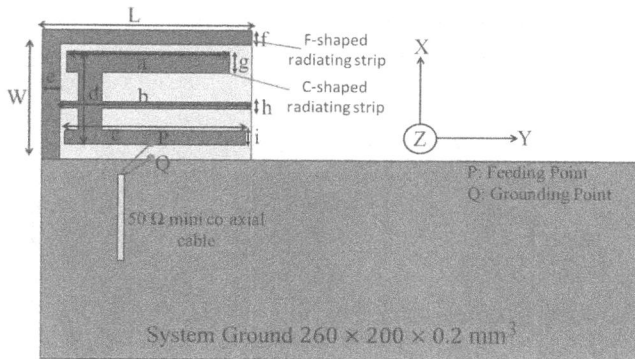

FIGURE 3.59   The proposed antenna attached on the top edge of system ground supporting 13-inch laptop display screen.

a unique structure consisting of an F-shaped strip and a C-shaped strip printed on an FR-4 substrate. Through coaxial feeding, the antenna produces three distinct bands, offering impedance bandwidths of 1.74–1.87 GHz in the lower band (f_l), 2.40–2.50 GHz in the medium band (f_m), and 5.12–6.06 GHz in the upper band (f_u). Remarkably, the antenna exhibits excellent radiation performance, including a gain of approximately 4–5 dBi and an efficiency above 80% within the operating bands. The simulated and measured results demonstrate strong agreement, affirming the suitability of the proposed antenna for GSM 1800/WLAN/WiMAX applications in laptop devices.

3.7.2   Proposed Antenna Geometry and Design

Figure 3.59 showcases the geometric layout of the proposed antenna, printed on a 0.4 mm thick FR-4 substrate with a dielectric constant of 4.3 and a loss tangent of 0.025. The dimensions of the antenna are provided in Table 3.17. Placed on the top edge of the system ground, the proposed antenna is designed to accommodate a 13-inch laptop with a system ground measuring $200 \times 260 \times 0.2$ mm³, composed of 91% brass material. As depicted in Figure 3.1, a straightforward C-shaped monopole strip and an F-shaped monopole strip are printed on a single FR-4 substrate.

TABLE 3.17
Dimensions of Proposed Antenna

Parameters	Values (mm)	Parameters	Values (mm)	Parameters	Values (mm)
L	17.5	c	15.1	g	01.5
W	09.0	d	06.5	h	00.5
a	13.6	e	01.5	i	01.0
b	16.0	f	01.0		

To ensure optimal performance, a small and low-loss 50 Ω miniature coaxial feed line is utilized, with its central conductor connected to point "P" on the lower part of the C-Shaped radiating strip, while its grounding sheath is soldered at point "Q" on the system ground. By embedding both the resonating C-shaped strip and the F-shaped strip on a single FR-4 substrate and employing a miniature coaxial cable for feeding, the antenna achieves resonance for multiband operations, covering GSM, WLAN, and WiMAX frequencies. The proposed antenna's design is simulated using CST MWS software.

3.7.3   WORKING MECHANISM OF PROPOSED ANTENNA

To gain a deeper understanding of the excitation of the triple bands in the proposed antenna, the simulated surface current (A/m) distribution at resonating frequencies of 1.8 GHz, 2.45 GHz, and 5.5 GHz for the f_l, f_m, and f_u bands, respectively, is presented and analyzed in Figure 3.60. From the current distribution observed in Figure 3.60A, the resonating path for the 1.8 GHz frequency is determined. By analyzing the current path, the resonating length for 1.8 GHz is calculated using the following equation:

$$L_1 = \frac{3}{4}L + \frac{W}{2} + \frac{a}{2} + \frac{d}{2} + \frac{3}{4}c \tag{3.24}$$

Using the above equation, the length of the resonance at 1.8 GHz is determined to be 39 mm. Notably, the obtained length corresponds to a quarter wavelength, as expressed in Equation (3.25):

$$f = \frac{c}{4L\sqrt{\varepsilon_{reff}}} \tag{3.25}$$

Here, c represents the velocity of free space, and ε_{reff} denotes the effective dielectric constant, which is derived from the relative permittivity of the substrate.

For the 2.45 GHz resonating band, the resonating path is identified from the current distribution depicted in Figure 3.60B, and the corresponding length is calculated using Equation (3.26):

$$L_2 = \frac{L}{2} + \frac{W}{2} + \frac{a}{4} + \frac{d}{2} + \frac{c}{2} \tag{3.26}$$

By applying this equation, the length of the resonance at 2.45 GHz is found to be 27.45 mm, approximately equal to a quarter wavelength as determined by Equation (3.25).

Similarly, for the 5.5 GHz resonating band, the resonating path is determined from the current distribution displayed in Figure 3.60C, and the corresponding length is calculated using Equation (3.27):

$$L_3 = \frac{L}{4} + \frac{W}{2} + \frac{d}{2} + \frac{c}{4} \tag{3.27}$$

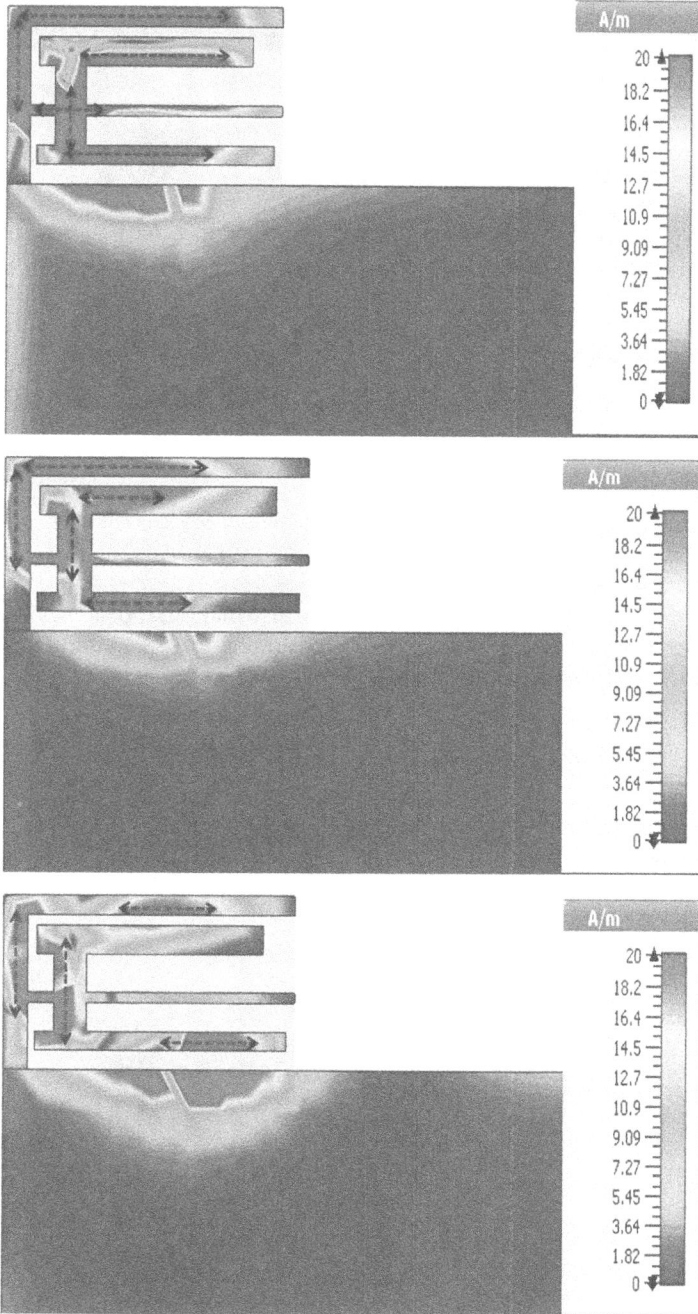

FIGURE 3.60   Simulated surface current (A/m) distribution of the proposed antenna.

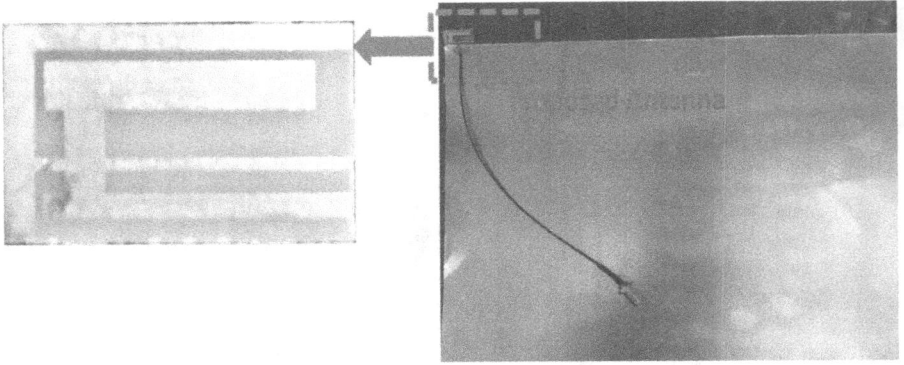

FIGURE 3.61   Fabricated prototype of proposed antenna mounted on the system ground.

Using this equation, the length of the resonance at 5.5 GHz is determined to be 15.9 mm, which is also approximately equal to a quarter wavelength as determined by Equation (3.25).

Through the analysis of the surface current distribution, it becomes evident that the proposed antenna operates in multiple bands (f_l, f_m, and f_u), making it well-suited for wireless operations in laptop computers.

3.7.4   PERFORMANCE CHARACTERIZATION OF PROPOSED ANTENNA

The design of the proposed antenna was validated by fabricating a prototype, as shown in Figure 3.61, and tested using a Rohde & Schwarz (9 KHz–16 GHz) network analyzer. The radiation performance, including radiation pattern, gain, and radiation efficiency, was measured in an $8 \times 4 \times 4$ m^3 calibrated anechoic chamber.

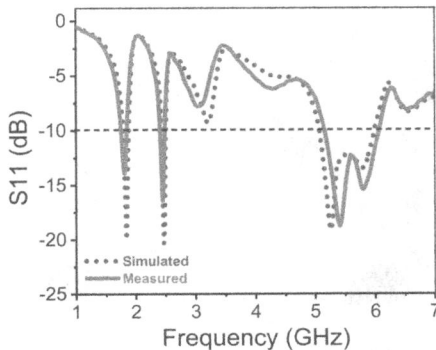

FIGURE 3.62   Measured and simulated reflection coefficient of the proposed antenna.

TABLE 3.18
Operating Bands of Proposed Antenna

	Operating Bands in GHz	
	Simulated	Measured
f_l	1.75–1.89	1.74–1.87
f_m	2.39–2.51	2.40–2.50
f_u	5.05–6.00	5.12–6.06

3.7.4.1   Reflection Coefficient (S_{11}) dB

Figure 3.62 illustrates the simulated and measured S11 for the proposed antenna. A good agreement between the simulated and measured S_{11} values is observed for the f_l and f_m bands. However, a slight deviation is observed at the resonant mode of 5.5 GHz in the f_u band, which can be attributed to soldering and fabrication tolerances. Three successful resonant modes were measured at approximately 1.83 GHz, 2.45 GHz, and 5.5 GHz, with S_{11} amplitudes better than 10 dB. The first resonance at 1.8 GHz covers the GSM 1800 band, the second resonance satisfies the 2.4 GHz Bluetooth/WLAN band, and the third resonance covers the 5.2/5.8 GHz WLAN and

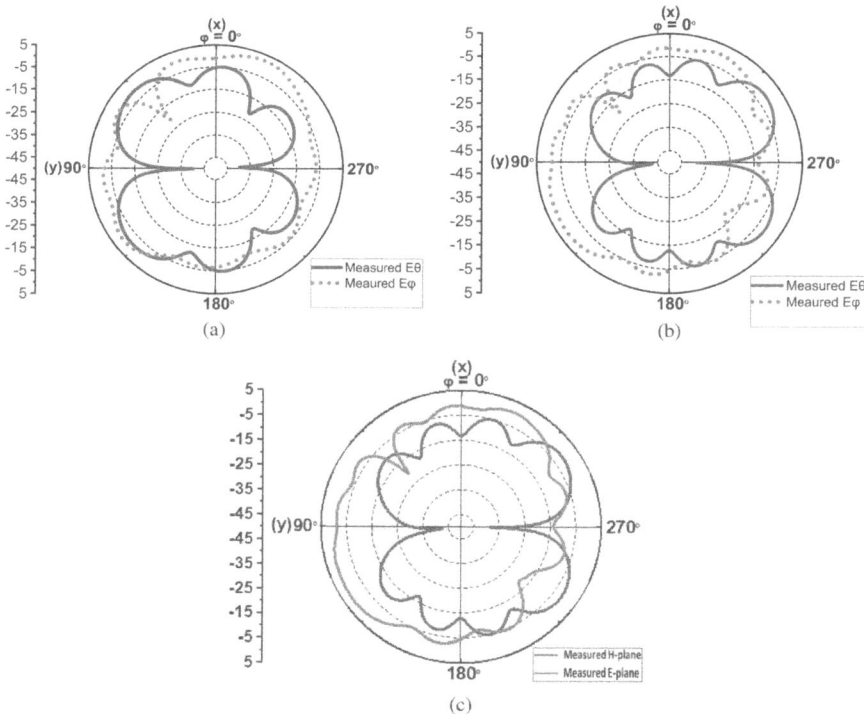

FIGURE 3.63 Measured 2D radiation pattern of proposed antenna at x-y plane (a) At 1.8 GHz (b) At 2.42 GHz (c) At 5.5 GHz.

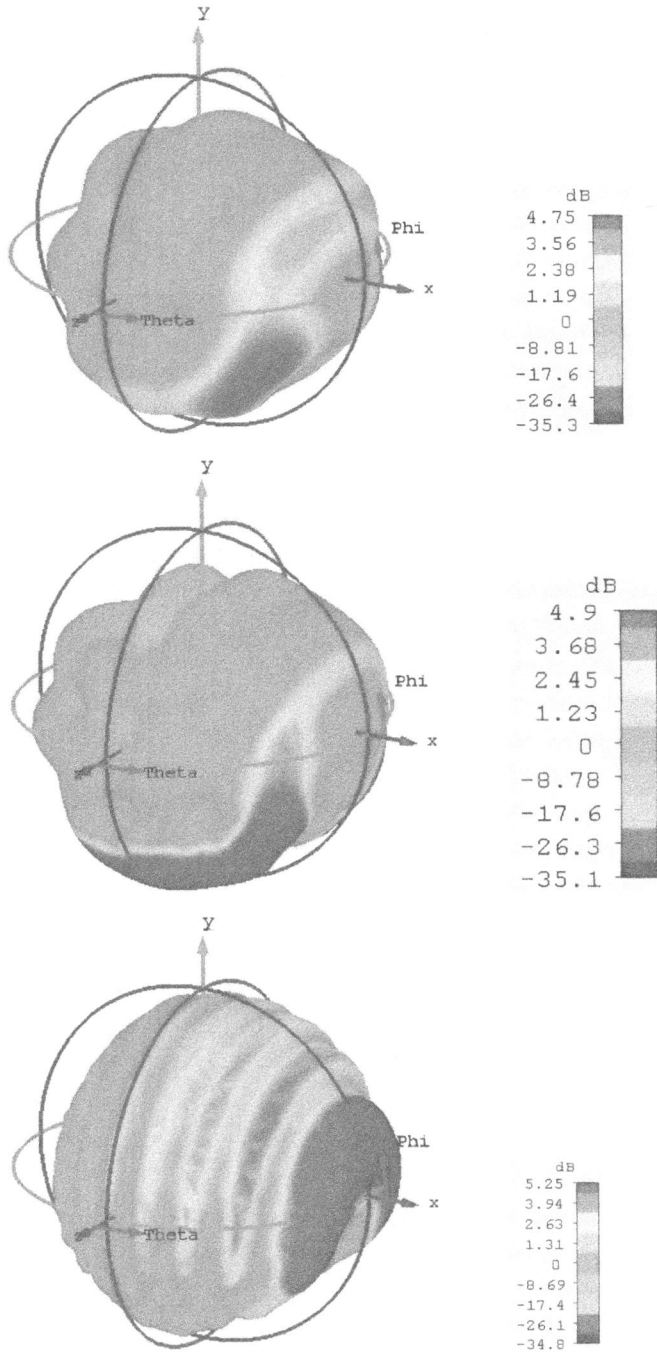

FIGURE 3.64 Simulated 3D radiation pattern of proposed antenna.

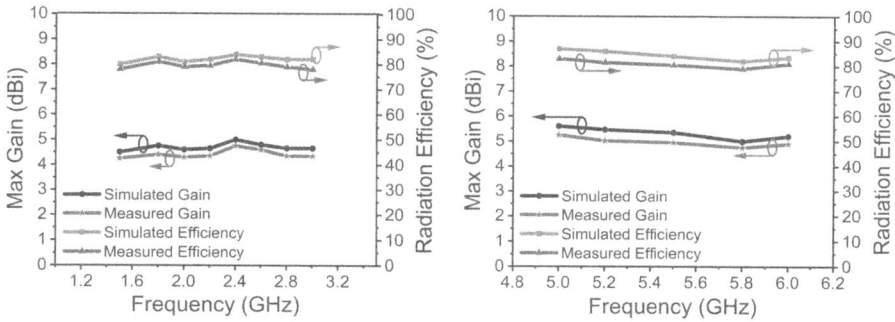

FIGURE 3.65 Measured and simulated gain and efficiency of the proposed antenna.

5.50 GHz WiMAX bands. Table 3.18 provides an overview of the operating bands of the proposed antenna.

3.7.4.2 Measured Radiation Patterns

The measured polar plots of the radiation pattern of the proposed antenna as a function of azimuth angle ($\varphi = 0°$) in the x-y planes are presented in Figure 3.63, while the simulated 3D radiation pattern is shown in Figure 3.64. In Figure 3.63, the E-plane and H-plane radiation patterns were measured at the resonant modes of 1.8 GHz, 2.42 GHz, and 5.5 GHz for the f_l, f_m, and f_u bands, respectively. The E-plane pattern for all resonant modes is nearly omnidirectional in the x-y plane, while the H-plane pattern exhibits a dipole pattern, forming a bi-directional radiation pattern in the x-y plane. This confirms the suitability of the proposed antenna for multiband applications in laptop computers.

3.7.4.3 Measured and Simulated Gain and Radiation Efficiency

The measured and simulated values of gain and efficiency for the proposed antenna across the three operating bands are depicted in Figure 3.65, with a comparison provided in Table 3.19.

TABLE 3.19
Comparison of Simulated and Measured Values of Gain and Radiation Efficiency of the Proposed Antenna

Covered bands (GHz)	Gain (dBi)		Efficiency (%)	
	Simulated	Measured	Simulated	Measured
f_l	4.50–4.75	4.25–4.40	80.00–83.00	78.20–81.10
f_m	4.60–5.00	4.35–4.75	82.69–84.98	79.93–82.60
f_u	5.00–5.45	4.75–5.25	82.59–87.98	79.15–83.50

From the analysis in Table 3.19, it is evident that there is minimal deviation between the simulated and measured values of both gain and radiation efficiency within the desired bands. This slight deviation can be attributed to fabrication tolerances in the construction of the proposed antenna. However, overall, the results indicate that the proposed antenna consistently achieves a stable gain of approximately 4–5 dBi and excellent efficiency exceeding 80% across all the operating bands. These findings validate the high-quality signal reception capability of the proposed antenna, which is crucial for wireless operations in laptop computers.

3.7.5 FEATURES OF THE PROPOSED ANTENNA

The proposed antenna exhibits the following notable features:

- **Simple structure with multiband application**: The antenna design is characterized by its straightforward structure, enabling multiband operation.
- **Low cost, less complexity, and fast time to market:** The antenna offers a cost-effective solution with reduced complexity, allowing for efficient production and shorter time to market.
- **No additional ground plane or reactive components required:** The antenna design eliminates the need for an additional ground plane, reactive components, or complex 3D structures to excite the desired bands.
- **Avoidance of expensive substrate:** By leveraging alternative materials, the proposed antenna design achieves its functionality without relying on expensive substrates.
- **Compact dimensions:** The antenna boasts a compact form factor, with dimensions of only $17.5 \times 9 \times 0.4$ mm³, making it suitable for space-constrained applications.

3.8 CONCLUSION

In conclusion, the successful design of a novel multiband monopole antenna for GSM, WLAN, and WiMAX wireless operations in laptop computers has been achieved. The proposed antenna exhibits a compact size of 17.5×9 mm², a simple structure, ease of fabrication, and operation within the 1.8 GHz GSM, 2.4/5.2/5.8 GHz WLAN, and 5.5 GHz WiMAX bands. It demonstrates nearly omnidirectional radiation characteristics, along with a consistent gain of more than 4 dBi and radiation efficiency exceeding 80%. These results affirm the practicality of the proposed antenna design for wireless operations in laptop computers.

REFERENCES

1. Chen, Z. N. *Antennas for Portable Devices*, John Wiley & Sons Ltd., 2007.
2. Pazin, L., Telzhensky, N. & Leviatan, Y. "Multiband flat-plate inverted-F antenna for Wi-Fi/WiMAX operation." *IEEE Antennas and Wireless Propagation Letters*, vol. 7, pp. 197–200, 2008.

3. Liu, H.-W., Lin, S.-Y. & Yang, C.-F. "Compact inverted-F antenna with meander shorting strip for laptop computer WLAN applications." *IEEE Antennas and Wireless Propagation Letters*, vol. 10, pp. 540–543, 2011.

4. Li, J., Wu, D. & Wu, Y. "Compact multi band inverted-F antenna for laptop operations." *5th IEEE International Symposium on Microwave, Antenna, Propagation*, Chengdu, China, pp. 29–31, 2013.

5. Liu, D., Flint, E. & Gaucher, B. "Integrated laptop antennas-design and evaluation." *IEEE Antennas and Propagation Society International Symposium*, San Antonio, TX, pp. 56–59, 2002.

6. Fujio, S. & Asano, T. "Dual band coupled floating element PCB antenna." *Proceedings of the IEEE Antennas and Propagation Society International Symposium*, Monterey, CA, vol. 3, pp. 2599–2602, 2004.

7. Lee, C.-T. & Wong, K.-L. "Uniplanar printed coupled-fed PIFA with a band-notching slit for WLAN/WiMAX operation in the laptop computer." *IEEE Transactions on Antennas and Propagation*, vol. 57, no. 4, pp. 1252–1258, 2009.

8. Liao, S.-J., Wong, K.-L. & Chou, L.-C. "Small-size uniplanar coupled-fed PIFA for 2.4/5.2/5.8 GHz WLAN operation in the laptop computer." *Microwave and Optical Technology Letters*, vol. 51, no. 4, pp. 1023–1028, 2009.

9. Chen, L.-Y. & Wong, K.-L. "2.4/5.2/5.8 GHz WLAN antenna for the ultrabook computer with metal housing." *2012 Asia Pacific Microwave Conference Proceedings*, Kaohsiung, Taiwan, pp. 322–324, 2012.

10. Chou, L.-C. & Wong, K.-L. "Uni-planar dual-band monopole antenna for 2.4/5 GHz WLAN operation in the laptop computer." *IEEE Transactions on Antennas and Propagation*, vol. 55, no. 12, pp. 3739–3741, 2007.

11. Abioghli, M. "Dual-band two layered printed antenna for 2.4/5GHz WLAN operation in the laptop computer." *IEICE Electronics Express*, vol. 8, no. 18, pp. 1519–1526, 2011.

12. Guo, L., Wang, Y., Gao, Y. & Shi, D. "A compact uniplanar printed dual-antenna operating at the 2.4/5.2/5.8 GHz WLAN bands for laptop computers." *IEEE Antennas Wireless Propagation Letters*, vol. 13, pp. 229–232, 2014.

13. Pazin, L., Telzhensky, N. & Leviatan, Y. "Multiband flat-plate inverted-F antenna for Wi-Fi/WiMAX operation." *IEEE Antennas and Wireless Propagation Letters*, vol. 7, pp. 197–200, 2008.

14. Sim, C.-Y.-D., Chien, H.-Y. & Lee, C.-H. "Uniplanar antenna design with adhesive ground plane for laptop WLAN operation." *IEEE Antennas and Wireless Propagation Letters*, vol. 13, pp. 337–340, 2014.

15. Naser-Moghadasi, M., Sadeghzadeh, R. A., Hafezi Fard, R., Yazdani Fard, S. O. & Virdee, B. S. "Highly compact meander line antenna using DGS technique for WLAN communication systems." *IEICE Electronics Express*, vol. 8, no. 10, pp. 722–729, 2011.

16. Jiang, H.-J., Kao, Y.-C. & Wong, K.-L. "High-isolation WLAN MIMO laptop computer antenna array." *2012 Asia Pacific Microwave Conference Proceedings*, Kaohsiung, Taiwan, pp. 319–321, 2012.

17. Liu, Y., Wang, Y. & Du, Z. "A broadband dual-antenna system operating at the WLAN/WiMAX bands for laptop computers." *IEEE Antennas Wireless Propagation Letters*, vol. 14, pp. 1060–1063, 2015.

18. Wang, Y.-Y. & Chang, S.-J. "A new dual-band antenna for WLAN applications." *IEEE Antennas and Propagation Society Symposium*, Monterey, CA, pp. 2611–2614, 2004.

19. Yang, J., Wang, H., Lv, Z. & Wang, H. "Design of miniaturized dual-band microstrip antenna for WLAN application." *Sensors*, vol. 16, pp. 1–15, 2016.

20. Bartwal, P., Gautam, A. K., Singh, A. K., Kanaujia, B. K. & Rambabu, K. "Design of compact multi-band meandered-line antenna for global positioning/wireless local area network/worldwide interoperability for microwave access band

applications in laptop/tablets." *IET Microwaves, Antennas & Propagation*, vol. 10, no. 15, pp. 1618–1624, 2016.

21. Sim, C.-Y.-D., Chen, C.-C., Zhang, X. Y., Lee, Y.-L. & Chiang, C.-Y. "Very small-size uniplanar printed monopole antenna for dual-band WLAN laptop computer applications." *IEEE Transactions on Antennas and Propagation*, vol. 65, no. 6, pp. 2916–2922, 2017.

22. Khan, R., Al-Hadi, A. A., Soh, P. J., Ali, M., Owais, O. & Islam, I. "Mutiband monopole antenna with minimized ground plane influence for portable devices." *IET Microwaves, Antennas & Propagation*, vol. 11, no. 13, pp. 1829–1835, 2017.

23. Su, S.-W., Chou, J.-H. & Liu, Y.-T. "Compact paper-clip-shaped wire antenna for 2.4GHz and 5.2GHz WLAN operation." *Microwave and Optical Technology Letters*, vol. 50, no. 10, pp. 2572–2574, 2008.

24. Lee, C.-T. & Su, S.-W. "Very-low-profile, stand-alone, tri-band WLAN antenna design for laptop-tablet computer with complete metal-backed cover." *2016 IEEE 5th Asia-Pacific Conference on Antennas and Propagation (APCAP)*, Kaohsiung, Taiwan, pp. 189–190, 2016.

25. Kim, T., Kim, S.-J., Byun, J., Harackiewicz, F. J., Park, M.-J., Chung, Y.-S. & Lee, B. "Internal dual-band WLAN antenna for laptop applications." *2010 IEEE Antennas and Propagation Society International Symposium*, Toronto, Canada, pp. 1–4, 2010.

26. Xu, Y., Jiao, Y.-C. & Luan, Y.-C. "Compact CPW-fed printed monopole antenna with triple-band characteristics for WLAN/WiMAX applications." *Electronics Letters*, vol. 48, no. 24, pp. 1519–1520, 2012.

27. Huang, H., Liu, Y., Zhang, S. & Gong, S. "Multiband metamaterial-loaded monopole antenna for WLAN/WiMAX applications." *IEEE Antennas and Wireless Propagation Letters*, vol. 14, pp. 662–665, 2014.

28. Hong, I.-P. & Lee, I.-G. "Design of a film antenna using a cloverleaf-shaped monopole structure for WiBro and WLAN." *IEICE Electronics Express*, vol. 9, no. 7, pp. 654–659, 2012.

29. Wong, K.-L. & Chou, L.-C. "Internal composite monopole antenna for WLAN/WiMAX operation in a laptop computer." *Microwave and Optical Technology Letters*, vol. 48, no. 5, pp. 868–871, 2006.

30. Kang, T.-W. & Wong, K.-L. "Very small size printed monopole with embedded chip inductor for 2.4/5.2/5.8 GHz WLAN laptop computer antenna." *Microwave and Optical Technology Letters*, vol. 52, no. 1, pp. 171–177, 2010.

31. Su, S.-W., Lee, C.-T. & Chen, S.-C. "Very-low-profile, triband, two-antenna system for WLAN notebook computers." *IEEE Antennas Wireless Propagation Letters*, vol. 17, no. 9, pp. 1626–1629, 2018.

32. Pozar, D. M. *Microwave Engineering*, 4th Edition, John Wiley, New York, 2012.

33. Mishra, D. K. *Radio Frequency and Microwave Communication Circuits Analysis and Design*, John Wiley and Sons, Inc., 2001.

34. Kumar, G. & Ray, K. P. *Broadband Microstrip Antenna*, Artech House, Inc., 2003.

35. Balanis, C. A. *Modern Antenna Handbook*, John Wiley & Sons, Inc., Canada, 2008.

4 Advanced Antenna Designs for WLAN and Wi-Fi 6E in Laptop Computers

4.1 INTRODUCTION TO WI-FI 6E BAND

The COVID-19 pandemic has precipitated an increased reliance on wireless broadband and internet-based services, encompassing online communication, education, entertainment, and video/audio conferencing calls to facilitate remote interactions and work-from-home arrangements. This unprecedented situation has also created a pathway for the adoption of digital learning [1, 2].

Amidst this landscape, Wi-Fi technology has emerged as an indispensable facilitator, placing users at the forefront of seamless communication. By offering wireless connectivity solutions, it caters to the escalating demands of our interconnected world.

4.1.1 Wi-Fi 6E Definition

Wi-Fi 6E presents an unprecedented opportunity by unlocking access to the expansive unlicensed 6 GHz frequency band, spanning from 5.925 GHz to 7.125 GHz. In the US and many other regions, this band operates with a remarkable 1.2 GHz bandwidth, while in the EU, it boasts nearly 0.5 GHz of spectrum. This allocation of spectrum alone triples the available capacity, with minimal advancements in standards.

Distinguished by the letter "E," Wi-Fi 6E signifies its extended capabilities, pushing the boundaries of Wi-Fi 6 to encompass the 6 GHz frequency band. This enables support for larger capacity, wider channels, uninterrupted spectrum, and multigigabit data rates, facilitating the seamless delivery of high-definition video content while minimizing interference. With the advent of Wi-Fi 6E, the rapid deployment of new APs and devices is made possible, further propelling the evolution of wireless connectivity.

4.1.2 Operating Bands of Wi-Fi 6E

While traditional Wi-Fi operates in two frequency bands, namely, 2.4 GHz and 5 GHz, the advent of Wi-Fi 6E introduces a groundbreaking third frequency band—6 GHz, as depicted in Figure 4.1. Wi-Fi 6E extends the impressive capabilities of Wi-Fi 6 into this new frequency band, unlocking enhanced efficiency, higher throughput, and elevated levels of security. By harnessing the power of the 6 GHz band, Wi-Fi 6E opens up a world of possibilities for improved wireless connectivity, empowering users with unrivaled performance and seamless experiences.

DOI: 10.1201/9781003331018-4

Band	Channels	Bandwidth
2.4 GHz	3 1	

60 MHz of Spectrum; 3 channels allocated

| 5 GHz | 25
12
6
2 | |

500 MHz of Spectrum; 25 channels allocated

| 6 GHz | 59
29
14
7 | |

Upto 1,200 MHz of Spectrum; 56 channels available (including up to 7 160 MHz channels)

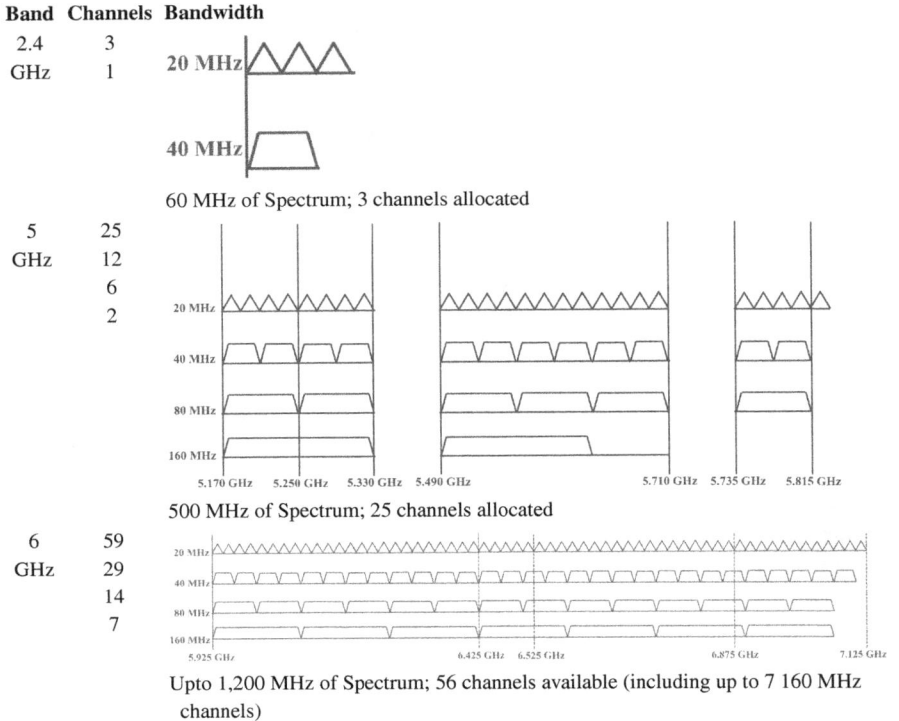

FIGURE 4.1 Wi-Fi 6E frequency spectrum.

4.1.3 NEED OF 6 GHz BAND

Building upon the success of Wi-Fi, the advent of Wi-Fi 6 has revolutionized wireless connectivity, enabling innovative applications such as unmanned aerial vehicles (UAVs or drones), automated guided vehicles, and industrial robots. Wi-Fi 6 has garnered significant recognition for its outstanding performance in indoor environments, including sports stadiums, large venues, offices, and hotels. Additionally, it has made significant strides in the automotive industry by supporting critical applications like infotainment, monitoring, maintenance, and manufacturer-provided upgrades.

However, the limitations of available spectrum, congestion in the 2.4 GHz band, and performance issues in the crowded 5 GHz band due to the proliferation of Wi-Fi devices operating within limited channels in close proximity have hindered the optimal performance of Wi-Fi 6. According to a study by the Wi-Fi Alliance in 2017, achieving the desired performance would require an additional 1.5 GHz of spectrum by 2025 [3], as illustrated in Figure 4.2. This realization prompted the Wi-Fi industry to seek a suitable spectrum to extend the capabilities of Wi-Fi 6, leading to the identification of the 6 GHz band (5.925 GHz to 7.125 GHz) as the ideal solution.

The 6 GHz band represents a unique opportunity in the Wi-Fi landscape, often referred to as a "greenfield band." This distinction arises from the fact that it is the

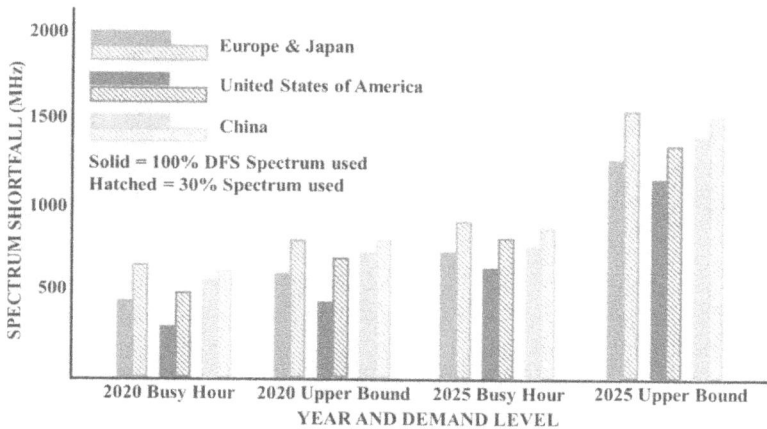

FIGURE 4.2   The estimated fall of spectrum availability in the near future.

first time this frequency range is being utilized for Wi-Fi activities, enabling the development of optimized protocols without the burden of backward compatibility requirements.

Devices operating within the 6 GHz band will be designated as "Wi-Fi 6E" devices, harnessing the benefits of expanded capacity and enhanced performance. This advancement unlocks the potential for new services characterized by higher data rates, reduced latency, and an exceptional quality of service. Moreover, Wi-Fi 6E ensures robust security measures, fortifying the protection of wireless communications in this frequency band.

4.1.4 DIFFERENCE BETWEEN WI-FI AND WI-FI 6E

Parameters	Wi-Fi 6	Wi-Fi 6E (6 GHz)
Operating bands	2.40 GHz, 5 GHz	2.40 GHz, 5 GHz, and 6 GHz
Features	Multiuser efficiencies, MU-MIMO to remove congestion.	Includes all features of Wi-Fi 6 plus: More spectrum in the 6 GHz band
	OFDMA to create carpool lanes to ride along smaller packets, such as voice data.	Wider channels of 160 MHZ bandwidth, suitable for high definition and virtual reality (VR).
	Target Wake Time (TWT) to allow APs to ring Internet IoT devices at greater intervals, minimize traffic, and extend battery life.	As Wi-Fi 6E capable devices operate in the 6 GHz band, there will be no interference from microwaves, etc.
	Use of Wi-Fi protected access 3 (WPA3) protocol to enhance guest access security.	
Benefits	Increased efficiencies to provide larger throughput with the same number of APs, suitable for dense areas and large numbers of IoT devices.	Support larger capacity, wider channel bandwidth support multigigabit data rate, suitable for high-definition video, AR) and VR.

	UNII - 5	UNII-6	UNII-7	UNII-8
Low Power Indoor AP	5 dBm/MHz			
Standard Power AP	36 dBm with AFC		36 dBm with AFC	
Client Devices	Mobile = 6 dB below AP/fixed indoor ("subordinate") = LP/fixed outdoor = AFC			
Very Low Power AP	Authorized Across the band, but rules not finalized			
	5.925 GHz	6.425 GHz 6.525 GHz	6.875 GHz	7.125 GHz

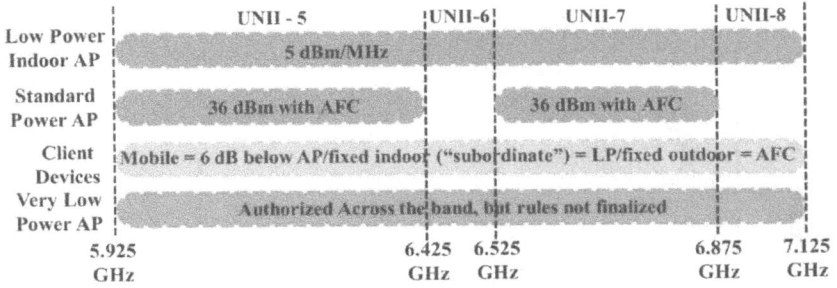

FIGURE 4.3 Wi-Fi 6E device classes.

4.2 SERVICE CLASSES SUPPORTED BY WI-FI 6E

To optimize performance while safeguarding the interests of licensed incumbents, the FCC has established four distinct device classes for Wi-Fi 6E operation. These classes, depicted in Figure 4.3, serve as a framework to ensure the coexistence of Wi-Fi 6E devices with existing licensed services. Let us delve into the details of each class in the subsequent subsections.

4.2.1 LOW POWER INDOOR (LPI)

Among the various classes defined for Wi-Fi 6E APs, LPI) class reigns supreme, as depicted in Figure 4.4. LPI APs are primarily deployed in residential and indoor enterprise environments. These APs are strategically positioned within buildings to effectively attenuate any power leakage beyond the designated frequency band. Consequently, they ensure seamless operations throughout the band, mirroring the power levels and coverage capabilities of existing Wi-Fi APs. This results in similar 6 GHz coverage as that of current 5 GHz Wi-Fi APs, delivering reliable connectivity within indoor spaces.

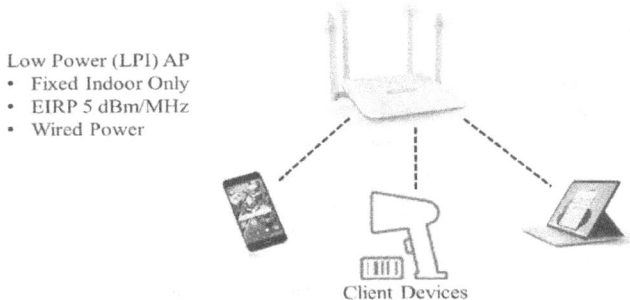

Low Power (LPI) AP
- Fixed Indoor Only
- EIRP 5 dBm/MHz
- Wired Power

Client Devices

FIGURE 4.4 Wi-Fi 6E low power indoor device class.

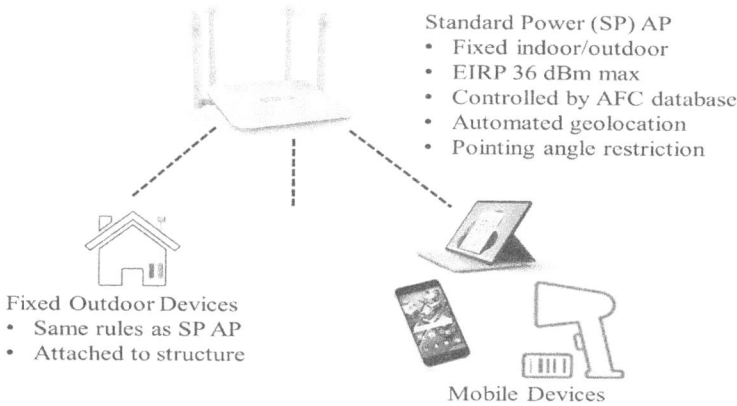

Standard Power (SP) AP
- Fixed indoor/outdoor
- EIRP 36 dBm max
- Controlled by AFC database
- Automated geolocation
- Pointing angle restriction

Fixed Outdoor Devices
- Same rules as SP AP
- Attached to structure

Mobile Devices

FIGURE 4.5   SP AP Wi-Fi 6E device class.

4.2.2 STANDARD POWER (SP) AP

The SP class, depicted in Figure 4.5, represents a Wi-Fi 6E device class that is versatile for both outdoor and indoor operations. In this class, the SP APs are carefully coordinated by an automated frequency coordination service (AFC). This coordination ensures that the 6 GHz Wi-Fi operations are conducted in a manner that minimizes interference with incumbent services, such as public safety, microwaves, satellite services, cellular backhaul, and broadcast services. By employing effective AFC mechanisms, SP APs enable reliable and uninterrupted Wi-Fi connectivity while maintaining harmonious coexistence with critical incumbent services.

4.2.3 CLIENT DEVICES AP

Figure 4.6 showcases the Wi-Fi 6E client device class, which operates in a similar manner to the lower sub-bands. Geographically, client devices are constrained by the coverage area of APs. Without an AP signal, devices are unable to establish connections and will refrain from transmitting. Thus, it is presumed that the AP is transmitting within authorized parameters, allowing the client to dynamically adjust its transmit power and channel selection in accordance with the AP's configuration.

Mobile Client
- Indoor/outdoor
- 6 dBm lower power
- Connected access point

Client Devices

FIGURE 4.6   Client devices.

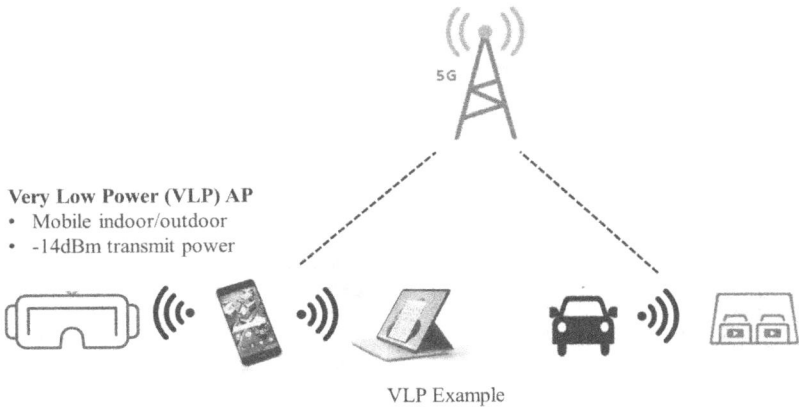

FIGURE 4.7   Very Low Power Wi-Fi 6E device class.

This mechanism ensures proper coordination and efficient utilization of the available spectrum resources between Wi-Fi 6E clients and APs, enabling seamless and reliable wireless communication.

4.2.4   Very Low Power AP

Figure 4.7 illustrates the Very Low Power class of Wi-Fi 6E, which caters to wearable or mobile APs and small cell deployments such as hotspots. These APs operate at sufficiently low power levels to ensure minimal interference with other wireless systems. By adhering to strict power constraints, very low power APs enable seamless and uninterrupted connectivity in environments where mobility and portability are crucial, without compromising the integrity and performance of neighboring wireless networks.

4.3   IMPORTANCE OF ANTENNA IN WIRELESS LOCAL AREA NETWORK (WLAN) AND WI-FI 6E STANDARDS

The advent of Industry 4.0, smart manufacturing, automated vehicles, and infotainment industries has brought about a growing reliance on WLAN and Wi-Fi 6E standards to enable their transformative applications. These industries are leveraging the combined power of these standards to enhance their operations. In the realm of smart cities, the deployment of both WLAN and Wi-Fi 6E is reinforcing real-time traffic monitoring, healthcare applications, and geofencing solutions, allowing for low-latency delivery of personalized alerts within specific geographic areas. The emergence of Wi-Fi 6E is also paving the way for next-generation standards that enable remote patient monitoring, high-definition video conferencing, and remote medical operations through the use of robots.

To deliver these wireless experiences and harness the full potential of these standards, antennas play a pivotal role. Antennas serve as critical components in ensuring robust signal strength and coverage, enabling the aforementioned applications to thrive. Considering this, the following section delves into insightful case studies that explore the design of antennas operating simultaneously in the WLAN and Wi-Fi 6E frequency bands.

4.4   CASE STUDY: A PLANAR INVERTED-F ANTENNA (PIFA) DESIGN WITH WLAN AND WI-FI 6E BAND FOR LAPTOP COMPUTER APPLICATIONS

A novel design of a compact PIFA has been developed to cater to the requirements of both WLAN and Wi-Fi 6E frequency bands in laptop computer applications. This innovative PIFA design, measuring a mere $18.2 \times 5 \times 0.4$ mm³, exhibits exceptional performance across the WLAN band (2.4/5 GHz) as well as the Wi-Fi 6E band (5.925–7.125 GHz) [4].

Through strategic modifications to the PIFA structure, including bending and slot-loading techniques, two resonant modes are excited at 2.4 GHz and 5.5 GHz. This ingenious design enables comprehensive coverage of the entire WLAN frequency bands (2.4/5 GHz) while maintaining optimal performance.

To expand its capabilities to include the Wi-Fi 6E band (5.925–7.125 GHz) without compromising the original PIFA's performance, a U-shaped parasitic element is strategically introduced behind the PIFA structure. This addition facilitates the excitation of an additional resonant mode at 6.5 GHz, thus achieving seamless coverage of the Wi-Fi 6E band.

4.4.1   OVERVIEW OF EXISTING STATE-OF-THE-ART TECHNOLOGY

Over the past decade, several innovative antenna designs have been proposed to enable WLAN operation in both the 2.4 GHz and 5 GHz bands for laptop computers [5–8]. These designs predominantly utilize popular antenna structures such as monopole and dipole configurations. Additionally, unique design techniques like open-loop structures [9] and modified inverted-F antennas (IFA) [10] have been explored to achieve dual-band functionality. However, these designs fall short in covering the upcoming Wi-Fi 6E frequency band (5.925–7.125 GHz) that extends beyond the operational range of WLAN 5 GHz [6–9].

To address this gap, recent research has focused on investigating antenna designs that can cover both the WLAN and Wi-Fi 6E bands [11–14]. For instance, in [11], a two-monopole slot antenna design was proposed to generate two resonant modes capable of covering both bands. However, this design exhibits a relatively large antenna height of 10 mm. Another design presented [12] utilizes a coupled-fed configuration, offering a low profile of only 3 mm but with a considerable length of 43 mm. In [13], a coupled-fed loop design measuring 30×6 mm² was studied, while [14] explored two sets of two-antenna MIMO designs.

In this paper, a novel approach is taken, introducing an antenna design with dimensions of only $18.2 \times 5 \times 0.4$ mm³. This design distinguishes itself from previous works by achieving a compact form factor, suitable for laptop computer applications, while covering both the WLAN and Wi-Fi 6E bands simultaneously.

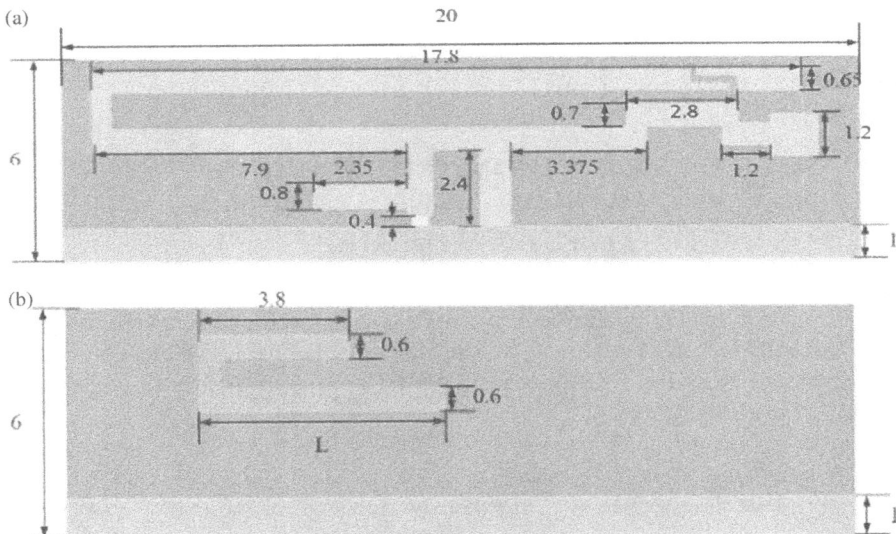

FIGURE 4.8 Geometry of proposed modified PIFA design: (a) Front and (b) back (with U-shaped parasitic element). $L = 6.2$ mm. Unit: millimeters.

4.4.2 ANTENNA GEOMETRY

The proposed PIFA geometry, which serves as the main radiator, is illustrated in Figure 4.8(a) and (b), showcasing its structural details. Fabricated on a 0.4 mm thick FR-4 substrate, the PIFA has a planar size of 20×6 mm^2. Additionally, an extended ground section, with a height of 1 mm, is connected to a metallic ground plane measuring $200 \times 300 \times 0.4$ mm^2.

Figure 4.8(a) showcases the initial modified PIFA with a folded structure, which successfully achieves operation within the WLAN band (2.4/5 GHz) by exciting two resonant modes. The loading of the open slot is introduced to finely adjust the low-frequency mode, ensuring precise matching with the WLAN 2.4 GHz band (2.4–2.48 GHz) without affecting the high band mode (5.15–5.925 GHz).

In Figure 4.8(b), the coupling effects between the PIFA and the U-shaped parasitic element result in the excitation of a mode at 6.5 GHz (tuning L), effectively covering the Wi-Fi 6E band. This innovative design integration enables the antenna to operate across both WLAN and Wi-Fi 6E frequency ranges, offering enhanced performance and compatibility.

4.4.3 RESULTS ANALYSIS

The proposed PIFA design was thoroughly evaluated through electromagnetic simulations using the Ansoft HFSS software. Figure 4.9(b) presents the simulated reflection coefficient of the proposed PIFA, showcasing its performance characteristics.

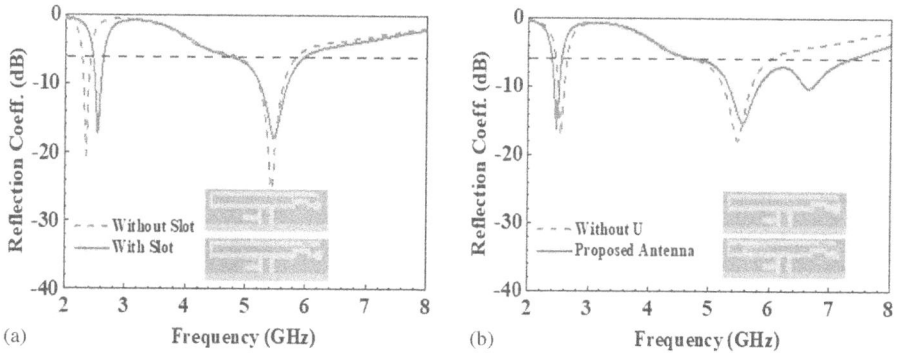

(a) Frequency (GHz)   (b) Frequency (GHz)

FIGURE 4.9 Simulated reflection coefficients: (a) With and without loading the open slot (the U-shaped parasitic element is absent in this case) and (b) with (reported antenna) and without loading the U-shaped parasitic element.

The simulation results indicate that the proposed PIFA achieves a 6 dB impedance bandwidth of 2.39–2.55 GHz for the low band and 4.78–7.31 GHz for the high band. These bandwidths align with the operational frequency ranges of the current WLAN (2.4/5 GHz) and Wi-Fi 6E (5.925–7.125 GHz) standards.

Figure 4.10(a–c) illustrates the simulated radiation patterns (normalized) of the proposed PIFA at three frequencies: 2.45 GHz, 5.5 GHz, and 6.5 GHz. These patterns are observed across the x-z plane, y-z plane, and x-y plane, respectively. In the

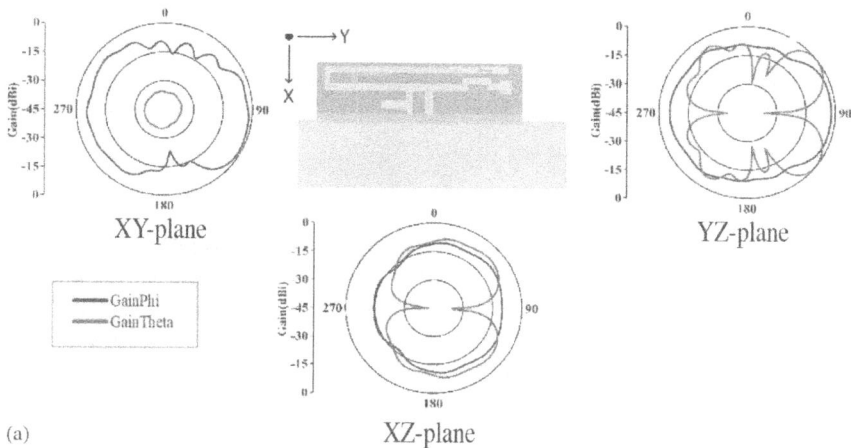

XY-plane   YZ-plane

XZ-plane

(a)

FIGURE 4.10 The radiation patterns of the PIFA at different modes: (a) 2.45 GHz, (b) 5.5 GHz, and (c) 6.5 GHz. (*Continued*)

(b)

(c)

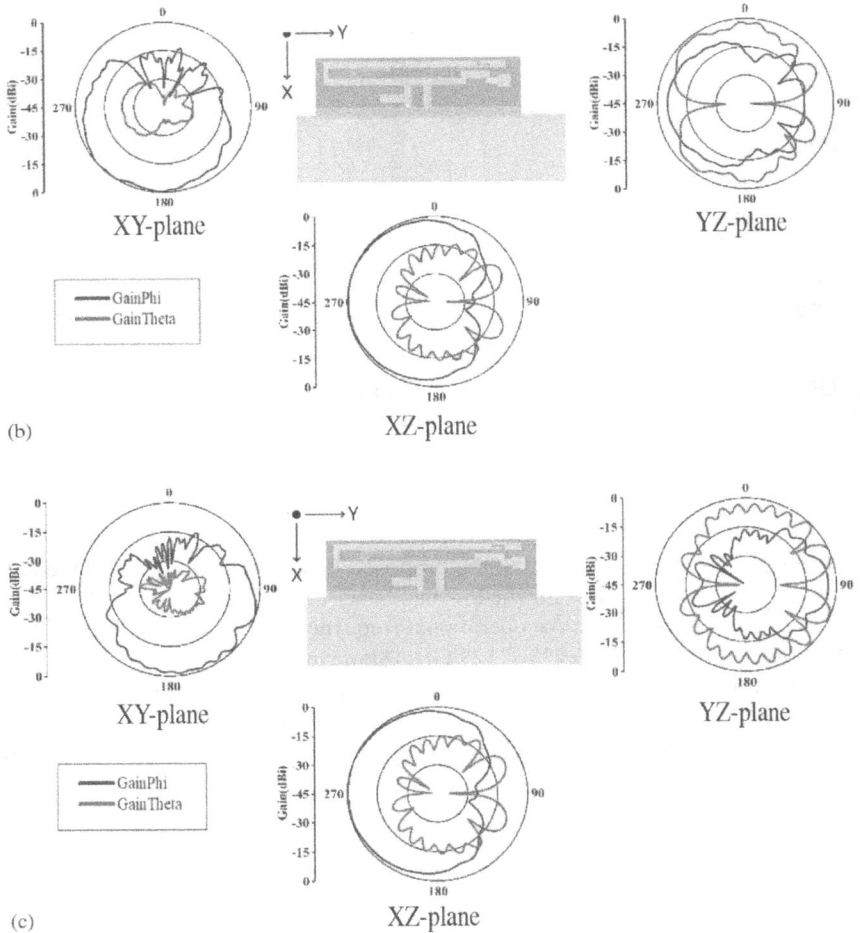

FIGURE 4.10 *(Continued)*

low band, the radiation pattern exhibits bi-directional characteristics. Conversely, the two high bands demonstrate broadside directional radiation, offering optimal coverage and signal propagation.

Figure 4.11 illustrates the simulated antenna gain and radiation efficiency of the proposed PIFA using the advanced electromagnetic simulation software (HFSS). In the lower frequency band (2.4–2.5 GHz), the PIFA demonstrates a commendable gain of approximately 1.5 dBi, coupled with a radiation efficiency exceeding 50%. Moving to the first higher frequency band (5.15–5.85 GHz), the PIFA achieves a higher gain range of approximately 2.8–3.2 dBi, accompanied by an improved radiation efficiency ranging between 75% and 80%. Lastly, within the second high

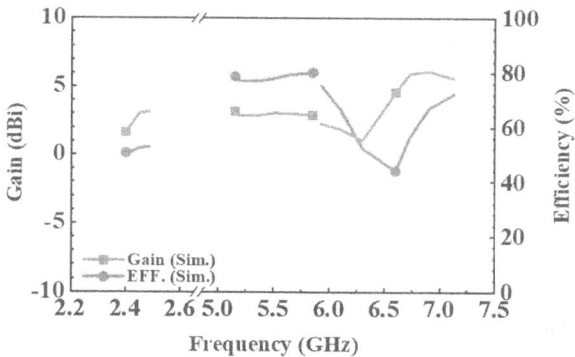

FIGURE 4.11   The simulated gain and radiation efficiency of the antenna.

frequency band (5.925–7.125 GHz) designated for Wi-Fi 6E applications, the PIFA exhibits simulated gains surpassing 1.8 dBi, accompanied by a radiation efficiency surpassing 44%.

4.4.4 Concluding Remarks

This section introduces a novel and compact modified PIFA design, specifically tailored for laptop computer applications operating within the WLAN frequency band (2.4/5 GHz) and the emerging Wi-Fi 6E frequency band (5.925–7.125 GHz). The proposed antenna incorporates a U-shaped parasitic element to effectively stimulate Wi-Fi 6E high-frequency operation. With its low-profile form factor, measuring no more than 5 mm, the antenna is exceptionally well suited for integration into laptop computers, where space constraints along the edge of the screen are prevalent.

4.5 CASE STUDY: LOW PROFILE LAPTOP ANTENNA DESIGN FOR WI-FI 6E BAND

A cutting-edge laptop antenna design catering to the Wi-Fi 6E frequency bands (2.4/5/6 GHz) is presented in recent research [15]. The antenna, boasting a compact and low-profile configuration of only 3.5 mm, utilizes a double-slit loaded modified inverted-L antenna (ILA) structure. This innovative design enables the antenna to efficiently excite multiple frequency modes, ensuring comprehensive coverage of the targeted 5 GHz and 6 GHz bands (5,150–7,125 MHz). Furthermore, to encompass the 2.4 GHz band (2,400–2,485 MHz), a C-shaped coupling strip is strategically integrated behind the modified ILA.

Notably, this state-of-the-art antenna exhibits exceptional attributes, such as a diminutive planar size of merely 22.3 mm × 3.5 mm, making it ideal for integration within the narrow screen bezels of contemporary laptops. Despite its compact form, the proposed antenna delivers commendable performance, as evidenced by the measured 10 dB bandwidths of 2.4–2.485 GHz and 4.74–7.26 GHz. Additionally, the antenna demonstrates excellent efficiency, surpassing 40% and 61% for the respective frequency bands.

4.5.1 Overview of Existing State-of-the-Art Technology

The WLAN operates under the IEEE 802.11 wireless communication standard. Initially, WLAN 2.4 GHz (2,400–2,485 MHz) was deployed, providing extensive coverage but with slower transmission speeds. Subsequently, WLAN 5 GHz (5,150–5,825 MHz) was introduced, offering faster transmission rates but reduced coverage. Recently, the Wi-Fi Alliance renamed the sixth-generation Wi-Fi standard from IEEE 802.11ax to Wi-Fi 6, while introducing Wi-Fi 6E as an extension to Wi-Fi 6. Wi-Fi 6E encompasses the 6 GHz unlicensed frequency band (5,925–7,125 GHz) alongside the WLAN 5/6 GHz bands, delivering the highest transmission speeds among the three frequencies but with the least coverage [16].

In recent years, numerous compact laptop antenna designs have been proposed, typically placed on the upper section of the laptop screen [5–8, 11–13, 17–29]. Among these designs, loop antennas have been employed in [17–19], with [17] utilizing a capacitive feeding structure for dual-band operation. Conversely, monopole antenna designs have been adopted [5, 7, 8, 20–25] due to their simplicity and smaller size. [5] extends the current path through coupling methods and modifies it into a meandering monopole antenna to generate multifrequency modes. Interestingly, [23] integrates two folded monopoles with a chip inductor to achieve a low-profile triple-band operation for a two-antenna system. In addition to loop and monopole designs [5, 7, 8, 17–25], dipole [6], open-slot [26], and IFA [27] designs are commonly utilized to induce multifrequency modes for WLAN 2.4/5 GHz coverage.

While the aforementioned laptop antenna designs cover WLAN 2.4/5 GHz bands, the inclusion of the Wi-Fi 6E frequency band has only recently been addressed [11–13, 28, 29]. In [11], two different monopole slots fed by a meandering feeding line are employed to excite low-frequency (f_L) and high-frequency (f_H) modes. However, the f_L mode exhibits low gain (-7 dBi) and efficiency (17%–20%), with a high antenna profile of 20 mm. Although [12] presents a low-profile (3 mm) laptop antenna covering the desired bands, it features a lengthy 43 mm antenna and requires a protrusion of the system ground to accommodate the feeding structure. [28] introduces two antenna pairs (four feeding ports) simultaneously to cover the bands of interest, but the WLAN and 5G NR antenna pairs have different sizes (38 mm × 5 mm and 35 mm × 5 mm, respectively). [13] presents a modified loop antenna with a widened portion and meandering section; however, it only covers the desired bands when considering the 6 dB impedance bandwidth and has a high profile of 6 mm and an antenna length of 30 mm. While [29] proposes a simple monopole design with a compact size (4 mm × 16 mm), it requires loading three chip inductors into the monopole.

To the best of the author's knowledge, a compact laptop antenna offering dual-band operation with a 10 dB impedance bandwidth effectively covering the desired f_L (2,400–2,485 MHz) and f_H (5,150–7,125 MHz) with a low profile (<4 mm) has not been found in the existing literature. Therefore, this section introduces a low-profile (3.5 mm) laptop antenna with a length of 22.3 mm, capable of covering the

Wi-Fi 6E band (2.4/5/6 GHz). This antenna exhibits impressive 10 dB impedance bandwidths for f_L (2.40–2.49 GHz) and f_H (4.71–7.26 GHz), along with excellent efficiency (>40%) across the two bands of interest.

4.5.2 ANTENNA GEOMETRY

Figure 4.12 illustrates the structure diagram of the proposed antenna, featuring a compact planar size of 22.3 mm × 3.5 mm. The antenna is fabricated on a 0.4 mm thick FR-4 substrate, with the modified ILA serving as the main radiator printed on the front side. On the opposite side, the C-shaped coupling strip (C-strip) is printed. The C-strip is a 3 mm wide parasitic element with asymmetric arm lengths of L_1 = 11.3 mm (shorter arm) and 12 mm (longer arm). The choice of these arm lengths is intentional to prevent complete cancellation of radiation along the Z-axis linear polarization. Each arm has a width of 1.25 mm and is partially coupled to the left section of the ILA.

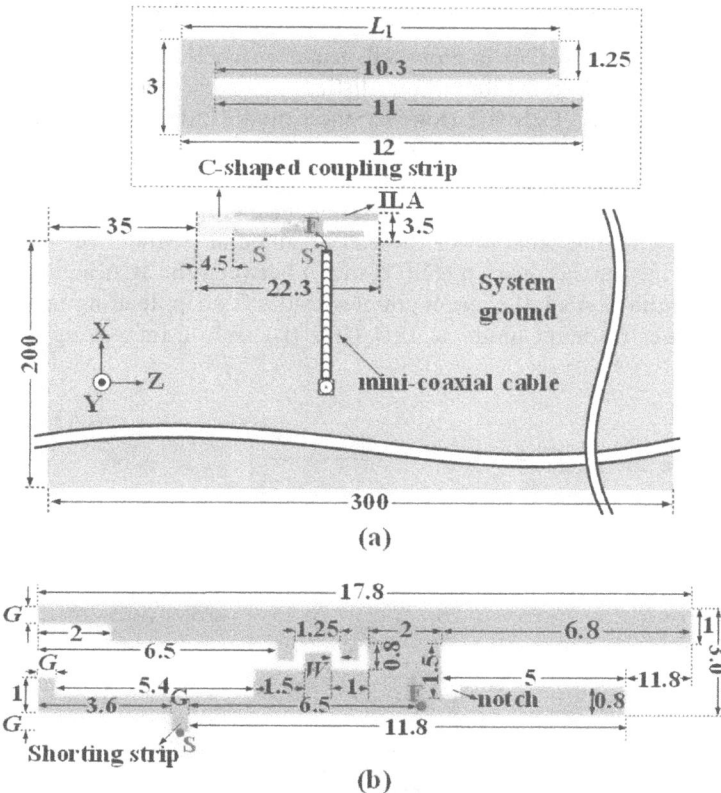

FIGURE 4.12 (a) Geometry and dimensions of the antenna composed of an inverted-L antenna (ILA) (printed at the front) and a C-shaped coupling strip (printed at the back). (b) Detailed dimensions of the ILA (B). L_1 = 11.3, W = 0.25, G = 0.5. Unit: millimeters.

The ILA itself is a modified double-slit loaded structure. On the left section, the slit is a step structure extended by a meandered narrow slit (width W = 0.25 mm, total length of 3.5 mm) that stretches toward the middle section of the ILA. On the right section, there is an inverted L-shaped slit measuring 6.8 mm in length. Additionally, a small notch (0.5 mm × 0.3 mm) is loaded at the left bottom corner of the slit to enhance impedance matching for the high-frequency band (fH). The ILA structure also incorporates a shorting strip (5 mm × 5 mm), located 3.6 mm from the left edge, and a feeding point (connected to a 50 Ω miniature coaxial cable) positioned 6.5 mm from the shorting strip.

To accommodate the proposed antenna, a gap space of 35 mm is reserved from the upper left corner of the system ground plane (300 mm × 200 mm). This gap space is typically allocated for the mechanical structure, aligning with the design requirements of narrow bezel laptop computers.

By employing this innovative design and configuration, the proposed antenna achieves a compact form factor and addresses the challenges of limited space in laptop computer applications.

Figure 4.13 provides insights into the design evolution process of the proposed antenna, accompanied by the corresponding reflection coefficient diagrams. Initially, a reference antenna (Ref. 1) is proposed, featuring an ILA without the extended meandering slit. Ref. 1 exhibits three distinct resonant modes at 5.4 GHz (f_1), 5.975 GHz (f_2), and 6.5 GHz (f_3). However, these modes only cover the desired high-frequency band (f_H, 5,150–7,125 MHz).

To expand the frequency coverage to include the desired low-frequency band (f_L, 2,400–2,485 MHz), a C-strip is introduced behind the left section of the ILA, resulting in the creation of the Ref. 2 antenna (without the meandering slit). Due to the partial overlap (11.075 mm²) between the ILA and the C-strip, energy from the ILA's left section couples to the C-strip, leading to the excitation of a lower resonant mode at 2.64 GHz (f_1). While increasing the shorter

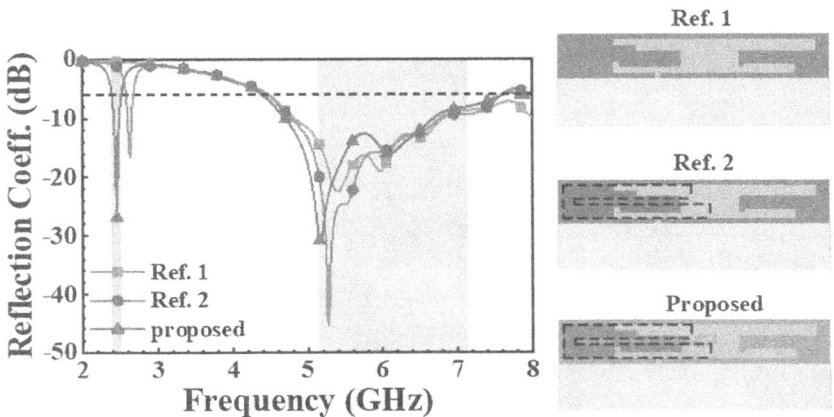

FIGURE 4.13 The design evolution of the antenna (including Ref. 1 and Ref. 2 antennas) and their respective simulated reflection coefficient.

arm length (L_1) of the C-strip could potentially further shift f_1 toward the lower spectrum, Figure 4.14A demonstrates that increasing L_1 to 13.6 mm yields an undesirable f_1 with a 6 dB bandwidth of 2.41–2.47 GHz, which does not cover the desired f_L.

To overcome the narrow bandwidth issue of f_1, an innovative approach is employed by extending a meandering slit from the initial left section slit of the Ref. 2 antenna, resulting in the proposed antenna configuration. Figure 4.13 illustrates that the extended meandering slit contributes to an increased current path for f_1, facilitating its shift from 2.64 GHz to 2.45 GHz in the lower spectrum.

Furthermore, Figure 4.14(b) demonstrates that the introduction of the meandering slit into the ILA significantly enhances the impedance matching of f_1 compared to the configurations depicted in Figure 4.13(a). By tuning the length L_1 from 10.3 mm to 12.3 mm, it is observed that at L_1 = 11.3 mm, f_1 exhibits a much-desired 6 dB bandwidth of 2.39–2.51 GHz.

Regarding the width (W) of the meandering slit, Figure 4.14(c) reveals that adjusting W from 0.1 mm to 0.4 mm slightly shifts f_1 from 2.48 GHz to 2.42 GHz, as further demonstrated in Figure 4.14(d). By setting W = 0.25 mm and L_1 = 11.3 mm, an optimized f_1 at 2.45 GHz is achieved with excellent matching. Notably, tuning L_1 or W has no impact on the resonant modes within the f_H range.

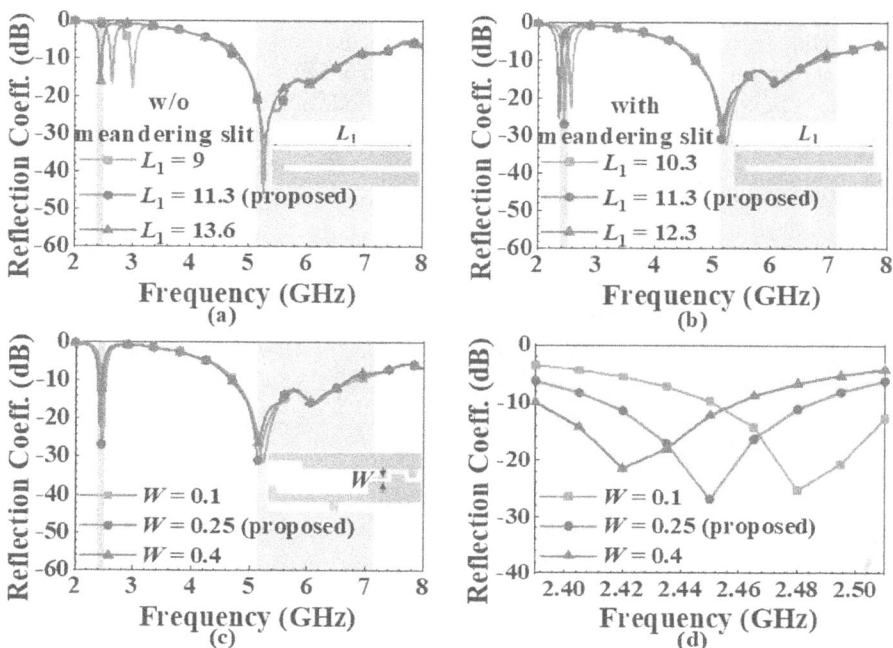

FIGURE 4.14 Simulated reflection coefficients: (a) tuning l_1 (without meandering slit), (b) tuning L_1 (with meandering slit), (c) tuning width w, and (d) f_1 only when tuning width W. Unit: millimeters.

The incorporation of the meandering slit into the ILA alters the coupling effects from the ILA to the C-strip, thereby significantly improving the impedance matching of f_1.

4.5.3   RESULTS AND DISCUSSION

Figure 4.15(a and b) provides a detailed front and back view of the fabricated proposed antenna, securely attached to the system ground constructed from a 0.8 mm thick FR-4 substrate. To facilitate a convenient connection between the proposed antenna and the system ground, a narrow width (4 mm) extended antenna ground is implemented and an adhesive copper tape is utilized to firmly affix the proposed antenna to the system ground. To ensure clarity and prevent any confusion, Figure 4.15(b) showcases the overlapped region between the ILA and the C-strip.

Figure 4.16 showcases the simulated and measured reflection coefficients of the proposed antenna, revealing the excitation of f_L primarily from f_1 and the combination of f_2 to f_4 contributing to f_H. The simulated results demonstrate that f_1 (or f_L) exhibits a 6 dB bandwidth of 4.8% (2.39–2.51 GHz), with a corresponding 10 dB bandwidth of 2.8% (2.41–2.48 GHz). On the other hand, f_H demonstrates a commendable simulated 6 dB bandwidth of 53.3% (4.41–7.67 GHz) and a corresponding 10 dB bandwidth of 35% (4.77–6.8 GHz).

Comparing the simulated and measured results presented in Figure 4.16, it becomes apparent that the impedance bandwidths observed in the measured results are slightly broader than those in the simulated results, potentially due to unexpected fabrication tolerances. Additionally, an unforeseen 5 GHz mode emerges in the measurement, possibly stemming from a higher order mode of f1. Moreover, the measured f_2 and f_3 are offset to higher frequencies by approximately 485 MHz and 310 MHz,

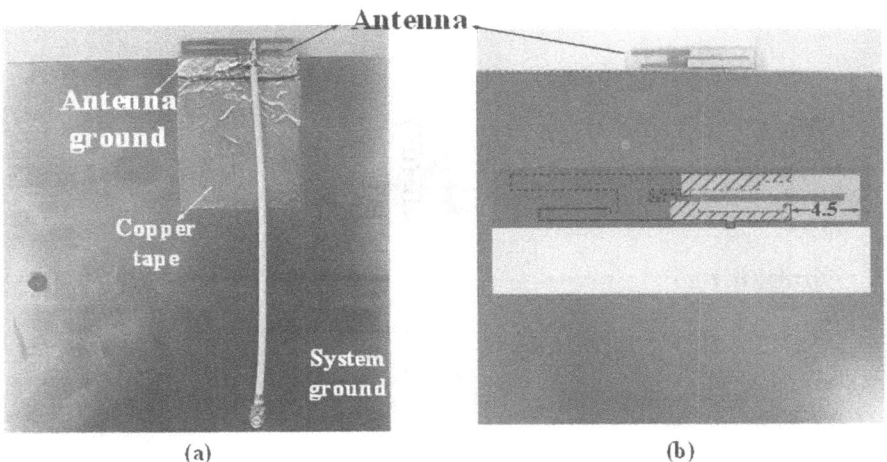

FIGURE 4.15   The fabricated proposed antenna prototype: (a) front and (b) back (showing the overlapped area of 11.075 mm²). Unit: millimeters.

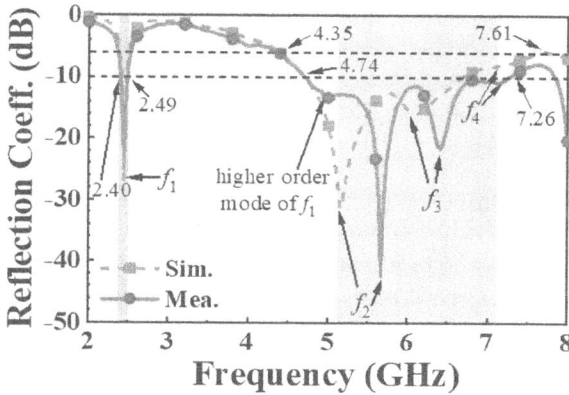

FIGURE 4.16   Simulated and measured reflection coefficients of an antenna.

respectively. It is noteworthy that the measured f_4 exhibits superior impedance matching compared to the simulated results at approximately 7.2 GHz.

The measured 10 dB bandwidths of f_L and f_H satisfactorily encompass the desired Wi-Fi 6E bands of interest, with f_L exhibiting a measured 6 dB bandwidth of 6.1% (2.37–2.52 GHz) and a corresponding 10 dB bandwidth of 3.6% (2.40–2.49 GHz). Additionally, f_H demonstrates a measured 6 dB bandwidth of 59.1% (4.35–8 GHz) and a corresponding 10 dB bandwidth of 42.6% (4.71–7.26 GHz)

To gain deeper insights into the four resonant modes and their corresponding current distributions, the simulated current paths using Ansoft HFSS are analyzed and visualized in Figure 4.17. Figure 4.17(a) illustrates that the excitation of f_1 at 2.45 GHz arises from the presence of the C-strip positioned behind the modified ILA, manifesting a half-wavelength current distribution path from point A to point B. Moving on to Figure 4.17(b), the current path of f_2 at 5.19 GHz is primarily concentrated along the lower left section of the ILA, extending from point C to point D. Notably, point D also serves as the shorting point of the ILA, and its current path spans approximately one-quarter wavelength.

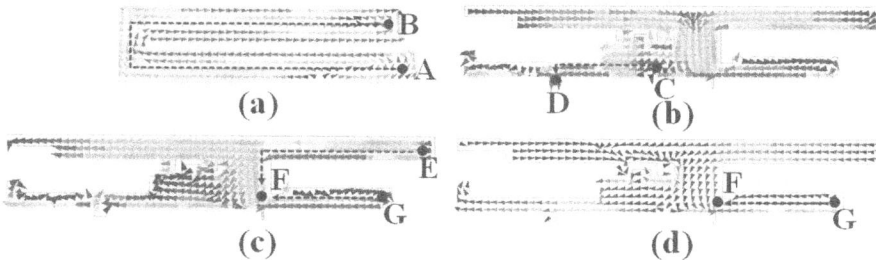

FIGURE 4.17   Simulated current vector distribution diagrams of the antenna at, (a) f_1 = 2.45 GHz, (b) f_2 = 5.19 GHz, (c) f_3 = 6.10 GHz, and (d) f_4 = 7.12 GHz.

In Figure 4.17(c), the current distribution path of f_3 at 6.1 GHz follows a half-wavelength pattern, traversing the inverted L-shaped slit loaded on the right section of the ILA, commencing from point E to point F, and subsequently from point G back to point F. On the other hand, Figure 4.17(d) depicts f_4 at 7.12 GHz as a less pronounced resonant mode, with its vector current amplitude appearing weaker compared to those observed between f_1 and f_3. Nevertheless, f_4 exhibits a quarter-wavelength current distribution path somewhere along the ILA, spanning between point G and point F.

These visualizations of the current paths provide valuable insights into the behavior and characteristics of each resonant mode, elucidating the mechanisms behind their excitation and propagation within the proposed antenna design.

Figure 4.18 shows the simulated and measured gain and antenna efficiency of the proposed antenna. Here, the measured maximum gain across the desired f_L and f_H was approximately 1.0–2.4 dBi and 2–4.6 dBi, respectively. As for the measured antenna efficiency, it was approximately 40%–45% for f_L, and 61%–78% for f_H.

The radiation patterns of the proposed antenna, ranging from f_1 to f_3, are depicted in Figure 4.20(a–c) for both the x-y plane and y-z plane. To streamline the presentation, the radiation pattern for f_4 is not included. As observed in Figure 4.19, there is a commendable agreement between the simulated and measured results.

Across the x-y plane, where f_1 to f_3 are plotted, the main beams of these three resonant modes are predominantly concentrated in the vicinity of the +X direction. On the other hand, when examining the radiation patterns across the y-z plane for f_1 to f_3, notable characteristics emerge. Specifically, these modes exhibit near broadside patterns for the E_ϕ component, representing the azimuthal electric field, while showcasing near conical radiation patterns for the E_θ component, denoting the polar electric field.

These findings shed light on the directional radiation properties of the proposed antenna at various resonant frequencies, underscoring its ability to focus the main beam in specific directions while generating distinctive radiation patterns in both the azimuthal and polar planes.

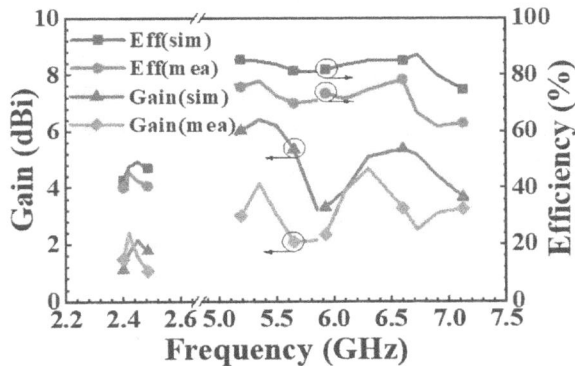

FIGURE 4.18 Measured and simulated gain and efficiency of the antenna.

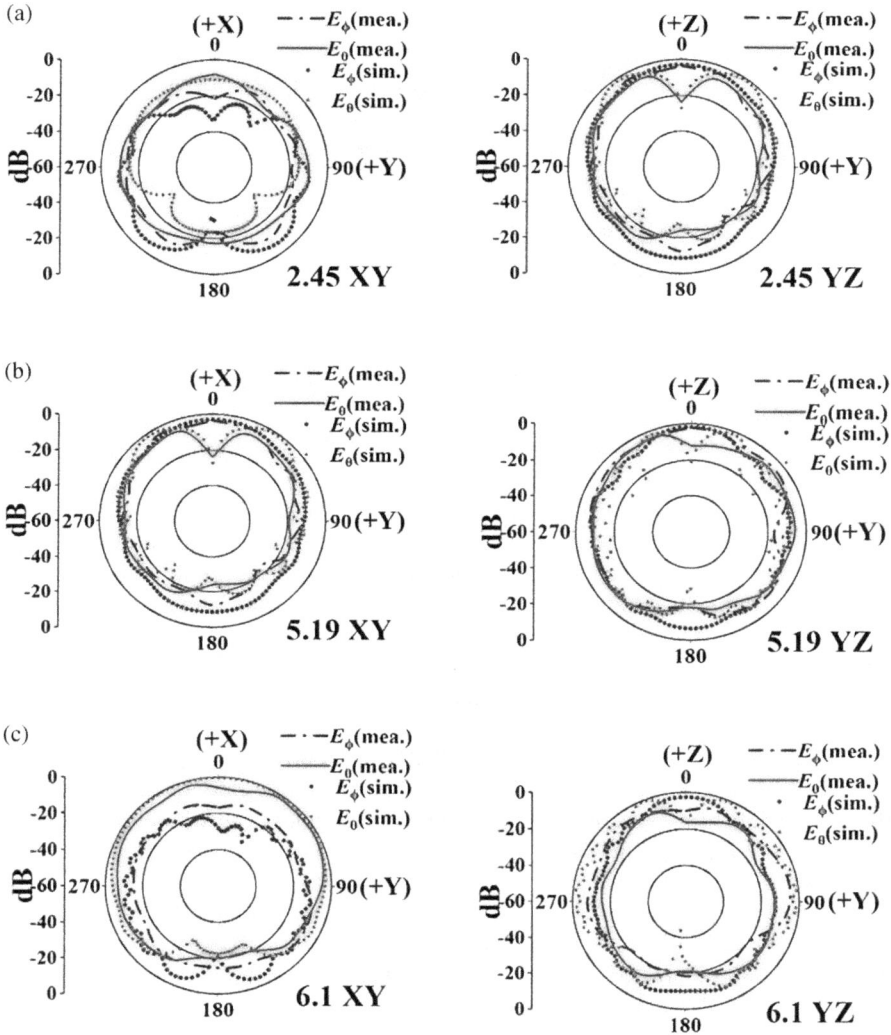

FIGURE 4.19 The simulated and measured radiation patterns (E_θ and E_ϕ) of the proposed antenna across the x-y plane and y-z plane: (a) $f_1 = 2.45$ GHz, (b) $f_2 = 5.19$ GHz, and (c) $f_3 = 6.10$ GHz.

To assess the performance of the proposed low-profile antenna when integrated into an actual laptop computer, Figure 4.20 illustrates the experimental setup. The proposed antenna was mounted at the top-right corner of a laptop cover, with a laptop screen positioned beneath it. The objective was to evaluate whether the antenna's S-parameter characteristics remained intact under these practical conditions.

Upon closer examination of Figure 4.20, it becomes evident that the laptop cover and screen have only a marginal impact on f_1, the lowest resonant frequency.

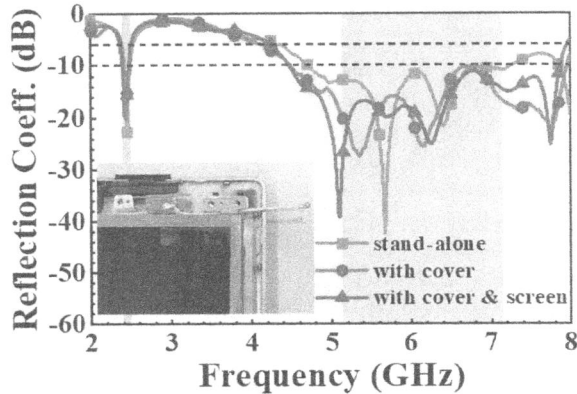

FIGURE 4.20 Measured reflection coefficients of the reported antenna mounted onto an actual laptop.

This implies that the presence of the laptop components has minimal effect on the antenna's ability to resonate at this frequency. Although f_2 to f_4 exhibit some deviations compared to the standalone configuration, the measured results, considering the laptop cover and screen, still manifest desirable 10 dB bandwidths. Specifically, the measured bandwidths for f_L (the lower frequency band) range from 2.4 GHz to 2.48 GHz, while f_H (the higher frequency band) demonstrates a bandwidth of 4.47 GHz to 7.86 GHz.

These findings reinforce the notion that the proposed antenna maintains its desirable performance characteristics even when integrated into a laptop computer, thereby establishing its suitability for real-world applications where compactness and robustness are paramount.

4.5.4 CONCLUDING REMARKS

The investigation into a novel low-profile dual-band laptop antenna has yielded successful results. To encompass the desired high-frequency range of the Wi-Fi 6E band (5,150–7,125 MHz), the antenna design incorporates a modified double-slit loaded ILA structure. Additionally, by employing a C-shaped coupling strip, the desired low-frequency band (2,400–2,485 MHz) is effectively obtained. The measured antenna efficiency across these two bands of interest surpasses 40%, confirming the excellent performance of the proposed design. Consequently, this antenna is well suited for modern narrow-bezel laptop applications, where compactness and high performance are essential considerations.

REFERENCES

1. De', R., Pandey, N. & Pal, A. "Impact of digital surge during Covid-19 pandemic: A viewpoint on research and practice." *International Journal of Information Management*, vol. 55, 102171, 2020. doi: 10.1016/j.ijinfomgt.2020.102171

2. Pokhrel, S. & Chhetri, R. "A literature review on impact of COVID-19 pandemic on teaching and learning." *Higher Education for the Future*, vol. 8, no. 1, pp. 133–141, 2021. doi: 10.1177/2347631120983481

3. Quotient Associates for the Wi-Fi Alliance, 2017. https://www.wi-fi.org/system/files/6_GHz_Wi-Fi_Connecting_to_the_future_202210.pdf

4. Sim, C.-Y.-D. *et al.* "A PIFA Design with WLAN and Wi-Fi 6E Band for Laptop Computer Applications." *2022 IEEE International Symposium on Antennas and Propagation and USNC-URSI Radio Science Meeting (AP-S/URSI)*, Denver, CO, USA, pp. 1808–1809, 2022. doi: 10.1109/AP-S/USNC-URSI47032.2022.9886623

5. Chien, H. Y., Sim, C. Y. D. & Lee, C. H. "Dual-band meander monopole antenna for WLAN operation in laptop computer." *IEEE Antennas and Wireless Propagation Letters*, vol. 12, pp. 694–697, 2013.

6. Sim, C. Y. D., Chien, H. Y. & Lee, C. H. "Dual-/triple-band asymmetric dipole antenna for WLAN operation in laptop computer." *IEEE Transactions on Antennas and Propagation*, vol. 61, no. 7, pp. 3808–3813, 2013.

7. Sim, C. Y. D., Chien, H. Y. & Lee, C. H. "Uniplanar antenna design with adhesive ground plane for laptop WLAN operation." *IEEE Antennas and Wireless Propagation Letters*, vol. 13, pp. 337–340, 2014.

8. Sim, C. Y. D., Chen, C. C., Zhang, X. Y., Lee, Y. L. & Chiang, C. Y. "Very small-size uniplanar printed monopole antenna for dual-band WLAN laptop computer applications." *IEEE Transactions on Antennas and Propagation*, vol. 65, no. 6, pp. 2916–2922, 2017.

9. Sim, C. Y. D., Chen, C. C., Li, C. Y. & Ge, L. "A novel uniplanar antenna with dual wideband characteristics for tablet/laptop applications." *International Journal of RF and Microwave Computer-Aided Engineering*, vol. 27, no. 9, e21145, 2017.

10. Sim, C. Y. D. & Lai, Y. N. "An inverted-F antenna design for WLAN/WiMAX dual-network applications." *International Journal of RF and Microwave Computer-Aided Engineering*, vol. 24, no. 5, pp. 523–528, 2014.

11. Han, T. Y., Hsieh, W. T., Jheng, K. H., Wang, S. H. & Sim, C. Y. D. "Design of laptop antenna for WLAN and Wi-Fi 6E applications." *2021 International Symposium on Antennas and Propagation (ISAP)*, Taipei, Taiwan, pp. 1–2, 2021.

12. Jhang, W. C. & Sun, J. S. "Small antenna design of triple band for WIFI 6E and WLAN applications in the narrow border laptop computer." *International Journal of Antennas and Propagation*, vol. 2021, pp. 1–8, 2021.

13. Su, S. W. & Wan, C. C. "Asymmetrical, self-isolated laptop antenna in the 2.4/5/6 GHz Wi-Fi 6E Bands." *2021 International Symposium on Antennas and Propagation (ISAP)*, Taipei, Taiwan, pp. 1–2, 2021.

14. Lee, C. T., Wan, C. C. & Su, S. W. "Multi-laptop-antenna designs for 2.4/5/6 GHz WLAN and 5G NR77/78/79 operation." *2020 International Symposium on Antennas and Propagation (ISAP)*, Osaka, Japan, pp. 421–422, 2021.

15. Sim, C.-Y.-D., Kulkarni, J., Wang, S.-H., Zheng, S.-Y., Lin, Z.-H. & Chen, S.-C. "Low-profile laptop antenna design for Wi-Fi 6E band." *IEEE Antennas and Wireless Propagation Letters*, vol. 22, no. 1, pp. 79–83, 2023. doi: 10.1109/LAWP.2022.3202697.

16. FCC, "Report and order and further notice of proposed rulemaking; In the matter of unlicensed use of the 6 GHz band (ET Docket No. 18–295); Expanding flexible use in mid-band spectrum between 3.7 and 24 GHz (GN Docket No. 17–183)," Apr. 2020, [Online]. Available: https://docs.fcc.gov/public/attachments/FCC-20-51A1.pdf

17. Su, S. W., Lee, C. T. & Chen, S. C. "Compact, printed, tri-band loop antenna with capacitively-driven feed and end-loaded inductor for notebook computer applications." *IEEE Access*, vol. 6, pp. 6692–6699, 2018.

18. Su, S. W. "Capacitor-inductor-loaded, small-sized loop antenna for WLAN notebook computers." *Progress In Electromagnetics Research M*, vol. 71, pp. 179–188, 2018.

19. Su, S. W. "Compact, self-isolated 2.4/5-GHz WLAN antenna for notebook computer applications." *Progress in Electromagnetics Research M*, vol. 83, pp. 1–8, 2019.

20. Abioghli, M. "Dual-band two layered printed antenna for 2.4/5 GHz WLAN operation in the laptop computer." *IEICE Electronics Express*, vol. 8, no. 18, pp. 1519–1526, 2011.

21. Chou, L. C. & Wong, K. L. "Uni-planar dual-band monopole antenna for 2.4/5 GHz WLAN operation in the laptop computer." *IEEE Transactions on Antennas and Propagation*, vol. 55, no. 12, pp. 3739–3741, 2007.

22. Wong, K. L., Chou, L. C. & Su, C. M. "Dual-band flat-plate antenna with a shorted parasitic element for laptop applications." *IEEE Transactions on Antennas and Propagation*, vol. 53, no. 1, pp. 539–544, 2005.

23. Su, S. W., Lee, C. T. & Chen, S. C. "Very-low-profile, triband, two-antenna system for WLAN notebook computers." *IEEE Antennas and Wireless Propagation Letters*, vol. 17, no. 9, pp. 1626–1629, 2018.

24. Su, S. W. "Very-low-profile, 2.4/5-GHz WLAN monopole antenna for large screen-to-body-ratio notebook computers." *Microwave and Optical Technology Letters*, pp. 1313–1318, 2018.

25. Su, S. W. "Very-low-profile, small-sized, printed monopole antenna for WLAN notebook computer applications." *Progress In Electromagnetics Research Letters*, vol. 82, pp. 51–57, 2019.

26. Lee, C. T., Su, S. W., Chen, S. C. & Fu, C. S. "Low-cost, direct-fed slot antenna built in metal cover of notebook computer for 2.4-/5.2/5.8-GHz WLAN operation." *IEEE Transactions on Antennas and Propagation*, vol. 65, no. 5, pp. 2677–2682, 2017.

27. Su, S. W., Lee, C. T. & Hsiao, Y. W. "Compact two-inverted-F-antenna system with highly integrated π-shaped decoupling structure." *IEEE Transactions on Antennas and Propagation*, vol. 67, no. 9, pp. 6182–6186, 2019.

28. Lee, C. T., Wan, C. C. & Su, S. W. "Multi-laptop-antenna designs for 2.4/5/6 GHz WLAN and 5G NR77/78/79 operation." *2020 International Symposium on Antennas and Propagation (ISAP)*, Osaka, Japan, pp. 421–422, 2020.

29. Su, S. W. "Compact, small, chip-inductor-loaded Wi-Fi 6E monopole antenna." *2021 IEEE International Symposium on Antennas and Propagation and USNC-URSI Radio Science Meeting (APS/URSI)*, Singapore, pp. 937–993, 2021.

5 Enhancing Laptop Connectivity with Multiple Input Multiple Output Antenna Design

5.1 IMPORTANCE OF MIMO SYSTEM

In today's world, advanced wireless communication systems must provide high speeds to fulfill the demands for increased communication data capacity being created by multimedia applications running on terminals. To achieve such high-speed communication, MIMO technology have been adopted by many wireless standards, such as Long-Term Evolution (LTE) and WLAN. In a MIMO system, multiple antennas send multiple signals in the same frequency bands and, as a result, enhance the communication performance. To achieve higher performance, the number of antennas in MIMO systems implemented in recent wireless equipment needs to be increased. For example, current WLAN routers adapting IEEE 802.11ax already have 4×4 MIMO antenna systems. However, wireless equipment must be compact from the viewpoints of installation, design, and harmony with the surroundings. Therefore, compact and multiantenna MIMO systems are needed to achieve both high performance and deploy ability. One simple approach for achieving a compact and multiantenna MIMO system is to implement miniaturized antennas within a small area. In this approach, it is necessary to miniaturize the antennas and suppress performance degradation caused by the nearby antenna.

Therefore, this chapter presents an arrangement of possible antenna arrays using a triple-band monopole antenna design of 0.05 mm thick with an overall dimension of 21×8 mm^2 for sub-6 GHz 5G and Wi-Fi 6 MIMO applications in the laptop computer [1, 2].

5.2 EXISTING STATE-OF-THE-ART TECHNOLOGY

There is a great demand for high-speed and better quality data transmission in wireless operations, which needs proper utilization of the available channel bandwidth. MIMO systems have received significant attention in sub-6 GHz 5G and Wi-Fi 6 operating systems because of their ability to provide high-speed data transmission, efficient utilization of spectrum, and security in the wireless communication system. Practically, a minimum of two antennas occupying a smaller area with high isolation are needed to produce a high-speed signal transmission and reception in a laptop computer. Literature reports many promising MIMO antenna arrays with multiband operations in the laptop computer.

DOI: 10.1201/9781003331018-5

A MIMO antenna based on a driven and shorted strip is proposed by [3, 4]. The antenna proposed in [3], uses a protruded ground plane and spiral open slot to reduce the mutual coupling between the antennas and has dimensions of 55 × 9 × 0.8 mm³. The antenna proposed in [4], has a smaller dimension of 35 × 9 × 0.8 mm³ but uses T-shape decoupling structure to improve the isolation between the two antennas, which provides isolation of around only −15 dB. The proposed antenna in [5] is based on an F-shaped strip and feeding pad technique. It has dimensions of 54 × 9 × 0.8 mm³ and uses a protruded ground with two symmetric branches as an isolating element. Even here, the isolation is approximately −15 dB. Generally, isolation better than -20 dB is preferred for optimum performance of the antenna.

MIMO antennas based on inverted-F technology are discussed in [6, 7]. The antenna [6] uses an Electronic Band Gap (EBG) structure to reduce the mutual coupling between the antennas, but has a very large dimension of 44.1 × 22 × 5.65 mm³. It also uses three kinds of substrate: FR-4, foam, and liquid crystal polymer Roger substrate, which makes the structure complex and expensive. The dual band antenna [7] uses an inverted T-shape slot and a meandering resonant slot as an isolation element and also has a large dimension of 52 × 77.5 × 1.6 mm³. Hence, both these proposed antennas are not suitable candidates for laptop computers.

MIMO antennas using a meandering strip are proposed in [8, 9]. The antenna proposed [8] has a large dimension of 12 × 18 × 1.6 mm³ and may not be suitable for laptop computers. Also, for mutual coupling between the antennas, the antennas are mounted orthogonally by keeping a distance of 15 mm between them. Even though it does not use any additional hardware to reduce mutual coupling, the isolation is not better than -20 dB in some of the operating bands. The antenna proposed [9] uses an additional ground plane as an isolation technique and has a larger dimension of 38 × 7 × 0.8 mm³ of each antenna element. Also, it measures the impedance bandwidth at −6 dB, which is practically not acceptable in laptop computers. It also uses an inductor, which makes the hardware complex and increases the power consumption of the antenna.

The parasitic loop and driven branch MIMO antenna [10] has dimensions of 40 × 9 × 0.8 mm³ and uses a protruded ground plane with a T-shaped slot to reduce the mutual coupling between the antennas. The patch-and-slot type antenna [11] with dimensions of 45 × 6 × 0.4 mm³ uses an additional two resonant slots and a ground pad as isolation between two antennas. Two identical antennas presented in [12] have a size of 5 × 42 × 0.8 mm³ and use a decoupling inductor to reduce the mutual coupling in the 2.4 GHz band. The use of inductor makes the hardware complex and also increases the power consumption of the antenna. These three antennas operate only in dual (2.4 GHz and 5 GHz) WLAN band.

Through the investigation of these antenna arrays formed for MIMO applications in the laptop computer, it is confirmed that the deployment of miniaturized multiple antennas with high isolation and excellent Radio Frequency (RF) performance, covering both sub-6 GHz 5G and Wi-Fi 6, pose challenges for researchers and antenna designers.

5.3   ANTENNA OVERVIEW

This section presents a triple-band monopole antenna design of 0.05 mm thick with an overall dimension of 21×8 mm^2 for sub-6 GHz 5G and Wi-Fi 6 applications in the laptop computer. The proposed antenna comprises three monopole radiating elements, namely, strip AD (inverted C), strip EG (inverted J) and strip FI (inverted U) along with two rectangular open-end tuning stubs namely "m" and "n" sized 1.5×0.9 mm^2 and 1.8×0.9 mm^2, respectively.

The proposed structure is compact, cost-effective, easy to integrate inside the laptop computers, and excites (2.4/5 GHz) bands, conforming to 802.11ax standards and 3.3 GHz bands for sub-6 GHz 5G. The proposed antenna design elucidates that it has measured -10 dB impedance bandwidth of 11.86% (2.22–2.50) GHz in a lower band (f_l), 6.83% (3.25–3.48) GHz in medium band (f_m), and 16.84% (5.00–5.92) GHz in upper band (f_u). The measured gain and radiation efficiency are above 3.65 dBi and 75%, respectively. The simulated and measured results are in good concurrence, which confirms the applicability of the proposed antenna for sub-6 GHz 5G and Wi-Fi 6 applications in the laptop computer.

The antenna is designed without using vias, reactive elements, or matching circuits for excitation of sub-6 GHz 5G and Wi-Fi 6 bands in the laptop computers. The design also does not require any additional ground plane for mounting the antenna. Further, the MIMO antenna is designed without using additional isolation elements to achieve high isolation. Here, the monopole antennas are placed at the top and bottom edge at a distance of 35 mm from left of system ground. It is important to note that the system ground itself acts as an isolating element and reduces the mutual coupling between the antennas.

5.4   ANTENNA GEOMETRY AND DESIGN OF SINGLE ANTENNA

The complete structure of the proposed monopole antenna for sub-6 GHz 5G and Wi-Fi 6 in the laptop computer is shown in Figure 5.1. It comprises of three monopole radiating elements, namely, strip AD (inverted C), strip EG (inverted J) and strip FI (inverted U) and two rectangular open-end tuning stubs, namely, "m" ($a_1 \times b_1$) and "n" ($a_2 \times b_2$). The proposed antenna is designed without using a lossy substrate and is made of only copper material, having a thickness of 0.05 mm. It is placed at a distance of L_{10} from the left corner on the top edge of the system ground. The system ground is made up of 91% brass and has dimensions of size $260 \times 200 \times 0.2$ mm^3 (supports a 13-inch laptop display screen). The antenna structure has a length of 21 mm and shows the height of only 8 mm above the system ground. The 8 mm height of the proposed antenna is promising for installation inside the small bezel of the laptop computer for wireless applications. The proposed monopole antenna is fed by using a 50 Ω miniature and low-loss coaxial cable whose central conductor and outer grounding sheath are connected at point P (the feeding point) on the lower edge of strip AD and at point Q (the grounding point) on the upper edge of the system ground, respectively. This feeding position makes the effective dielectric constant of all radiating elements equal to 1 and, hence, contributes to attaining the desired bands of the proposed antenna.

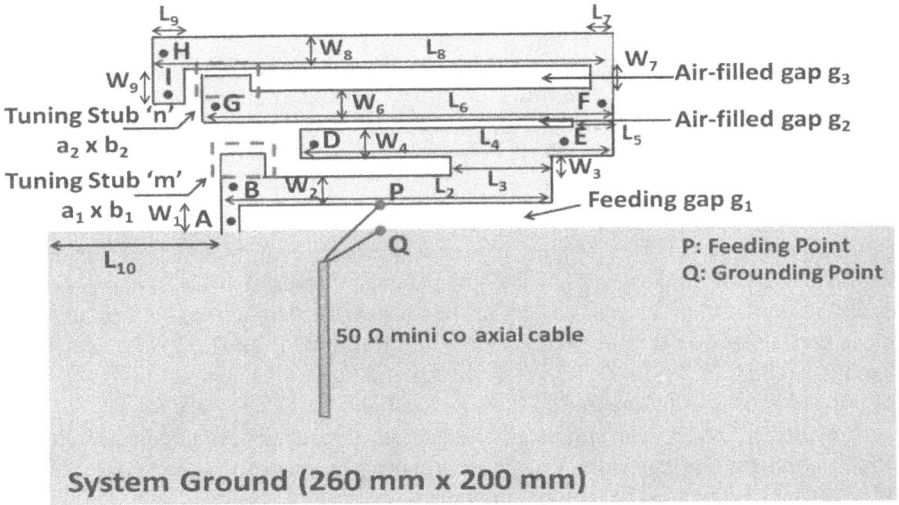

FIGURE 5.1   Complete structure of the proposed antenna.

5.5 WORKING MECHANISM OF PROPOSED ANTENNA

The evolution of the proposed antenna and its reflection coefficient (S_{11}) characteristic at different stages is shown in Figure 5.2. The main objective of the proposed antenna is to obtain a triple-band with wider bandwidth to meet the requirement of sub-6 GHz 5G and Wi-Fi 6 bands. In this regard, the design starts with a simple monopole strip as shown in Figure 5.2A. The antenna consists of an inverted-C-shaped strip AD and a system ground. To achieve resonance at about 5.5 GHz of f_u, the strip AD is designed in such a way that its total length is approximately equal to half of the wavelength at a resonant mode of 5.5 GHz. The strip AD is connected to the system ground at point "A" and fed by coaxial feed line at points "P" and "Q," which introduces a feeding gap g_1 at the feeding points as shown in Figure 5.2A. The impedance matching between strip AD and the coaxial feed line is achieved by adjusting the position of feed points using the CST Microwave Studios (MWS) software. Strip AD, proper feeding position, and the correct dimension of g_1, successfully generate the desired f_u band, with impedance bandwidth spanned in the range of 5.15–5.91 GHz as shown in Figure 5.2A.

To obtain the resonance mode at 3.35 GHz of f_m, strip EG is designed so that its total length is approximately equal to a quarter wavelength at a resonant mode of 3.35 GHz. As shown in Figure 5.2B, the strip EG is placed above the strip AD and connected at point "E" without disturbing the feeding position. This introduces air-filled gap g_2, which neutralizes the inductive reactance of strip EG by producing equal amount of capacitance and helps to achieve impedance matching across f_m. With the introduction strip EG, accurate dimension of g_2, and proper feeding position, the structure successfully generates the f_m band with impedance bandwidth spanned in the range of 3.23–3.46 GHz. At the same time, it shifts the f_u band toward a higher frequency, as impedance mismatch is created at 5.5 GHz as shown in Figure 5.2B.

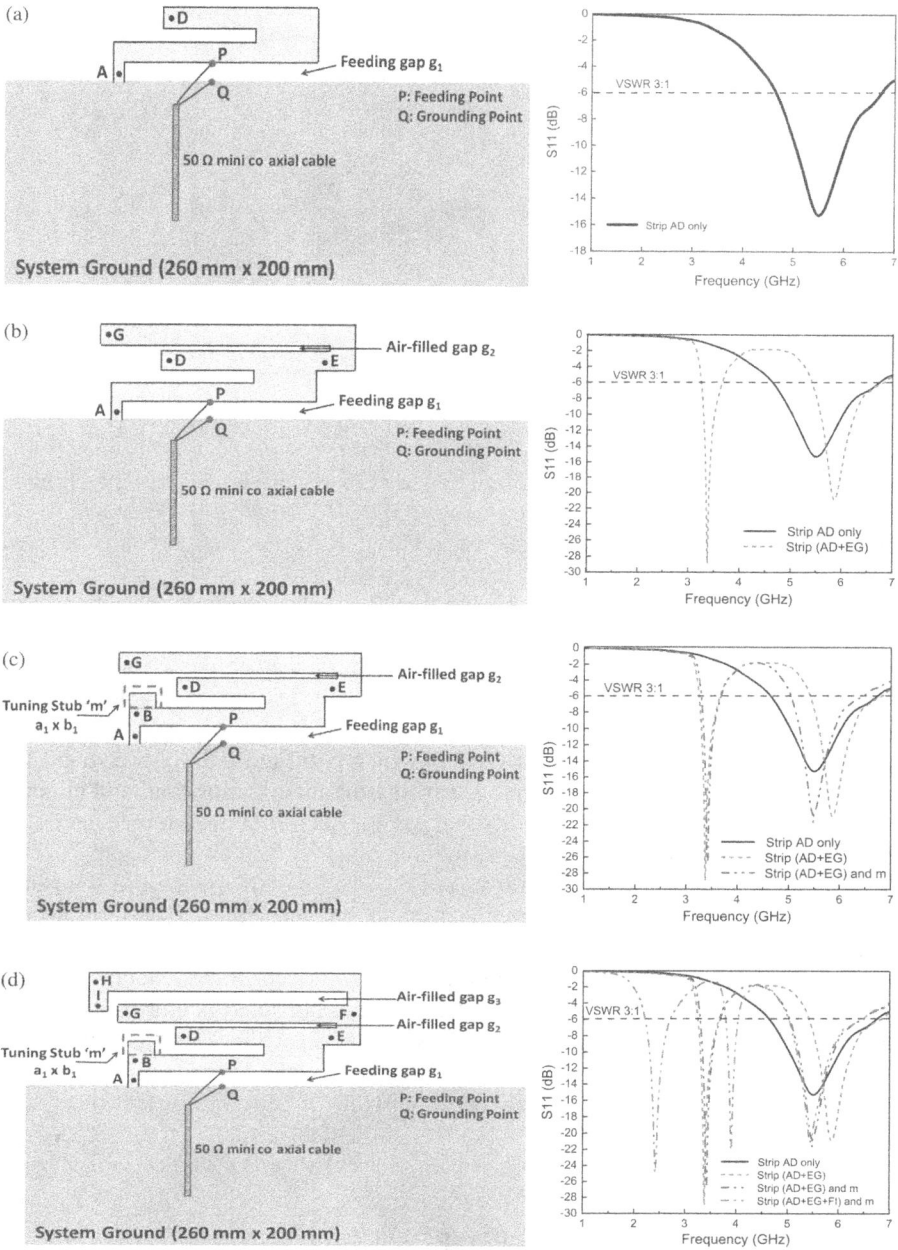

FIGURE 5.2   Design and working process of the proposed monopole antenna: step 1 (A), step 2 (B), step 3 (C), step 4 (D), and step 5 (E). (*Continued*)

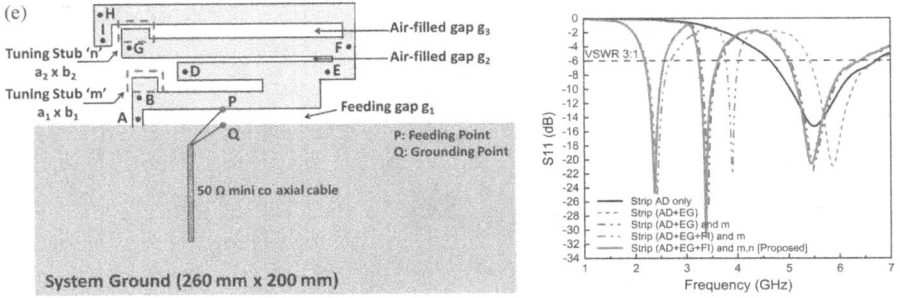

FIGURE 5.2 *(Continued)*

To tune the f_u for obtaining the required band and to achieve impedance matching in the f_u band, a rectangular open-end tuning stub "m" of length (a_1) and height (b_1) is added at point "B" to strip AD as shown in Figure 5.2C. Furthermore, it is inferred that the tuning stub "m" does not affect the f_m band.

Further, to achieve resonant mode at about 2.4 GHz and cover the f_l band, the total length of strip FI is chosen to be a quarter wavelength long at the resonant mode of 2.4 GHz. To limit the dimension of the proposed antenna, the strip FI is folded at point "H" to form an inverted-U shape and, because of this folding, the required quarter wavelength of strip FI reduces to 25.5 mm (about 0.2λ), where λ is the free space wavelength at 2.4 GHz. As shown in Figure 5.2D, the strip FI is attached above the strip EG without disturbing the feeding position, and coupled at point "F," which introduces an air-filled gap g_3. Therefore, impedance matching occurs at 2.4 GHz due strip FI, proper feeding position, and air-filled gap g_3. This generates the required f_l band with impedance bandwidth spanned in the range of 2.26–2.50 GHz without affecting the f_u band, but shifts the f_m band toward the higher frequency due to more capacitive reactance produced by air-filled gap g_3, as shown in Figure 5.2D. To mitigate this increase in capacitive reactance, a second rectangular open-end tuning stub "n" of length (a_2) and height (b_2) is added at point "G," as shown in Figure 5.2E. This stub acts as "blockage" for additional capacitive coupling produced by the air-filled gap g_3 and tunes the f_m band. It is noted that the addition of rectangular tuning stub "n" does not affect the f_l and f_u bands.

5.6 INPUT IMPEDANCE OF PROPOSED ANTENNA

To further gain the understanding of the operating principle and impedance matching of the proposed antenna structure, the curve of input impedance Z_{in} versus frequency is depicted in Figure 5.3. It is visualized that there are exactly three anti-resonances adjacent to three resonating modes at 2.4 GHz, 3.35 GHz, and 5.5 GHz. It

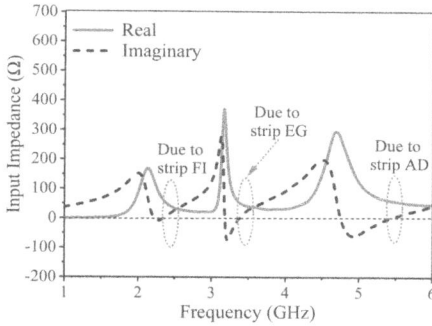

FIGURE 5.3   Input impedance versus frequency of the proposed antenna.

is also observed that at these three resonances, the input impedance of the proposed antenna is about 50 Ω and reactance is almost equal to 0 Ω. This results in excellent impedance matching across the operating bands of proposed antenna. These characteristics lead to maximum power transfer and, hence, a broader impedance bandwidth across the operating bands of the proposed antenna.

5.7   SURFACE CURRENT DISTRIBUTION OF PROPOSED ANTENNA

To further understand the design of the proposed antenna, the surface current distribution at resonances of 2.4 GHz, 3.35 GHz, and 5.5 GHz is shown in Figure 5.4. At 5.5 GHz, the maximum distribution of current flows through the strip AD. This indicates that strip AD contributes to generate the f_u band successfully.

FIGURE 5.4   Surface current distribution of the proposed antenna at 5.5 GHz (A), at 3.35 GHz (B), and at 2.45 GHz (C). (*Continued*)

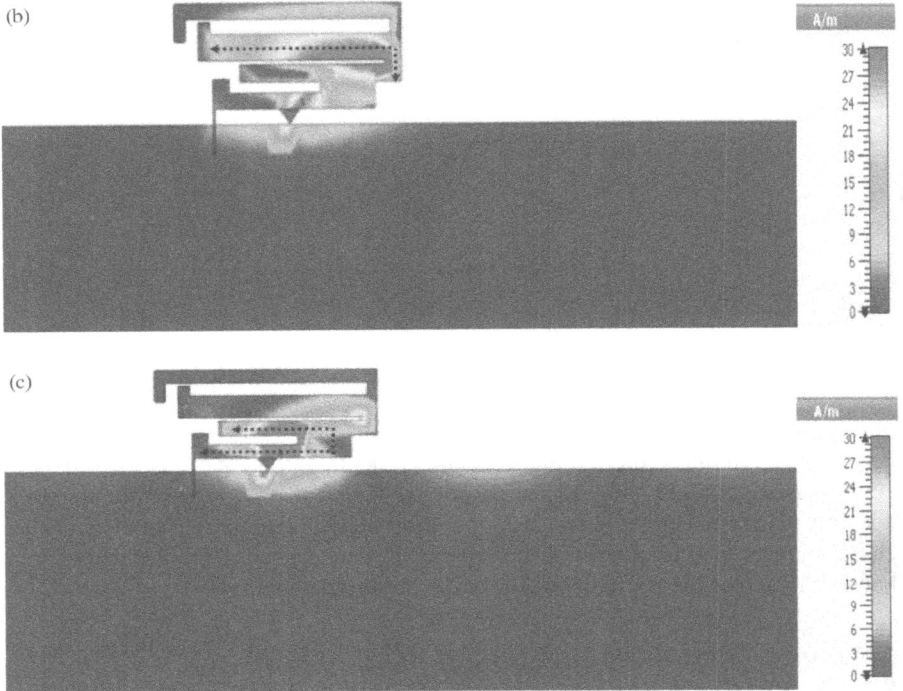

FIGURE 5.4 (*Continued*)

At 3.35 GHz, the maximum distribution of current is only strip EG, which indicates that it generates the f_m band successfully. Further, at 2.4 GHz, the maximum distribution of current is observed through strip FI, which indicates that the strip FI contributes in producing the 2.4 GHz band. Almost zero current flowing through the stub "n" clearly indicates that it acts as a blockage for additional capacitive coupling produced by air-filled gap g_3 and, hence, contributes in achieving the impedance matching at 2.4 GHz.

5.8 PARAMETRIC STUDY OF PROPOSED ANTENNA

Due to its very low profile and ultra-thin and compact size of the antenna, the mirror performance of the proposed antenna is very responsive to the variation of geometrical parameters. To illustrate, the influence of essential parameters on operating bands f_l, f_m, and f_u, a parametric analysis is carried out in this section. The important parameters selected for analysis are rectangular tuning stub "m" of size ($a_1 \times b_1$), rectangular tuning stub "n" of size ($a_2 \times b_2$) and L_{10} for mounting the proposed antenna on the system ground. It also includes the analysis of "Effect of different sizes of system ground" on the amplitude of S_{11} (dB) and impedance bandwidth

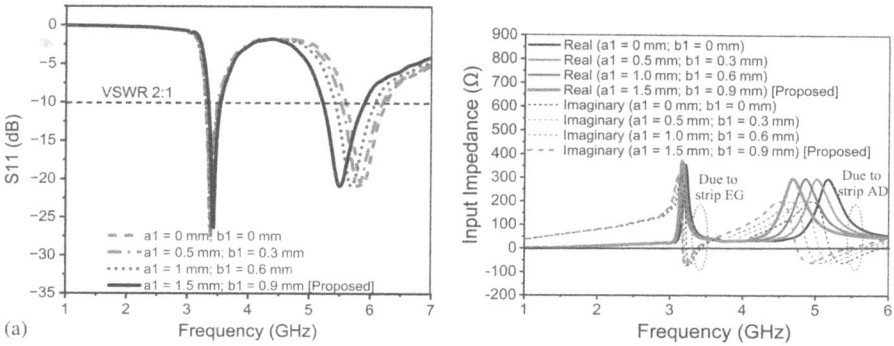

FIGURE 5.5 (A) Simulated reflection coefficient and (B) input impedance (Ω) as a function of tuning stub "m" ($a_1 \times b_1$).

across operating bands of the proposed antenna to check its suitability for other portable devices.

5.8.1 EFFECTS OF A RECTANGULAR TUNING STUB "M" ($A_1 \times B_1$) ON THE PROPOSED ANTENNA

The effects of a rectangular open-end tuning stub "m" of length (a_1) and height (b_1) on impedance bandwidth and the input impedance of the proposed antenna over the f_u band is studied in Figure 5.5. From Figure 5.5A, it is noted that as the value of a_1 increases from 0 mm to 1.5 mm (in step increments of 0.5 mm), and b_1 increases from 0 mm to 0.9 mm (in step increments of 0.3 mm), the f_u band shifts toward a lower frequency as it attenuates the capacitive reactance produced by air-filled gap g_2. With the aid of Figure 5.5B, it is observed as the value of a_1 and b_1 increases, there is a smooth variation of the input impedance from 68 Ω toward 50 Ω, and reactance also becomes 0 Ω at a resonant mode of 5.5 GHz. Hence, it is observed that at $a_1 = 1.5$ mm and $b_1 = 0.9$ mm, the input impedance of the antenna is equal to 50 Ω and reactance is equal to 0 Ω at the desired resonant mode of 5.5 GHz as shown in Figure 5.5B. This condition leads to impedance matching at 5.5 GHz and successful generation of the f_u band with the impedance bandwidth spanned in the range of 5.15–5.91 GHz in the presence of strip EG. Hence, the optimized size of "m" is 1.5×0.9 mm^2.

5.8.2 EFFECTS OF A RECTANGULAR TUNING STUB "N" ($A_2 \times B_2$) ON THE PROPOSED ANTENNA

The effects of rectangular tuning stub "n" of length (a_2) and height (b_2) on the impedance bandwidth and input impedance of the proposed antenna across the f_m band is studied in Figure 5.6. As the value of a_2 is increased from 0 mm to 1.8 mm (in step increments of 0.6 mm), and b_2 increased from 0 mm to 0.9 mm (in step increments of 0.3 mm), the f_m band shifts from the resonant mode of 3.85 GHz toward a resonant

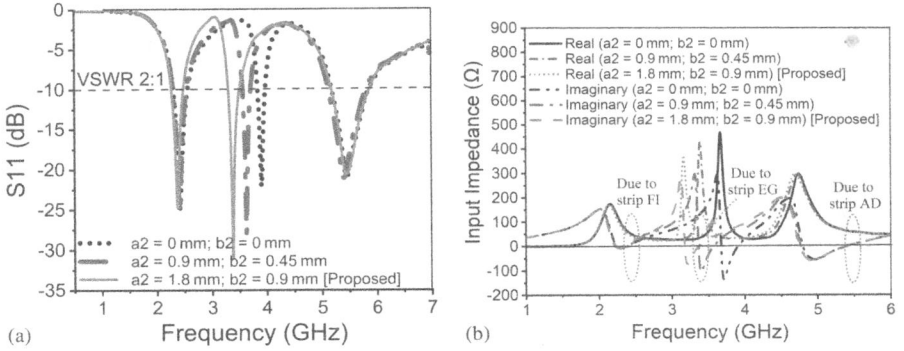

(a)   Frequency (GHz)   (b)   Frequency (GHz)

FIGURE 5.6 (A) Simulated reflection coefficient and (B) input impedance (Ω) as a function of tuning stub "n" ($a_2 \times b_2$).

mode of 3.35 GHz with impedance bandwidth spanned in the range of 3.23–3.46 GHz as shown in Figure 5.6A. Also, as seen in Figure 5.6B, for $a_2 = 1.8$ mm and $b_2 = 0.9$ mm, the input resistance is equal to 50 Ω and capacitive reactance is equal to 0 Ω at the desired resonance of 3.35 GHz. This leads to impedance matching between the antenna and coaxial feed and generation of the desired f_m band. From Figure 5.6, it is also noted that the f_l and f_u bands remain unaffected. Therefore, the optimized size of "n" is 1.8×0.9 mm^2.

5.8.3 EFFECTS OF VARYING L_{10} ON THE PROPOSED ANTENNA

The effects of varying L_{10} from 0 mm to 105 mm (in the step increments of 35 mm) for mounting the proposed antenna on system ground over f_l, f_m, and f_u bands are analyzed in Figure 5.7. It is visualized that at $L_{10} = 0$ mm, the resonant mode 2.4 of f_l

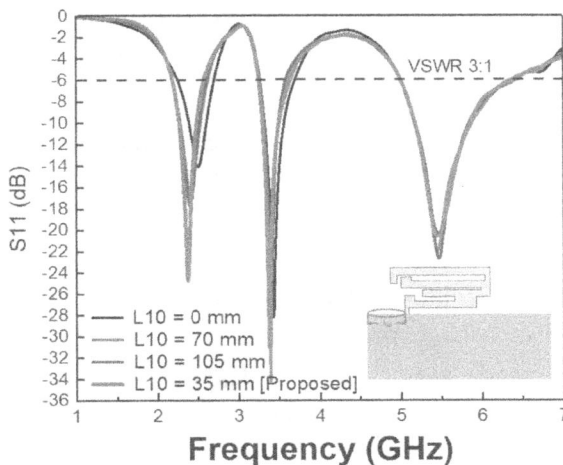

Frequency (GHz)

FIGURE 5.7 Simulated reflection coefficient of proposed antenna as a function of L_{10}.

shifts toward a higher frequency with reduced bandwidth because of degradation in impedance matching, while the resonant mode 3.35 GHz of f_m and 5.5 GHz of the f_u bands have negligible effect. At $L_{10} = 35$ mm, all the desired bands of the proposed antenna are obtained with the required bandwidth and Voltage Standing Wave Ratio (VSWR) less than 2. Further, increase in L_{10} affects the bandwidth and the amplitude of S_{11} (dB) of all the desired bands along with minor shifting of the resonant modes. This effect can be visualized at $L_{10} = 105$ mm, as shown in Figure 5.7. Hence, from this study, the optimum value of $L_{10} = 35$ mm is selected for mounting the proposed antenna to achieve the optimal performance on the system ground.

5.8.4 EFFECTS OF DIFFERENT SIZES OF SYSTEM GROUND ON THE PROPOSED ANTENNA

To study the effect of different sizes of the system ground on the proposed antenna, few standard sizes such as "5-inch" (120 mm × 70 mm), "10-inch" (200 mm × 150 mm) and "15-inch" (300 mm × 190 mm) of display screens are compared with the selected 13-inch" (260 mm × 200 mm) system ground across the three operating bands f_l, f_m , and f_u. Figure 5.8 confirms that the proposed antenna performs equally well for different sizes of the system ground without any effect on the impedance bandwidth and amplitude of S11 (dB) of the proposed antenna.

5.8.5 DIMENSIONS OF PROPOSED ANTENNA

From the previously discussed design and parametric studies, it is confirmed that the proposed antenna operates with a simulated impedance bandwidth spanned in the range of 2.26–2.5 GHz in the lower band f_l, 3.23–3.46 GHz in medium band f_m, and 5.15–5.91 GHz in upper band f_u. Hence, the proposed antenna covers both the

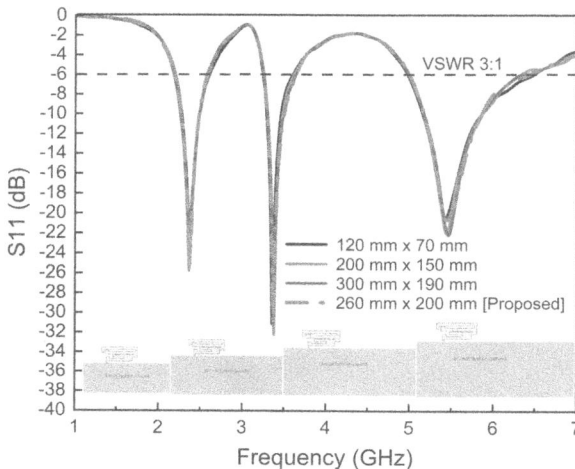

FIGURE 5.8 Effect of different sizes of the system ground (display ground) on the performance of the proposed antenna.

TABLE 5.1

Parametric and Optimized Values of the Proposed Antenna

Parameter	Value (mm)	Parameter	Value (mm)	Parameter	Value (mm)
L_1	0.2	L_8	21	W_6	1.8
$L_2 = L_4$	15	$W_1 = W_7 = W_8$	1	W_9	1.5
L_3	5	W_2	1.10	a_1	1.5
L_5	1.5	W_3	0.9	b_1	1.8
L_6	19	W_4	1.2	$a_2 = b_2$	0.9
$L_7 = L_9$	1	W_5	3.2	L_{10}	35

sub-6 GHz 5G and Wi-Fi 6 bands. The parametric values from this parametric study and optimized values of other parameters using optimization techniques in CST MWS are cataloged in Table 5.1.

5.9 TWO-ANTENNA MIMO SYSTEM USING PROPOSED MONOPOLE ANTENNA

Possible antenna arrays formed by using the proposed monopole antenna for the sub-6 GHz 5G and Wi-Fi 6 band are studied in this section and shown in Figure 5.9.

In the first antenna array (case I), two antennas, Antenna 1 and Antenna 2, are mounted at the top edge of the same system ground and at a distance of 155 mm apart. Antenna 1 is the proposed antenna as shown in Figure 5.1 whereas Antenna 2 is an exact replica of Antenna 1. The simulated S parameters (dB) S_{11}, S_{21} and S_{22} of case I are shown in Figure 5.10. In this case, no effect on impedance bandwidth is observed for S_{11} and S_{22}, but the amplitude of S_{11} (dB) is higher across the f_l and f_u

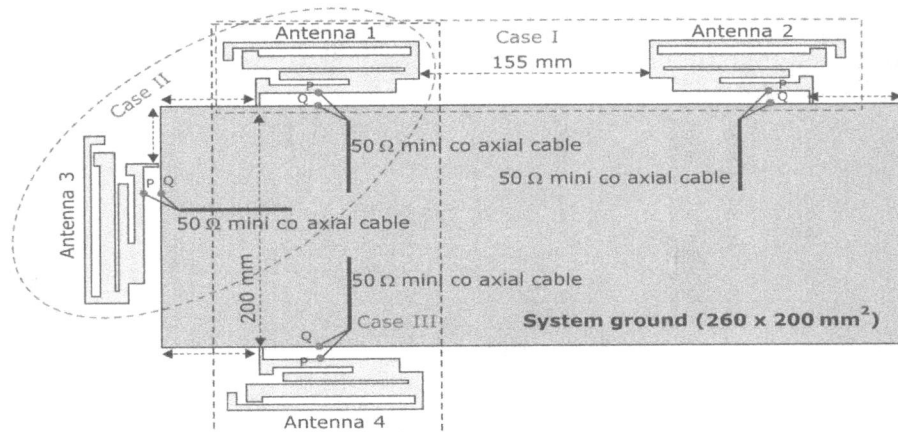

FIGURE 5.9 Possible antenna array cases for multiple input multiple output applications mounted on the system ground.

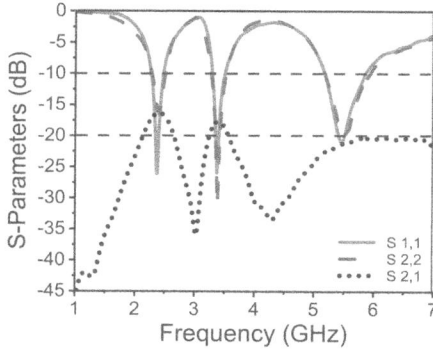

FIGURE 5.10 Simulated S parameters for MIMO antenna array of case I.

bands, whereas the amplitude of S_{22} (dB) is higher across the f_m band. Hence, in this case, there is a small mismatch in terms of amplitude of S_{11} (dB) and S_{22} (dB). Also, the required S_{21} is not below -20 dB for the f_l and f_m bands. Hence, this case is not considered.

The second possible antenna array (case II), as shown in Figure 5.9, is further studied. In this case, Antenna 1 and Antenna 3 are placed adjacent to each other at a distance of 35 mm from the top left corner of system ground. Antenna 3 is an exact replica of Antenna 1 and is rotated anticlockwise by 90° before placing it on the vertical left edge of the system ground. The simulated parameters are shown in Figure 5.11.

It is observed that there is a great mismatch between S_{11} (dB) and S_{22} (dB). From S_{22} of Antenna 3, it is seen that the impedance bandwidth of the f_l and f_u bands becomes reduced and does not conform the required bandwidth. Also, the amplitude of S_{21} (dB) is approximately -10 dB in the f_l band and -13 dB in the f_m band, which is not suitable for receiving excellent RF performance of the MIMO system. Therefore, this array (case II) is not attractive for forming a MIMO system in the laptop computer.

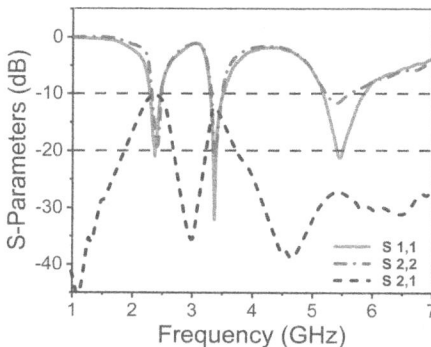

FIGURE 5.11 Simulated S parameters for the antenna array of case II.

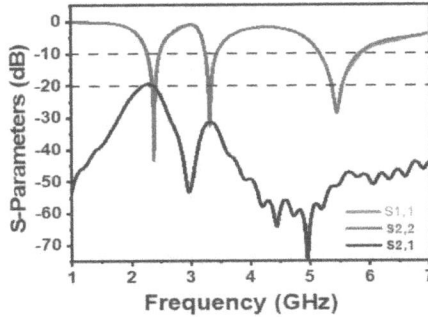

FIGURE 5.12   Simulated S parameters for the proposed MIMO antenna array of case III.

Another possible array (case III) was formed by using Antenna 1 and Antenna 4, as analyzed in Figure 5.9. In this case, Antenna 4 is same as Antenna 1 and is rotated 180° along the x-axis. It is then placed at the bottom edge of the system ground at a distance of 200 mm from Antenna 1. Figure 5.12 shows the simulated S parameters (dB) of case III. Here, it is observed that amplitudes of S_{11} (dB) and S_{22} (dB) are the same across all the operating bands. Also, isolation between antennas or S_{21} (dB) is achieved as -20 dB, -35 dB, and -45 dB across the f_l, f_m, and f_u bands, respectively, as the system ground itself acts as an isolating element between Antenna 1 and Antenna 4. The obtained isolation values are good for achieving optimal performance of the MIMO system. Hence, from the simulated results observed in Figure 5.12, case III was a good candidate for MIMO and was immediately taken for fabrication, the results of which are discussed later in the chapter.

5.9.1   Results and Discussion for Proposed MIMO Array of Case III

To validate the simulated results of case III, the prototype of the proposed antenna array is fabricated as shown in Figure 5.13. Antenna 1 and Antenna 4 were tested using Rohde & Schwarz (9 KHz–16 GHz) network analyzer. The radiation performance, including radiation patterns, gain, and radiation efficiency of proposed case III were tested in an anechoic chamber of size $8 \times 4 \times 4$ m³.

All the parameters were measured under the condition that while measuring Antenna 1, Antenna 4 was terminated with a 50 Ω load and vice-versa.

5.9.1.1   Reflection Coefficient

Figure 5.14 shows the comparison of simulated and measured S-parameters (dB) of the proposed MIMO antenna array and provides only S_{11} and S_{21} due to the analogy with S_{22} and S_{12}, respectively. The simulated -10 dB impedance bandwidth of Antenna 1 and Antenna 4 are 10.08% (2.26–2.50 GHz), 6.87% (3.23–3.46 GHz), and 13.74% (5.15–5.91 GHz), whereas the measured values are 11.86% (2.22–2.5GHz), 6.83% (3.25–3.48 GHz), and 16.84% (5.00–5.92 GHz) in the f_l, f_m, and f_u bands, respectively. The simulated and measured isolation of S_{21} (dB) between Antenna 1 and Antenna 4 are better than -20 dB across all the operating bands, as shown in Figure 5.14.

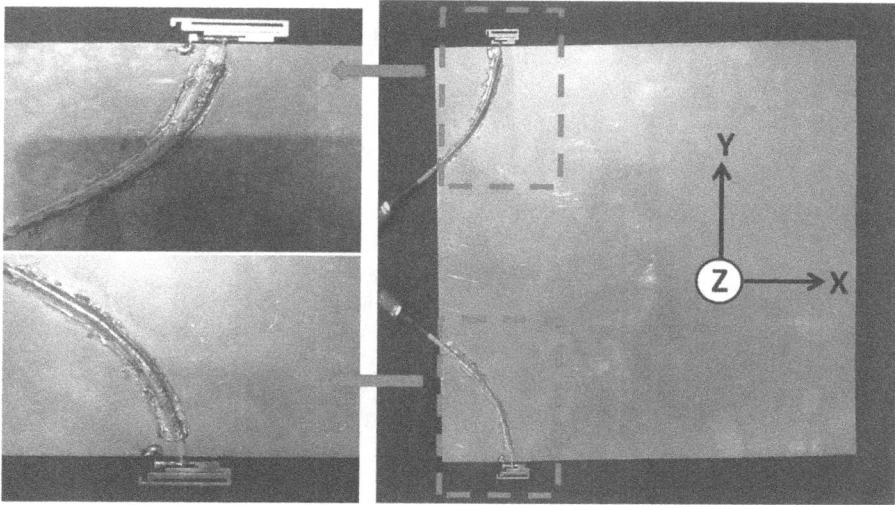

FIGURE 5.13   Fabricated prototype of proposed MIMO antenna for laptop computers.

5.9.1.2   Measured Radiation Pattern

The normalized measured radiation patterns of proposed case III are shown in Figure 5.15. Here, E_θ and $E_\varnothing$ of the proposed MIMO antenna array are shown at the measured resonant mode of 2.39 GHz, 3.37 GHz, and 5.5 GHz of f_l, f_m, and f_u, respectively. $E_\varnothing$ in both antennas, namely, Antenna 1 and Antenna 4, at all resonant modes is nearly omnidirectional, whereas E_θ contributes dipole pattern forming bi-directional radiation. This confirms the applicability of the antenna array for MIMO applications in the laptop computer.

5.9.1.3   Simulated and Measured Gain and Radiation Efficiency

Figure 5.16 shows simulated and measured gain and radiation efficiency of the proposed antenna array (case III). Figure 5.16 presents the radiation efficiency of only

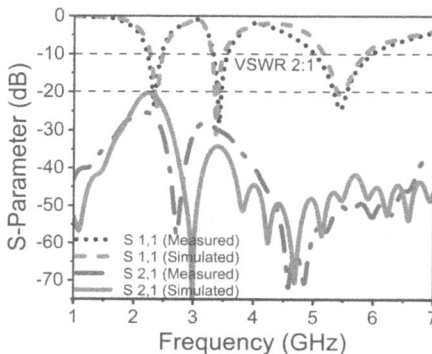

FIGURE 5.14   Simulated and measured S parameters of the proposed MIMO antenna.

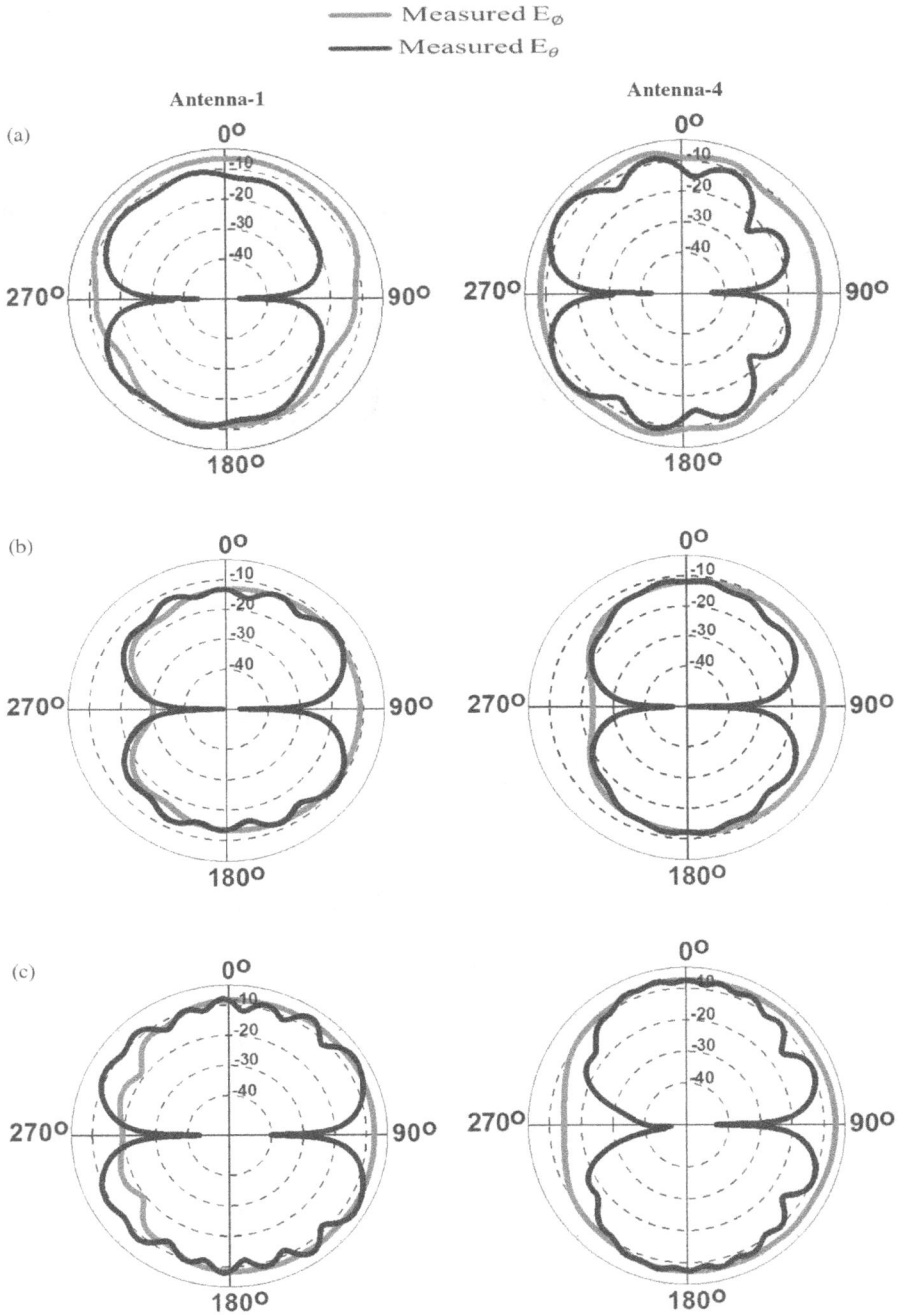

FIGURE 5.15 Measured radiation pattern of the proposed MIMO antenna at 2.42 GHz, (a), 3.37 GHz (b), and 5.5 GHz (c).

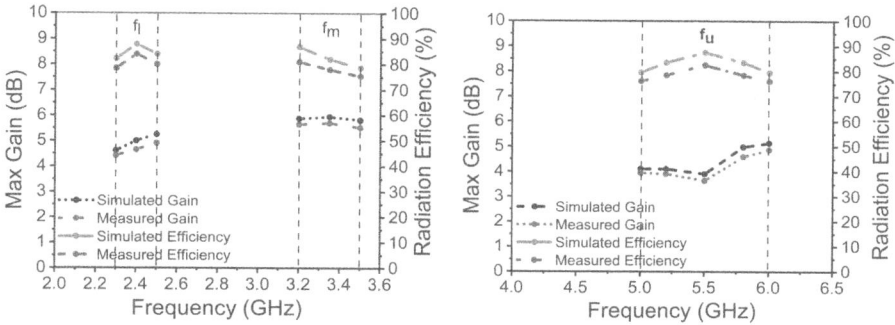

FIGURE 5.16   Simulated and measured gain and efficiency of the proposed MIMO antenna.

Antenna 1 due to the analogy with Antenna 4. The values of simulated and measured gain and efficiency are as shown in Table 5.2. A small deviation in simulated and measured values of gain and efficiency is observed, which may be due to fabrication tolerances.

5.9.1.4   Envelope Correlation Coefficient (ECC)

In a MIMO system, envelope correlation coefficient (ECC) is a very important parameter to evaluate channel capacity and cross-correlation performances between two antennas. It can be calculated using the formula [11] as follows:

$$ECC = \frac{\left|S_{11}^{*}S_{12} + S_{21}^{*}S_{22}\right|^{2}}{\left(1 - \left|S_{11}\right|^{2} - \left|S_{21}\right|^{2}\right)\left(1 - \left|S_{22}\right|^{2} - \left|S_{12}\right|^{2}\right)} \tag{5.1}$$

where * indicates the complex conjugate. Figure 5.17 shows the simulated and measured ECC using S-parameters. The implemented MIMO antenna demonstrates optimal performance, as evidenced by simulated and measured ECC values consistently below 0.004 across all operating bands. Remarkably, this achievement was reached without the need for any additional isolation techniques between the two antennas.

TABLE 5.2
Comparison of Simulated and Measured Values of Gain and Efficiency (Case-III)

Covered Bands (GHz)	Gain (dBi)		Efficiency (%)	
	Simulated	Measured	Simulated	Measured
f_l	4.61–5.25	4.42–4.90	82.24–87.86	78.5–83.9
f_m	5.82–5.95	5.52–5.70	78.69–86.88	75.43–81.0
f_u	3.92–5.14	3.64–4.87	79.59–87.58	76.15–81.5

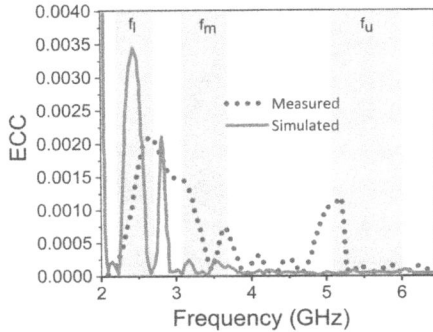

FIGURE 5.17   Simulated and measured ECC for the proposed MIMO antenna.

5.9.2   EXPERIMENTAL ANALYSIS OF PROPOSED ANTENNA

To study the applicability of the proposed MIMO antenna array, Figure 5.18A shows the mounting of the proposed MIMO antenna array inside a L412 ThinkPad Lenovo laptop computer. Figure 5.18B shows the measured reflection coefficient S_{11} (dB) versus frequency (GHz) result of the proposed array. Only S_{11} results are shown due to analogy with S_{22}. It is observed that there is a negligible effect on the resonant mode of f_l and f_m, whereas the f_u band was slightly shifted toward the higher frequency due to the material effect of the laptop cover, and the measured operating bandwidth is spanned as 5.30–6.25 GHz. The obtained results are also verified by mounting the proposed MIMO antenna array in an HP Pavilion 15 laptop computer; the obtained results are same as obtained in the ThinkPad Lenovo laptop. Hence, due to analogy, the results are not shown. However, the desired f_u band can be recovered by slightly tuning the a_1 and b_1 of tuning stub "m."

5.9.3   PERFORMANCE COMPARISON OF PROPOSED
MIMO ARRAY WITH EXISTING WORK

To prove the potency of the proposed MIMO array, a comparison including size, type of antenna, covered operating bands, and isolating technique has been performed in Table 5.3. The proposed MIMO antenna array has been compared with recently reported MIMO array for wireless operations in the laptop computer.

5.9.4   FEATURES OF PROPOSED ANTENNA

- The proposed antenna is simple in structure, miniaturized, ultra-thin and cost-effective, unlike [6, 8], and suitable for next-generation laptops.
- The proposed antenna operates in sub-6 GHz 5G and Wi-Fi 6 bands and an array is formed without using an isolating element, reactive element, or any additional ground plane, unlike all the reported literature.

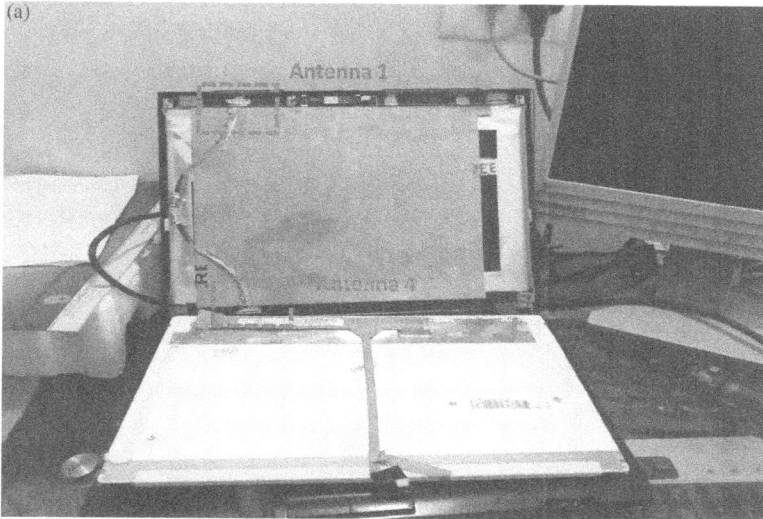

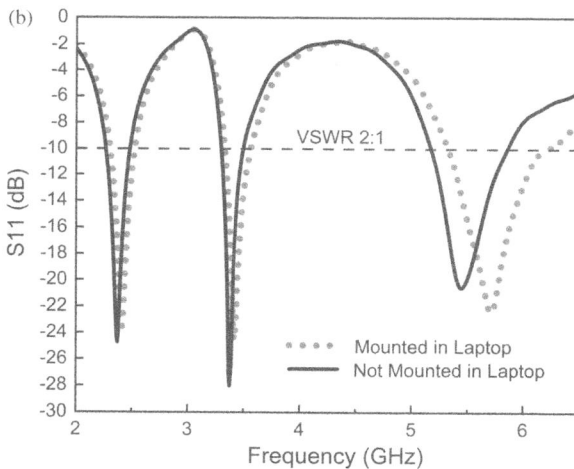

FIGURE 5.18   The proposed antenna installed inside practical laptop (A) and a comparison of measured S_{11} results when the proposed antenna is installed inside the laptop and when it is not installed in the laptop (B).

- The ECC of the proposed antenna array is below 0.004 over all the operating band of interest.
- The antenna operates in a triple-band unlike [3, 7, 10–12].
- The configuration of the antenna array is simple and enhances the data rate in MIMO applications.
- The isolation between antennas is better than −20 dB, unlike [4, 5, 8, 12].
- The proposed antenna array is the thinnest compared to reported state-of-the-art technology.

TABLE 5.3

Performance Comparison with Recently Reported MIMO Array

Ref.	Size (mm³)	Operating Bands (GHz)	Type of Antenna	Isolation Technique
[3]	55 × 9 × 0.8	2.4/5.2/5.5/5.8	Driven strip and shorted strip	Protruded ground plane and spiral open slot
[4]	35 × 9 × 0.8	2.4/3.5/5.2/5.5/5.8	Driven strip and shorted strip	Protruded T-shape decoupling structure
[5]	54 × 9 × 0.8	2.4/2.5/3.5/5.2/5.5/5.8	F-shaped strip and feeding pad	Protruded ground plane and decoupling structure of two branches
[6]	44.1 × 22 × 5.65	1.8/2.4/3.4/5.2	Planar Inverted-F	EBG structure
[7]	52 × 77.5 × 1.6	2.4/5.2/5.5/5.8	Inverted-F	Inverted-T and meandering branch
[8]	12 × 18 × 1.6	2.3/2.4/3.5/5.2/5.5/5.8	Meandering line slit	Orthogonal mounting and space of 15 mm
[9]	38 × 7 × 0.8 (each antenna element)	1.8/2.3/2./5.2/5.5/5.8	Meandering and driven strip	Additional ground plane
[10]	40 × 9 × 0.8	2.4/5.2/5.5/5.8	Parasitic loop and driven branch	Additional ground plane with inverted-T slot
[11]	45 × 6 × 0.4	2.4/5.2/5.5/5.8	Patch and slot antenna	Protruded ground plane and two resonant slots
[12]	42 × 5 × 0.8	2.4/5.2/5.5/5.8	Monopole antenna	Decoupling inductors
Proposed	21 × 8 × 0.05 (Each antenna element)	2.3/2.4/3.3/5.2/5.5/5.8	Monopole strips	No additional isolation elements are used. System ground itself acts as an isolating element.

5.9.5 Concluding Remarks

The triple-band monopole antenna design operating in sub-6 GHz 5G and Wi-Fi 6 for MIMO applications in the laptop computer is verified successfully. The proposed antenna has a very small size of 21×8 mm^2, a simple structure, is easy to fabricate, and operates in three sub-6 GHz 5G and Wi-Fi 6 bands. Additionally, owing to excellent RF performance and the small size of the proposed antenna, an antenna array formed by using the same antenna for MIMO system shows that it has excellent gain and efficiency well above 3.65 dBi and 75%, respectively, isolation between two antennas is better than -20 dB, and the ECC is below 0.004 over all the operating bands of interest. Hence, the proposed antenna and the antenna array formed are promising and good candidates for MIMO application in the laptop computer.

REFERENCES

1. Kulkarni, J. "Multiband triple folding monopole antenna for wireless applications in the laptop computers." *International Journal of Communication Systems*, vol. 34, no. 8, e4776, 2021.
2. Kulkarni, J. S. & Seenivasan, R. "Design of a novel triple-band monopole antenna for WLAN/WiMAX MIMO applications in the laptop computer." *Circuit World*, vol. 45, no. 4, pp. 257–267, 2019.
3. Chen, S.-C., Sze, J.-Y. & Chuang, K.-J. "Isolation enhancement of small-size WLAN MIMO antenna array for laptop computer applications." *Journal of Electromagnetic Waves and Applications*, vol. 31, no. 3, pp. 323–334, 2017.
4. Chen, W.-S., Shu, J. H. & Sim, C.-Y.-D. "Small-size, WLAN/5G MIMO antenna for laptop computer applications." *2017 Sixth Asia-Pacific Conference on Antennas and Propagation (APCAP), IEEE*, Xi'an, China, pp. 1–3, 2017.
5. Deng, J. Y., Li, J. Y., Zhao, L. & Guo, L. "A dual-band inverted-F MIMO antenna with enhanced isolation for WLAN applications." *IEEE Antennas and Wireless Propagation Letters*, vol. 16, pp. 2270–2273, 2017.
6. Guo, L., Wang, Y., Gao, Y. & Shi, D. "A compact uniplanar printed dual-antenna operating at the 2.4/5.2/5.8 GHz WLAN bands for laptop computers." *IEEE Antennas Wireless Propagation Letters*, vol. 13, pp. 229–232, 2014.
7. Liu, Y., Wang, Y. & Du, Z. "A broadband dual-antenna system operating at the WLAN/WiMAX bands for laptop computers." *IEEE Antennas and Wireless Propagation Letters*, vol. 14, pp. 1060–1063, 2015.
8. Ojaroudi, N., Ghadimi, N., Mehranpour, M., Ojaroudi, Y. & Ojaroudi, S. "A new design of triple-band WLAN/WiMAX monopole antenna for multiple-input/multiple-output applications." *Microwave and Optical Technology Letters*, vol. 56, no. 11, pp. 2667–2671, 2014.
9. Soliman, A. M., Eisheakh, D. M., Abdallah, E. A. & ElHennawy, H. "Design of planar inverted-F antenna over uniplanar EBG structure for laptop MIMO applications." *Microwave and Optical Technology Letters*, vol. 57, no. 2, pp. 277–285, 2015.
10. Su, S.-W., Lee, C.-T. & Chen, S.-C. "Very-low-profile, triband, two-antenna system for WLAN notebook computers." *IEEE Antennas and Wireless Propagation Letters*, vol. 17, no. 9, pp. 1626–1629, 2018.
11. Su, H.-L., Huang, B.-W., Lang, H.-R., Tsai, J. Y. & Lee, J. H. "Uniplanar multi-band MIMO antennas for laptop computer applications." *2017 IEEE International Conference on Antenna Innovations & Modern Technologies for Ground, Aircraft and Satellite Applications (iAIM), IEEE*, Bangalore, India, pp. 1–4, 2017.
12. Wong, K.-L., Jiang, H. J. & Kao, Y. C. "High-isolation 2.4/5.2/5.8 GHz WLAN MIMO antenna array for laptop computer applications." *Microwave and Optical Technology Letters*, vol. 55, no. 2, pp. 382–387, 2012.

6 Innovative Antenna Design Techniques for Next-Generation Wireless Communication

6.1 INTRODUCTION

Over the past decade, there has been a significant acceleration in the miniaturization of electronic devices. This trend demands the integration of smaller components in each new generation of devices. This pressure extends to antennas, which must also become smaller without sacrificing performance as this is an essential terminal component in all types of electronic devices. The well-designed antenna facilitates signal transmission from one place to any other place anywhere on Earth. If the antenna is not well designed with a reduced size, this results in an inefficient electronic system with a reduced range of communication. Therefore, the design of antennas compact in nature is a challenge. This chapter discusses the case studies on compact and multiband antennas for modern wireless applications.

6.2 CASE STUDY: WIDEBAND COPLANAR WAVEGUIDE (CPW) FED OVAL-SHAPED MONOPOLE ANTENNA FOR Wi-Fi 5 AND Wi-Fi 6 APPLICATIONS

A wideband coplanar waveguide (CPW)-fed monopole antenna designed for Wi-Fi 5 and Wi-Fi 6 applications is reported in Ref. [1]. The reported antenna (main radiator) has a designed footprint of only $20 \times 8.7 \times 0.4$ mm^3, which is composed of an oval-shaped ring radiator with three concentric rings and a double-T structure loaded with a J-shaped slot. The main novelty of this work is that the measured wideband operation of 34.5% (5.15–7.29 GHz) is contributed by only a single resonance at 6.2 GHz, conforming to the bandwidth requirement of Wi-Fi 5 (5.15–5.85 GHz) and Wi-Fi 6 (5.925–7.125 GHz). Furthermore, the proposed antenna also exhibits good radiation characteristics, including a gain of around 2.25 dBi, a radiation efficiency above 80%, a total efficiency of above 70%, and omnidirectional radiation patterns with a low magnitude of cross-polarization throughout the bands of interest.

6.2.1 EXISTING STATE-OF-THE-ART TECHNOLOGY

Presently, there is a greater demand for higher data rates due to the increasing demand in places such as multifamily housing, offices, airports, entertainment venues, and other high-density environments. Not only is there a need to improve the

DOI: 10.1201/9781003331018-6

data rate, but it is also vital to improve the capability of a network when several devices are connected to it simultaneously. A new next-generation Wi-Fi standard known as Wi-Fi 6 (802.11ax) has been developed in this journey of continuous innovation. Wi-Fi 6 is becoming very popular because it can provide a maximum data rate of up to 9.6 Gb/s while maintaining strong connections, even when more devices are added. Wi-Fi 6 standards also improve the reliability, efficiency, flexibility, security, and scalability of the network, thus allowing greater network performances. Therefore, antennas that maintain compatibility with the existing standard IEEE 802.11ac (also known as Wi-Fi 5) along with conforming to the bandwidth requirement of Wi-Fi 6 simultaneously are of greater demand for next-generation wireless communications.

Several antenna design techniques that exhibit multiband or wideband operations have been reported in the literature [2–17] for wireless applications. Amid these designs, the dielectric resonator antennas (DRAs) are very popular, as they offer high dielectric strength and no inherent conductor loss, which results in better radiation efficiency. An example of such a design has been investigated in Ref. [2], in which a DRA antenna with dimensions of $14 \times 30 \times 8$ mm^3 has been designed for multiband operation in the WiMAX (3.4–3.69 GHz), 5G (3.3–3.6 GHz), and WLAN (5.72–5.80 GHz) bands. In Ref. [3], even though the artificial magnetic conductor (AMC) applied can increase the gain of the antenna, as well as achieving multiband operations in the 2.13–2.87 GHz, 3.22–4.75 GHz, and 5.54–5.86 GHz bands, the planar size of the antenna has a larger area of 37.26×37.26 mm^2. To maintain a compact size antenna for multiband operations, the design method known as complementary split ring (CSR) is widely used in antenna design. In Ref. [4], CSR has been applied to achieve multiband operations across the 842–947 MHz, 2.3–2.46 GHz, and 4.98–5.16 GHz bands. To further achieve small sizes with omnidirectional patterns, a half-wave dipole fed by microstrip balun that has dimensions of $50 \times 10 \times 1$ mm^3 operating in the Wi-Fi bands of 2.24–2.7 GHz and 4.73–5.6 GHz has been studied in Ref. [5]. To realize a flexible antenna for UMTS (1.9 GHz) and WiMAX (5.8 GHz) applications, Ref. [6] has reported a flexible paper-based printed antenna having dimensions of 35×30 mm^2. To achieve very small dimensions of 17.5×8 mm^2, Ref. [7] has presented a simple multifrequency monopole antenna operating in 2.4 GHz, 3.3 GHz, and 5.8 GHz bands. Notably, the antenna designs reported in Refs. [2–7] do not cover the complete Wi-Fi 5 bands. Even though the patch antenna design in [8] has a size of $34.6 \times 33.05 \times 1.57$ mm^3 and can cover the entire Wi-Fi 5 band, the design has used an expensive Roger substrate to fabricate the antenna. Thus, in Refs. [9–12], several unique antenna designs with very wide bandwidth for specific industry applications are introduced using low-cost FR-4 substrate. In Ref. [9], a multiband shark-fin shorted monopole antenna (of dimensions $57 \times 40 \times 1.6$ mm^3) for automotive applications has been investigated. Even though it can cover the 0.81–5.84 GHz band, the reflection coefficients measured are across the -6 dB threshold, which is not a standard practice as far as wireless devices are concerned. To improve the reflection coefficient from -6 dB to -10 dB, as well as to cover multiband operations at wireless standards 3.3/5.0/5.8/6.6/9.9/15.9 GHz, the antenna structure ($44 \times 39 \times 1.6$ mm^3) in Ref. [10] has used the split-ring resonator type. However, its hexa-band characteristics can only be achieved by loading a PIN diode located at a reclined L-shaped slot in the ground plane, where the use of the diode can further increase the manufacturing complexity of the designs. To reduce the manufacturing complexity, the antennas in Ref. [11, 12] present a simple monopole antenna operating in the 2.4/5 GHz WLAN bands with dimensions of 44×33 mm^2 and 42×30 mm^2, respectively. As wireless devices such as smartphones, laptops, notebooks, and tablets play a very crucial role for the end-user in day-to-day life, several antennas designed for laptop/portable devices are reported in Ref. [13–18]. Unfortunately, all the antennas reported are unable to cover the frequency band of Wi-Fi 6.

Through the investigation of the aforementioned existing state-of-the-art technology, it is obvious that the need to design an antenna with Wi-Fi 6 and Wi-Fi 5 operating bands simultaneously is essential for future wireless devices. Therefore, this section presents a novel, compact, low-cost, simple structure, and easy-to-manufacture CPW-fed wideband printed oval-shaped monopole antenna for Wi-Fi 5 and Wi-Fi 6 applications. The proposed antenna is built on a low-cost FR-4 substrate, and the main radiator is comprised of an oval-shaped ring radiator with three concentric rings and a double T-shaped structure loaded with a J-slot. The complete structure is fed using the CPW technique, and a single resonance is excited at around 6.2 GHz, which covers the 10 dB impedance bandwidth (5.15–5.85 GHz) of Wi-Fi 5 and (5.925–7.125 GHz) of Wi-Fi 6 bands. Moreover, to validate the suitability of the proposed antenna for 802.11ax technology, simulated experimentation of an 8×8 multiuser multiple input multiple output (MU-MIMO) antenna array has also been carried out. It is found that the array exhibits the same impedance bandwidth as presented by a single proposed antenna along with an isolation of larger than 15 dB throughout the bands of interest.

6.2.2   DESIGN AND ANALYSIS OF PROPOSED ANTENNA

Figure 6.1 depicts the structural layout of the proposed CPW-fed antenna with detailed dimensions. The proposed antenna is printed on a 0.4 mm thick rectangular FR-4 substrate of size 20×8.7 mm^2 with a dielectric constant (ε_r) of 4.3 and a loss tangent ($\tan\delta$) of 0.025. The simulation is carried out using Computer Simulation Technology Microwave Studio (CST MWS) software. As mentioned earlier, the

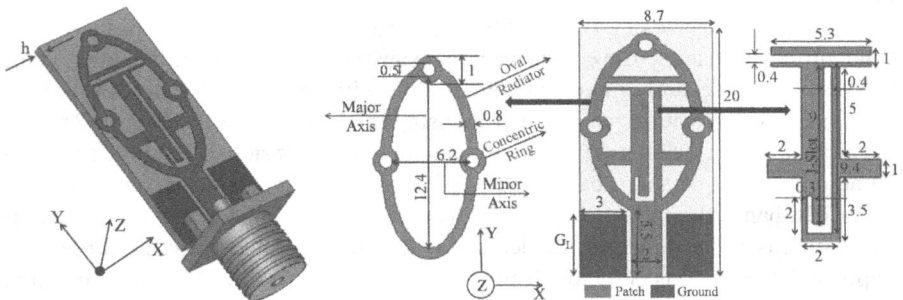

FIGURE 6.1 Detailed structural layout with dimensions of antenna (all dimensions are in millimeters).

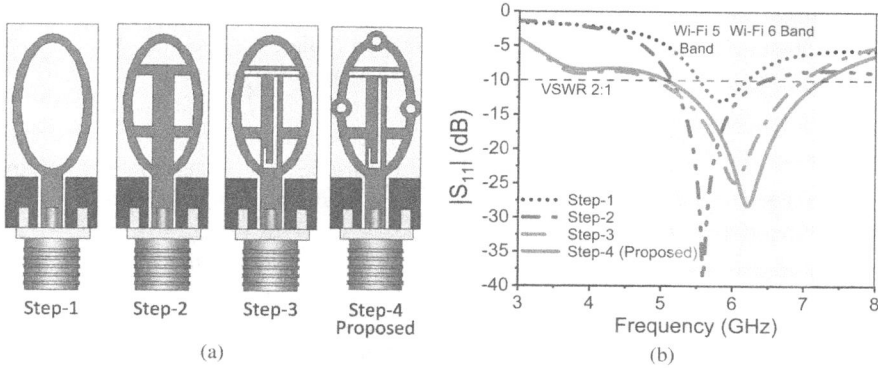

FIGURE 6.2   (a) Design steps of the antenna; (b) stepwise simulated reflection coefficient $|S_{11}|$ (dB) of the proposed antenna.

main radiator is an oval-shaped ring structure that has three concentric rings (inner radius of 0.5 mm and outer radius of 1 mm) loaded along the left, right, and top sections. The minor axis and major axis lengths of the oval-shaped ring are 6.2 mm and 12.4 mm, respectively, sharing a common width of 0.8 mm. A double-T structure is loaded and connected to the inner circumference of the oval-shaped ring structure, and it comprises two vertical structures (the top vertical structure has a size of 5 mm × 2 mm, and the bottom vertical structure has a size of 3.5 mm × 2 mm), each supporting a horizontal structure (the top horizontal one has a size of 5.3 mm × 1 mm, and the bottom horizontal one has a size of 6 mm × 1 mm). Notably, a narrow horizontal slot of size (5.3 mm × 0.4 mm) is loaded to the top horizontal structure, while a bending narrow slot is loaded across the two vertical structures and the bottom horizontal structure, thereby forming the J-slot, as shown in Figure 6.1.

To give a better insight into the reported antenna, a stepwise design is illustrated in Figure 6.2 and explained as follows. Initially, to induce a resonance at around 6.2 GHz, an oval-shaped ring structure radiator is designed on a 0.4 mm thick FR-4 substrate and fed using the CPW-fed technique, in which the feeding line and the two symmetrical ground planes of the CPW are of sizes 5.5 × 2 mm² and 5 × 3 mm², respectively. Here, major axis of the oval-shaped ring radiator is considered to be a quarter-wavelength long at the resonating frequency 6.2 GHz and it is calculated by using Equations (6.1) and (6.2) [18], whereas the minor axis length of the oval-shaped ring radiator is half the length of the major axis one.

$$\epsilon_{eff} = \frac{\epsilon_r + 1}{2} + \frac{\epsilon_r - 1}{2}\left(\frac{1}{\sqrt{1 + 12\dfrac{h}{W}}}\right) \tag{6.1}$$

$$L = \frac{c}{4f_r\sqrt{\epsilon_{eff}}} \tag{6.2}$$

In these equations, L denotes the length of the major axis, h is the height of the FR-4 substrate, W is the width of the oval shaped radiator, ϵ_{eff} is the effective permittivity, c is the speed of light (3×10^8 m/s), and f_r is the resonating frequency at 6.2 GHz. The design steps (denoted as Step 1 to Step 4) layout is illustrated in Figure 6.2a, and its corresponding reflection coefficient magnitude ($|S_{11}|$) dB characteristics are depicted in Figure 6.2b. From Figure 6.2b, it is seen that the deployment of Step 1 (a simple oval-shaped ring radiator) can induce a resonance at around 6 GHz, but it has a narrow 10 dB impedance bandwidth of 5.6 – 6.2 GHz, thus, unable to cover the entire Wi-Fi 5 and Wi-Fi 6 bands. To further improve the impedance matching (bandwidth) of this excited resonance, Step 2 is improvised, in which a double-T structure is loaded within the inner circumference of the oval-shaped ring structure, without disturbing the compactness of the proposed antenna. From its corresponding $|S_{11}|$ curve, a well-improved 10 dB impedance bandwidth of 5.20–6.40 GHz is seen to be achieved. However, such improvement in matching by Step 2 is still inadequate to cover the two bands of interest. Therefore, to further enhance the operating bandwidth, Step 3 is introduced, in which the J-slot is embedded into the double-T structure. Notably, this embedded J-slot in Step 3 has aided in increasing the impedance bandwidth of the antenna to 5.00–6.80 GHz, as depicted in Figure 6.2b, but the bandwidth requirement of Wi-Fi 6 (5.925–7.125 GHz) remains unsatisfied. Hence, via further studies from the simulation, it is realized that the loading of three concentric rings along the left, right, and top sections of the oval-shaped ring radiator (now denoted as Step 4 of the proposed antenna) can excite a very wide 10 dB impedance bandwidth of 5.08–7.23 GHz, and it can very well cover the two bands of interest ranging from 5.15 GHz to 7.125 GHz.

To further comprehend the design principle of the proposed wideband antenna, its corresponding surface current distribution (A/m) at the resonance frequency of 6.2 GHz is plotted in Figure 6.3a. Here, it is visualized that an equal amount of current is flowing along the oval-shaped ring radiator and J-slot as well as the three

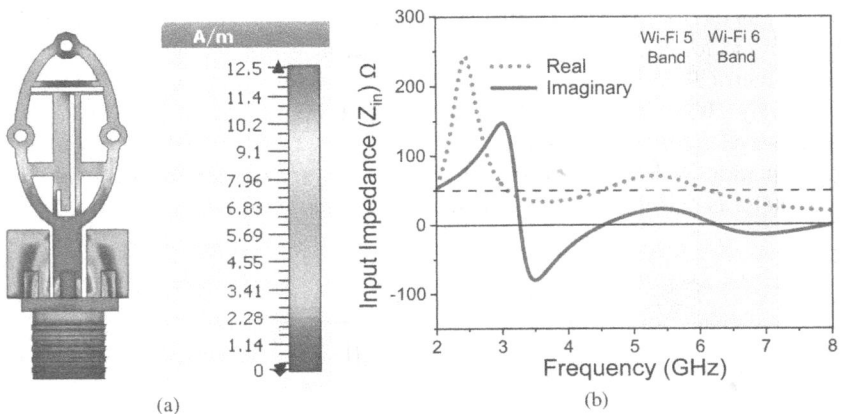

FIGURE 6.3   (a) Surface current distribution (a/m) of the proposed antenna (b) input impedance (Ω) versus frequency (GHz) curve of the proposed antenna.

concentric rings, forming a near half-wavelength distribution. To validate the imped-ance matching characteristics, Figure 6.3b depicts the input impedance (Ω) versus frequency (GHz) curve of the proposed antenna. In this figure, it is clearly seen that at the resonance frequency of 6.2 GHz, the real part of the impedance is approxi-mately equal to 50 Ω, whereas the imaginary part is approximately equal to 0 Ω. Furthermore, across the two bands of interest, the variations of the real and imagi-nary impedances are very slow (moderate) as well, which confirms that the proposed antenna has achieved a good impedance matching and satisfies the requirement of maximum power transfer theorem.

6.2.3 Parametric Analysis of the Antenna

Owing to the compact size, the mirroring performance of the proposed antenna is sensitive to the variation of geometrical parameters. To illustrate the influence of essential parameters on the performance of antenna across the desired Wi-Fi 5 (5.15–5.85 GHz) and Wi-Fi 6 (5.925–7.125 GHz) bands, a vital parametric analysis has been carried out. Here, the parameters selected are height "h" of the substrate and length "G_L" of the symmetric ground plane, and the studies are performed by keeping all other parameters unchanged.

6.2.3.1 Influence of Substrate Height "h" on the Performance of the Proposed Antenna

To examine the influence of substrate height "h" on the performance of the proposed antenna, "h" is varied from 0.2 mm to 0.8 mm in step increments of 0.2 mm, as illustrated in Figure 6.4. In this figure, it is observed that "h" is mainly affecting the resonance and bandwidth of the proposed antenna. By increasing "h" from 0.2 mm to 0.8 mm, the resonance (initially at 6.5 GHz) will shift linearly toward the lower frequency spectrum at near 5.6 GHz, while the operating bandwidth is reduced by approximately half. Therefore, "h" of 0.4 mm is chosen as the optimized value, as it

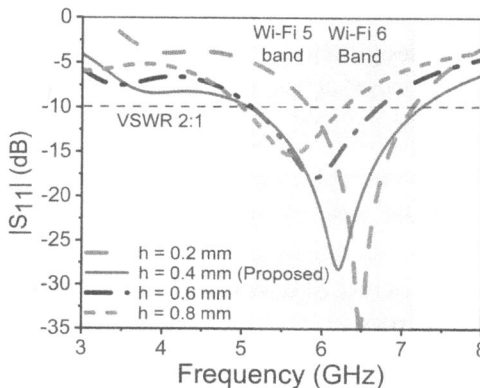

FIGURE 6.4 Influence of substrate height "h" on the performance of the antenna.

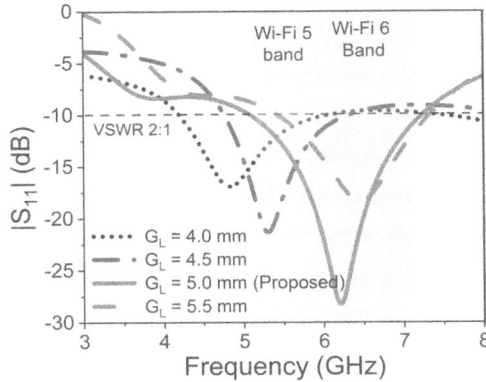

FIGURE 6.5 Influence of ground plane length "G_L" on the impedance bandwidth of the proposed antenna.

fulfills the desired bandwidth requirement as well as maintains a low profile for the proposed antenna.

6.2.3.2 Influence of Symmetric Ground Plane Length "G_L" on the Performance of the Proposed Antenna

Figure 6.5 illustrates the influence of ground plane length "G_L" on the impedance bandwidth of the proposed antenna. The parametric study of this parameter is very essential as it plays a very important role for impedance matching purpose. As depicted in Figure 6.5, as "G_L" increases from 4 mm to 5.5 mm in step increments of 0.5 mm, the resonance shifts linearly from a lower frequency (approximately 4.8 GHz) to a higher frequency (approximately 6.5 GHz) spectrum. Here, "G_L" of 5 mm is chosen as the optimized value because the desired impedance bandwidth with resonance at 6.2 GHz is obtained.

6.2.4 Validation, Results, and Discussion of the Proposed Antenna

To validate the simulated design of the proposed antenna (based on dimensions cataloged in Figure 6.1), the proposed antenna was fabricated as shown in Figure 6.6, and its scattering characteristics were validated by using the Rohde & Schwarz ZVH8 (100 KHz–8 GHz) vector network analyzer (VNA). As for its radiating characteristics, such as radiation patterns, gain, and radiation efficiency, they were measured using an anechoic chamber of size $8 \times 4 \times 4$ m^3.

6.2.5 Reflection Coefficient Magnitude ($|S_{11}|$) dB Characteristic of Proposed Antenna

Figure 6.7 shows the simulated and measured $|S_{11}|$ of the proposed antenna. A good agreement between the two results is observed across the Wi-Fi 5 and Wi-Fi 6 bands. Even though there is a slight deviation between the simulated and

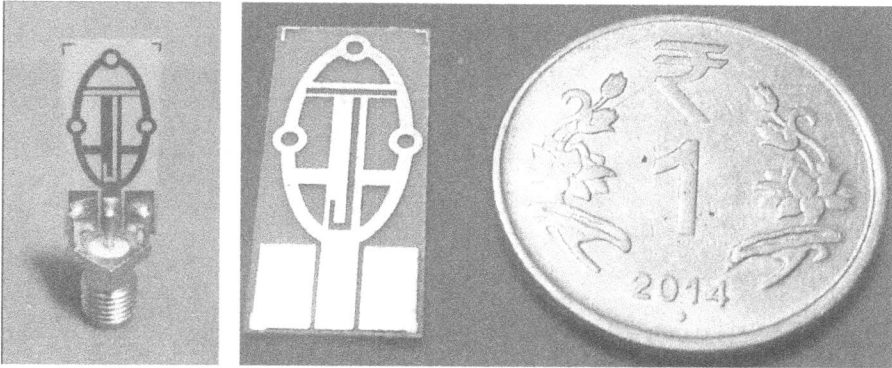

FIGURE 6.6 Fabricated prototype of the antenna with SMA-Type -10 connector.

measured $|S_{11}|$ results, this may be due to fabrication tolerances, soldering, or inaccuracies in dimensions while fabricating the proposed antenna.

Notably, from Figure 6.7, it is confirmed that both simulated as well as measured $|S_{11}|$ curves cover the desired operating bandwidth (simulated 5.08–7.23 GHz and measured 5.15–7.29 GHz) requirements of Wi-Fi 5 and Wi-Fi 6 bands with $|S_{11}| < -25$ dB at near 6.2 GHz (resonance). Therefore, the proposed antenna has achieved excellent impedance matching throughout the bands of interest.

6.2.6 Radiation Patterns of the Proposed Antenna

The simulated and measured two-dimensional (2D) polar radiation patterns of the proposed antenna plotted across the E-plane (x-y plane) and H-plane (y-z plane) at the resonance frequency of 6.2 GHz are shown in Figure 6.8. In this figure, the proposed antenna has exhibited good conventional monopole radiation patterns,

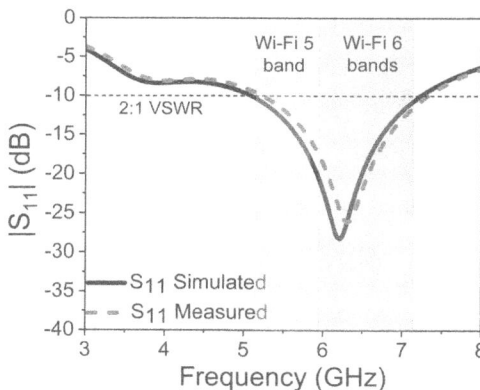

FIGURE 6.7 Measured and simulated reflection coefficient $|S_{11}|$ (dB) of the proposed antenna.

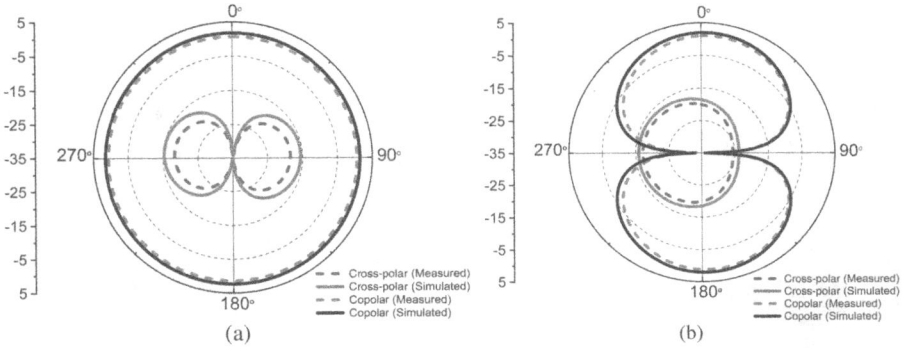

FIGURE 6.8 Simulated and measured 2D polar pattern of the proposed antenna at 6.2 GHz in the (a) E-plane and (b) H-plane.

in which the copolar pattern is an omnidirectional pattern, and its corresponding cross-polar pattern is a bi-directional pattern (eight-shaped). Figure 6.9 depicts its corresponding simulated three-dimensional (3D) radiation patterns at 6.2 GHz. It is observed that the 3D patterns are co-relating with the 2D patterns with negligible side lobes. This validates that the proposed antenna exhibits very good omnidirectional radiation patterns that ensure the good strength of received and transmitted signals for wireless applications.

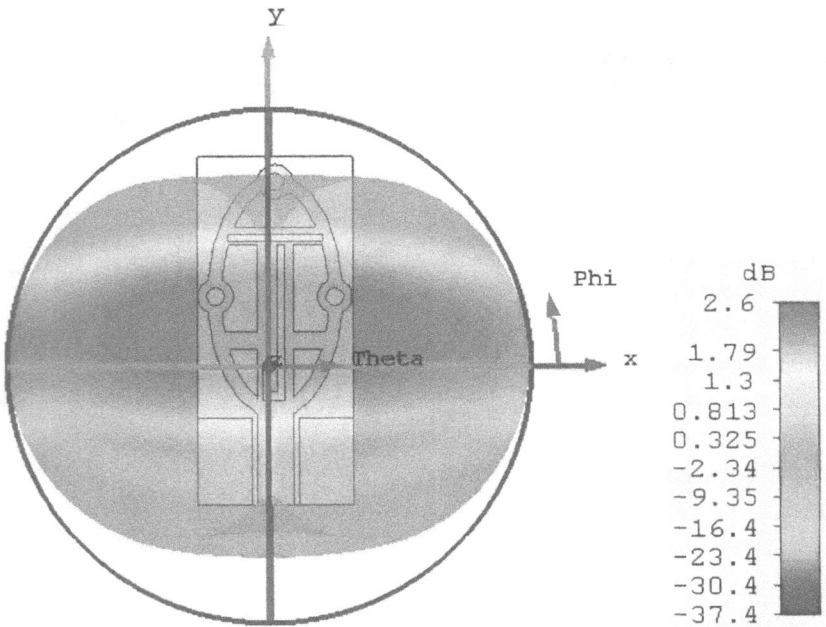

FIGURE 6.9 Simulated 3D radiation pattern of the antenna at 6.2 GHz.

6.2.7 GAIN AND EFFICIENCY OF THE PROPOSED ANTENNA

The simulated and measured gain and radiation efficiency of the proposed antenna across the two bands of interest are shown in Figure 6.10a. In this figure, it is observed that the simulated values are slightly higher than the measured ones, which could be due to fabrication tolerances, soldering, or slight inaccuracies in dimensions while fabricating the proposed antenna. Nevertheless, Figure 6.10a affirms that the proposed antenna emanates a constant measured gain of approximately 2.25 dBi and excellent radiation efficiency greater than 80% throughout the bands of interest.

The simulated and measured total efficiency of the proposed antenna is computed using formula (6.3) and shown in Figure 6.10b.

$$\text{Total Efficiency} = \text{Radiation Efficiency} \times \left(1 - \mid S_{11} \mid^2\right) \quad (6.3)$$

In Equation (6.3), the values of $\mid S_{11} \mid$ and radiation efficiency are obtained from Figures 6.7 and 6.10a. From Figure 6.10b, it is observed that the total efficiency is well above 70% throughout the operating band. The above results ensure that the proposed antenna has good signal reception quality, which is essential for wireless operations to achieve better performance of wireless communication systems.

6.2.8 SIMULATED ANALYSIS OF THE PROPOSED MU MIMO ANTENNA

To validate the suitability of the proposed antenna for the MU-MIMO IEEE 802.11ax standard, an array of 8×8 antenna elements with a size of 83.6×45 mm^2 is simulated using CST MWS simulator, as shown in Figure 6.11a.

Here, the antenna elements are deployed horizontally with an edge-to-edge distance of 2 mm, whereas the vertical distance is maintained at approximately 3 mm. This deployment of 8×8 antenna elements offers better isolation of larger than

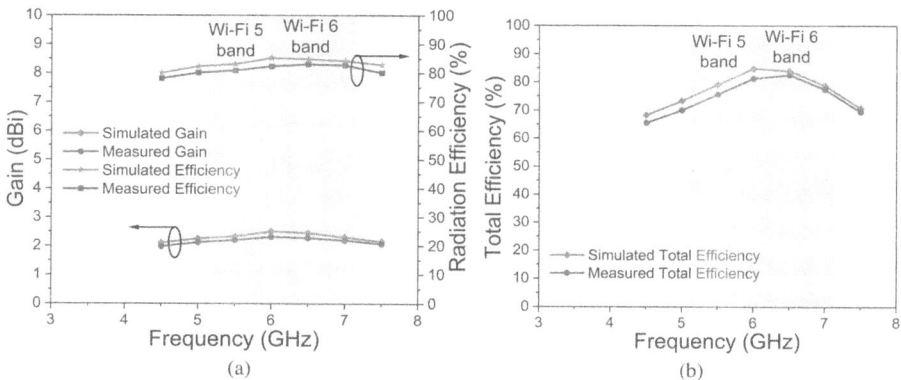

FIGURE 6.10 (a) The simulated and measured gain and radiation efficiency and (b) simulated and measured total efficiency.

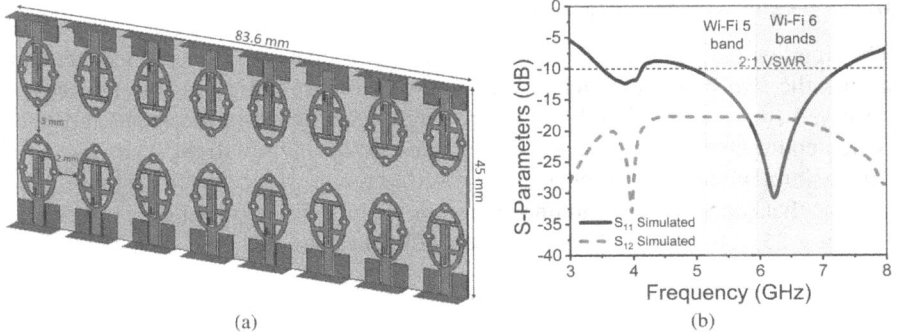

FIGURE 6.11 (a) An 8 × 8 antenna array and (b) simulated $|S_{11}|$ and $|S_{12}|$ of the antenna array.

20 dB without impacting the impedance matching and impedance bandwidth at 6.2 GHz. For brevity, only $|S_{11}|$ (dB) and isolation $|S_{12}|$ (dB) curves are shown in Figure 6.11b. In this figure, it is ascertained that the simulated 8 × 8 MIMO antenna array exhibits the same bandwidth of 2.15 GHz (5.08–7.23 GHz) as offered by the proposed antenna. Therefore, because of better performance and a compact size of 83.6 × 45 mm², the MU-MIMO antenna is a potentially good candidate for IEEE 802.11ax standards applications.

6.2.9 PERFORMANCE COMPARISON OF THE PROPOSED ANTENNA

To validate the potency of the proposed antenna, the performances (antenna size, operating bands, type of substrate used, gain, radiation efficiency, and feeding technique) of the proposed antenna are compared with the recently published pioneering state-of-the-art technology, and they are shown in Table 6.1. It is noteworthy that the proposed antenna has some obvious advantages over these reported ones, such as small size, cost effectiveness, and better radiation performances.

Based on this performance comparison, some of the salient and important features of the proposed antenna are as follows:

- The proposed antenna is compact and has smaller dimensions as compared to [2–6], and [8–12].
- Unlike [2, 8] the proposed antenna does not require an expensive substrate.
- The proposed antenna is very thin and can be used in next-generation wireless devices (with slim characteristics) as it has less thickness (low in profile) as compared to [2–5], and [7–12].
- The proposed antenna has higher gain and efficiency as compared to [5–7], and [10].
- It does not require a 3D structure [2], additional ground plane [9], or reactive component [10] for the excitation of required bands.

TABLE 6.1

Performance Comparison of Proposed Antenna with Recent Pioneering State-of-the-art Technology

Ref.	Dimension (mm³)	Operating Bands (GHz)	Substrate	Gain (dBi)	Efficiency (%)	Feeding Technique
[2]	$14 \times 30 \times 8$	3.3/5.2/5.8	RT Duroid and Foam	4.0	80	CPW
[3]	$37.26 \times 37.26 \times 1.6$	2.4/3.3/5.8	FR-4	4.93	NG	CPW
[4]	$18 \times 22 \times 1.6$	0.9/2.4/5.2	FR-4	NG	NG	Microstrip
[5]	$50 \times 10 \times 1$	2.4/5.2	FR-4	2.09	NG	Microstrip
[6]	$35 \times 30 \times 0.12$	1.9/5.8	Flexible paper	0.512	55	CPW
[7]	$17.5 \times 8 \times 0.8$	2.4/3.3/5.8	FR-4	1.5	80	Microstrip
[8]	$34.6 \times 33.05 \times 1.57$	5	RT Duroid	7	95	Coaxial
[9]	$57 \times 40 \times 1.6$	0.8/1.7/2.4/5	FR-4	6.7	NG	Coaxial
[10]	$44 \times 39 \times 1.6$	3.3/5/6.6/9.9	FR-4	1.98	78	Microstrip
[11]	$43 \times 33 \times 1.6$	2.4/3.8/5	FR-4	5.6	NG	Microstrip
[12]	$42 \times 30 \times 1.6$	2.4/5	FR-4	4	90	Microstrip
[Reported]	$20 \times 8.7 \times 0.4$	5/6	FR-4	2.25	80	CPW

Abbreviation: NG = Not Given

- The proposed antenna is cost-effective as it does not use expensive RT-Duroid substrate unlike [2, 8], additional ground plane unlike [9], reactive components [10] and also has less fabrication cost due to its compact size unlike [2–6] and [8–12].

6.2.10   CONCLUDING REMARKS

The design of a novel, cost effective, and wideband printed CPW-fed oval-shaped ring monopole antenna has been successfully studied. The antenna is easy to fabricate and has a compact size of $20 \times 8.7 \times 0.4$ mm³. Based on the excitation of only a single resonance at 6.2 GHz, a wide impedance bandwidth of 34.5% (5.15–7.29 GHz) was measured, and it can cover the desired Wi-Fi 5 (5.15–5.85 GHz) and Wi-Fi 6 (5.925–7.125 GHz) bands. Besides demonstrating good monopole radiation patterns (omnidirectional), the reported antenna also demonstrates a constant gain of approximately 2.25 dBi, the radiation efficiency of above 80%, and total efficiency of above 70% throughout the operating band. It is worth mentioning that the proposed antenna is also suitable to be used as an element of 8×8 MU-MIMO antenna for IEEE 802.11ax technologies. Therefore, owing to the simple structure, cost effectiveness, space efficiency, and good performances across the Wi-Fi 5 and Wi-Fi 6 bands, the proposed antenna is a good candidate to support the 802.11ac and 802.11ax standards for wireless applications in next-generation wireless devices.

6.3 CASE STUDY: MIMO ANTENNA FOR SUB-6 GHZ 5G/IEEE802.11AC/AX/C-BAND/X-BAND WIRELESS AND SATELLITE APPLICATIONS

A tapered symmetrical coplanar waveguide (S-CPW)-fed monopole antenna is studied and reported in Ref. [19]. To achieve multiband characteristics, the radiating element of this monopole antenna is loaded with multiple narrow slots and multiple slotted stubs (MSS). The designed slot-loading monopole is further transformed into a two-antenna MIMO type with a gap distance of only 0.12λ (at 5 GHz) and, thus, it has a small overall size of $32 \times 20 \times 0.8$ mm^3. By deploying five concentric ring elements between the two adjacent antenna elements, desirable isolation of better than 20 dB is yielded. As the low-band and high-band operation of the proposed two-antenna MIMO are 81.08% (3.3–7.8 GHz) and 40% (8.0–12.0 GHz), respectively, it can therefore satisfy the sub-6 GHz 5G New Radio (NR) n77/78/79, IEEE 802.11ac/ax, X-band/C-band wireless and satellite applications. Furthermore, it has shown a desirable gain of above 3 dBi and radiation efficiency greater than 69% throughout the two bands of interest.

6.3.1 OVERVIEW OF EXISTING STATE-OF-THE-ART TECHNOLOGY

The main challenging part of designing a compact size MIMO antenna is to obtain a very high isolation of >20 dB between the two adjacent antenna elements without affecting the scattering, radiation, and diversity performances. Therefore, several decoupling techniques have been reported in recent years for compact size planar MIMO antennas with multiband or broadband operations [20–40]. In Ref. [20], a two-antenna MIMO with frequency reconfigurable characteristics is designed using radio frequency microelectromechanical systems (RF-MEMS) switches. This antenna has a designed footprint of $32 \times 98 \times 1$ mm^3, and it can switch among the 0.6 GHz, 1.8 GHz, 2.4 GHz, 3.5 GHz, and 5.5 GHz bands with isolation >15 dB. However, the use of an RF-MEMS switch makes the antenna design more complex. To reduce the complexity, Ref. [21] has proposed two symmetrically located spider-shaped antennas (of size $37 \times 56 \times 1.6$ mm^3) that can operate in the Wi-Fi/WiMAX/Bluetooth and C-band applications. However, it does not cover the entire 5 GHz band. A dual band, dual-port antenna operating in the 2.4/5 GHz band with a dimension of $46 \times 20 \times 1.6$ mm^3 is proposed in Ref. [22], but the isolation is only around 12 dB. Therefore, a Wang-shaped triple-band MIMO antenna with high isolation of 31 dB is reported in Ref. [23], but this antenna has occupied a large area of 70×52 mm^2. To achieve smaller dimensions for the MIMO antenna, Ref. [24] has designed a two-antenna MIMO with a planar size of only 24×20 mm^2. Here, a single complementary split-ring resonator (S-CSRR) is loaded into the radiating element to enhance the performance, as well as operating in the X- and Ku-band. In Ref. [25], a two-port MIMO monopole antenna with a small size of $27 \times 16 \times 0.8$ mm^3 is reported. This antenna is a single-band operation (4.85–7.32 GHz), and it has shown good isolation of >15 dB [26]. To achieve dual-band operation with higher isolation, a dual-band MIMO antenna (size $27 \times 21 \times 1.6$ mm^3) using orthogonal polarization has been reported [26]. Even though it can yield a very high isolation of

>22.5 dB, the two operating bands are rather narrow at 4.15% (5.19–5.41 GHz) and 4.81% (7.30–7.66 GHz). To further achieve triple-band operation, a MIMO antenna with a compact size of 20×14.75 mm² has been reported [27]. The antenna element is composed of an asymmetric coplanar strip (ACS) feedline along with an inverted L-shaped slot and a meander line to operate in the WLAN (2.4/5.2/5.8 GHz) and WiMAX (2.5/3.5/5.5 GHz) applications. Notably, all the MIMO antennas reported in Ref. [20–27] do not apply any decoupling structure between adjacent antenna elements. Hence, they will suffer from low isolation if a wider operating bandwidth is implemented.

The easiest way to reduce the mutual coupling of any two adjacent antenna elements is to apply the spatial diversity method. The advantage of this method is that the MIMO antenna does not require any decoupling structure, such as the one reported in Ref. [28] that has a gap distance of 36 mm (>1/4 wavelength) between the two adjacent antennas and good isolation of >30 dB is achieved. However, applying the spatial diversity method will result in occupying more space and, hence, an increased dimension. To reduce the mutual coupling without occupying much gap distance between the antenna elements, several MIMO antenna designs with different decoupling structures such as the U-shaped slots [29], T-shape stub [30], comb-shaped structures [31], modified T-shape stub at ground plane [[32], and defected ground structures [33–36] have been reported. However, the volume size of [29–36] is between ($20 \times 35 \times 0.8$ mm³, 560 mm³) and ($50 \times 50 \times 7$ mm³, 17500 mm³), and none of these MIMO design exhibit very wide operational bandwidth (>80%) and high isolation >20 dB for modern wireless applications with physical volume size of below or near 500 mm³.

To achieve high isolation without increasing the overall volume size of the MIMO antenna, recent decoupling techniques such as the loading of electromagnetic bandgap (EBG) structure [37, 38] and metamaterials [39, 40] have been widely used. In Ref. [37], the fractal EBG technique is applied to enhance the isolation to 24.67 dB, but the two-antenna MIMO has a volume size of $38.2 \times 95.94 \times 1.6$ mm³ and it covers only a narrow single ISM band from 2.39 GHz to 2.48 GHz. Even though the volume size of the antenna in Ref. [38] ($55 \times 28 \times 1.6$ mm²) is much less than that of ef. [37] and it has exhibited wider operating bandwidth of 64.42% (2.01–3.92 GHz), one can see that it has exhibited isolation of only >15 dB after applying the EBG structure. In Ref. [39], the textile-based antenna has applied a unique metamaterial inspired decoupling network to reduce the mutual coupling between adjacent antenna elements. The two bands of interest, in this case, are 56% (1.34–3.92 GHz) and 37.4% (4.34–6.34 GHz), and isolation of >18 dB was measured. However, this antenna has a very large dimension of $100 \times 60 \times 1$ mm³. To achieve a small volume size of $47.5 \times 40 \times 1.6$ mm³, authors in Ref. [40] have proposed a single band, two-antenna MIMO monopole that has applied a metamaterial split-ring resonator unit to improve the isolation. However, the antenna is a single-band operation with 10 dB impedance bandwidth of 12.3% (3.35–3.78 GHz) and the isolation is only15 dB.

From the study of literature [20–40], it is observed that the reported MIMO antennas can yield high isolation but with a narrow operational band or vice-versa. Furthermore, many of them require a very large volume size. Therefore, in this chapter, a compact volume size ($32 \times 20 \times 0.8$ mm³, 512 mm³) two-antenna monopole MIMO with a dual wideband operation (3.3–7.8 GHz and 8.0–12.0 GHz) as well as high isolation (>20 dB) for the wireless and satellite applications is proposed. The proposed two-antenna MIMO functions in the sub-6 GHz 5G NR n77 (3.3–4.2 GHz)/n78 (3.3–3.8 GHz)/n79 (4.4–5.0 GHz), standard C-band uplink (3.7–4.2 GHz)/downlink (5.92–6.425 GHz), extended C-band uplink (3.4–3.7 GHz)/ downlink (6.425–6.725 GHz), 802.11ac (5.15–5.85 GHz)/802.11ax (5.92–7.125 GHz), and X-band radar (8.0–12.0 GHz). Therefore, an integration of sub-6 GHz 5G, IEEE 802.11ac/ax, X-band, and C-band into a single antenna along with MIMO configuration would be a good candidate for future wireless and satellite applications. In this case study, the geometry and evolution mechanism of the single monopole element and its characteristics are explained along with the working principles. The performance of the MIMO antenna and diversity performance of the MIMO antenna are also analyzed. To give a better understanding of the reported MIMO antenna, pioneering technology is compared. Finally, concluding remarks are also drawn.

6.3.2 DESIGN LAYOUT OF SINGLE MONOPOLE ANTENNA ELEMENT

The design layout and geometry of the single monopole antenna element that will be further applied for realizing the proposed two-antenna monopole MIMO are depicted in Figure 6.12. Here, the radiating section (rectangular radiator) of the monopole antenna is loaded with two narrow L-shaped slots, an inverted-U narrow slot, and two symmetrical MSS (deployed on the right and left side of the rectangular radiator). To achieve better impedance matching, a two-step CPW feeding line (of width 1.5 mm and 2 mm) is applied, and the tapered structure that is loaded at the bottom of the rectangular radiator (links to the feeding line of the CPW) is for yielding wider operating bandwidth. As seen in Figure 6.12, the two coplanar ground planes have an area of

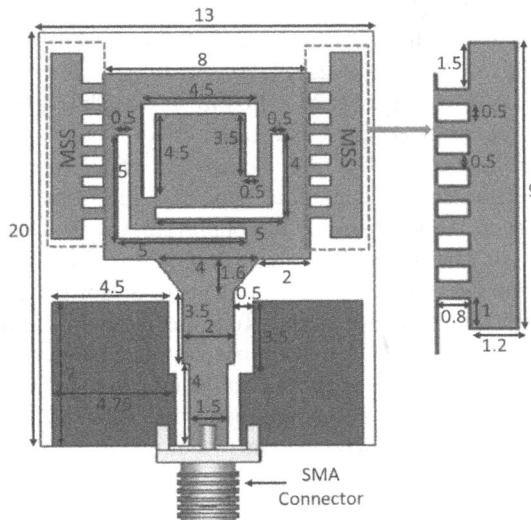

FIGURE 6.12 Layout and geometry of the single monopole antenna element.

4×7 mm^2 each, and it is slightly truncated with an air gap of 0.5 mm due to the two-step CPW feeding line. The antenna structure is printed on a 0.8 mm thick FR-4 substrate with relative permittivity (ε_r) of 4.3 and loss tangent (tan δ) of 0.025. The overall planar size of this single monopole antenna element is 20×13 mm^2.

6.3.3 ANTENNA EVOLUTION MECHANISM

To comprehend the excitation of the two wide operational bandwidths from the single monopole antenna element, this section explains the antenna evolution mechanism including the stepwise design and its associated reflection coefficient curve.

6.3.3.1 Step 1: Design of Rectangular Radiator

The antenna design begins with a two-step CPW-fed rectangular radiator with a tapered structure at the bottom, as illustrated in Figure 6.13a, and it is denoted

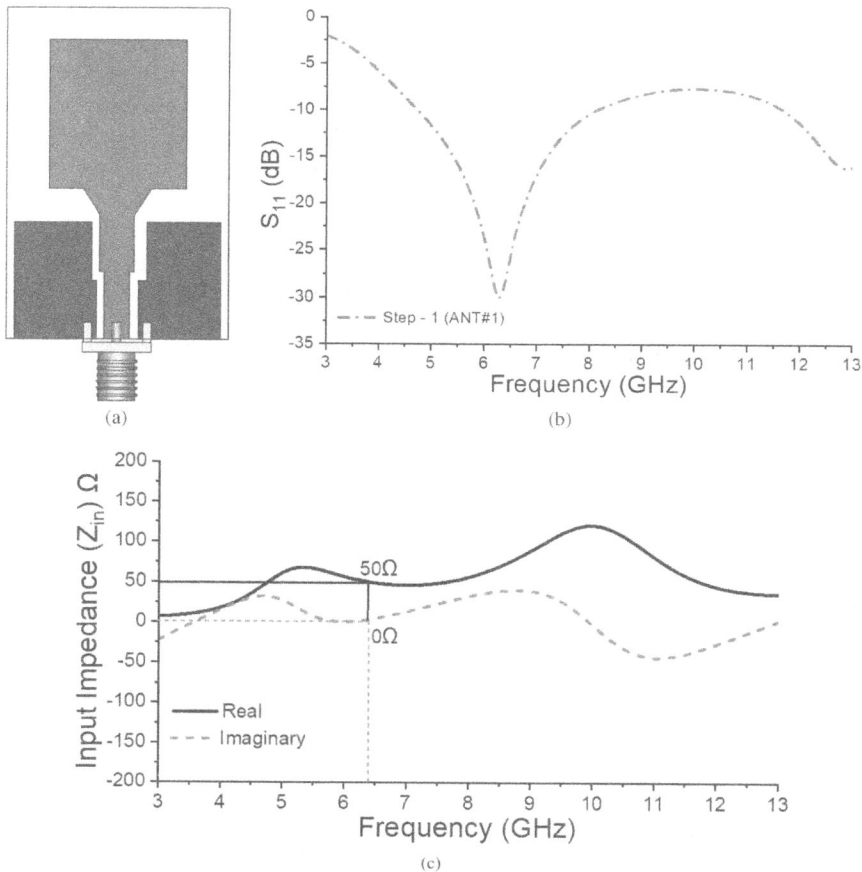

FIGURE 6.13 Step 1 of the single monopole antenna element, (a) ANT#1 structure, (b) S11, and (c) input impedance.

as ANT#1. Here, ANT#1 was analyzed and numerically simulated using CST Microwave Studio® (CST MWS) software. The width and length of the rectangular radiator are optimized using CST MWS.

From Figure 6.13b, it is clearly observed that ANT#1 has successfully induced a wideband operation with a 10-impedance bandwidth of 4.74–8.10 GHz (centered at 6.3 GHz). Thus, it can meet the wideband demand for X- and C-band applications. Figure 6.13c validates that the incorporation of rectangular patch, truncated ground planes, and tapered structure offer equal amounts of inductive and capacitive reactance at a resonating frequency of 6.3 GHz, hence, achieving good impedance-matching throughout the operating band (4.74–8.10 GHz).

6.3.3.2   Step 2: Deployment of MSS on Right and Left Side of ANT#1

As seen in Figure 6.13c, the frequency band generated by ANT#1 is not enough to cover the standard and extended C-band. Therefore, to shift the frequency band toward the lower spectrum and to match the load impedance of 100 Ω to the 50 Ω feed line, a series of open-circuit MSS are embedded into the left and right side of ANT#1 (which forms ANT#2), without disturbing the feeding arrangement of ANT#1, as shown in Figure 6.14a. Notably, there are six small slots (each has a size of 0.8×0.5 mm^2) in each MSS (see Figure6.12), and these slots act as capacitors and behave like an open circuit to block the high inductive reactance across the operating frequency range.

From the reflection coefficient curve of Figure 6.14b, it is visualized that the incorporation of the two symmetrical MSS into ANT#1 (that forms ANT#2) can shift the previous frequency mode (6.3 GHz) toward the lower spectrum at approximately 5.3 GHz, and a good 10 dB impedance bandwidth of approximately 3.4–7.0 GHz (low band) is achieved. Meanwhile, ANT#2 can also generate another high-band operation with 10 dB impedance bandwidth of 9.4–11.1 GHz. Therefore, ANT#2 is able to operate in the IEEE 802.11ac, C-band, and partial X-band applications.

To comprehend the contributions of the MSS that achieve good impedance-matching across the two bands of interest, Figure 6.14c depicts the input impedance Z_{in} (Ω) diagram of ANT#2. Here, one can clearly see that good impedance-matching has been achieved throughout the two operating bands of 3.4–7.0 GHz and 9.4–11.1 GHz, as their resistive impedances are very much closer to 50 Ω and their corresponding reactive impedances are near 0 Ω. This validates that the two symmetrical MSS can aid in achieving good impedance-matching across the low band and exciting a new high band.

6.3.3.3   Step 3: The Loading of L-Shaped Slots and Inverted U-Shaped Slot

To further integrate the sub-6 GHz NR 5G bands along with the IEEE 802.11 ac/ax, C-band, and X-band into a single antenna simultaneously, ANT#3 (as shown in Figure 6.15a) is developed by further loading three narrow slots, namely, two L-shaped slots and an inverted U-shaped slot. When these slots are loaded into the rectangular radiator, they become interlocked with each other and aid in reducing the capacitive reactance. Moreover, these slots also help to widen the previous frequency band by forcing the current distribution on the surface of the radiator to flow for a longer time, as well as diverting the current paths to flow in various directions with various velocities.

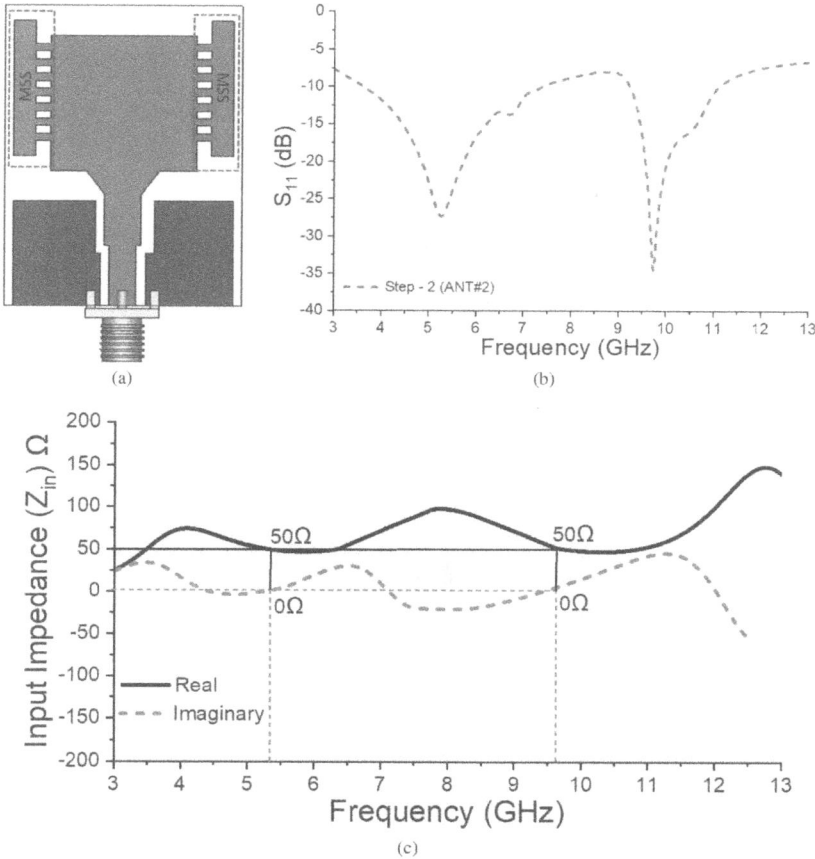

FIGURE 6.14 Step 2 of the single monopole antenna element, ANT#2 structure (a), S11 (b), and input impedance (c).

This results in the merging of all the currents coming from various directions with different velocities and offers wide bandwidth, which can be seen in Figure 6.15b that plots the current distributions at 5 GHz for ANT#1 and ANT#3. To illustrate the wideband characteristics of ANT#3, its corresponding reflection coefficient is plotted in Figure 6.15c. It is clearly seen that ANT# 3 is able to yield a dual-bandwidth operation spanning the range of 81.08% (3.3–7.8 GHz) and 40% (8.0–12.0 GHz) at the resonant frequency of 5.0 GHz and 10.5 GHz, respectively. Figure 6. 15d shows the input impedance diagram of ANT#3. Here, one can see excellent impedance matching throughout the two bands of interest, in which the resistive and reactive impedances are around 50 Ω and near 0 Ω across the two bands, respectively. Therefore, ANT#3 is considered for further analysis and will be applied for MIMO applications. The stepwise configuration of ANT#1 to ANT#3 and their associated operating bands are summarized in Table 6.2.

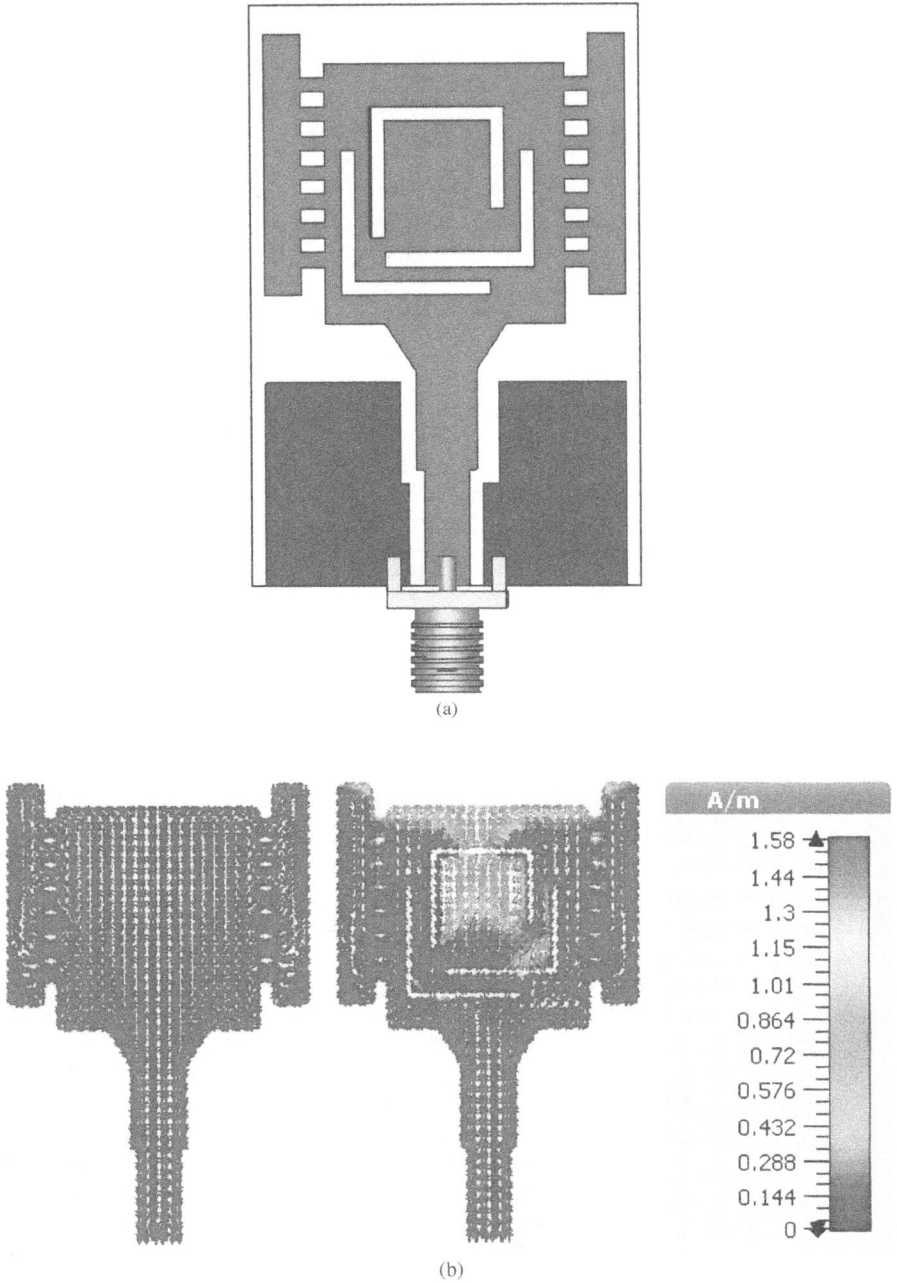

FIGURE 6.15   Step 3 of single monopole antenna element, (a) proposed monopole antenna element, (b) current distributions at 5 GHz, (c) S11, (d) input impedance. (*Continued*)

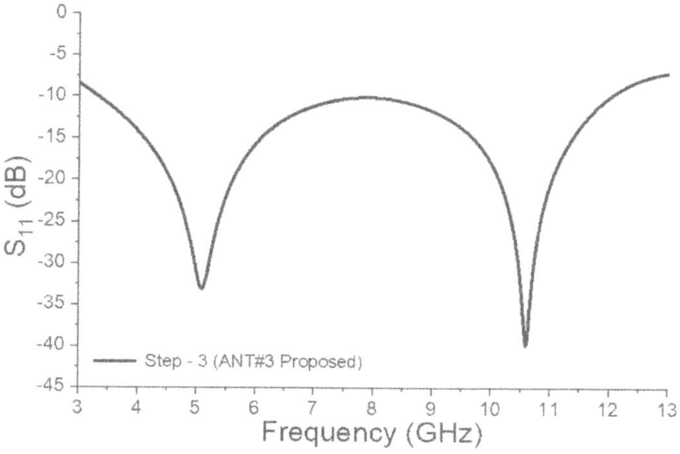

FIGURE 6.15   (*Continued*)

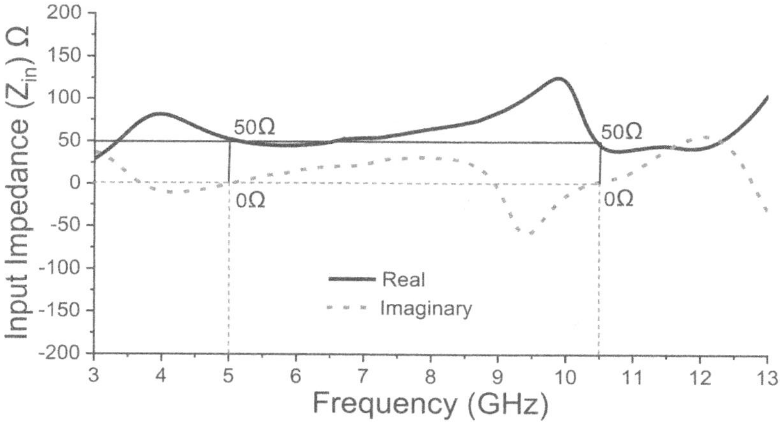

TABLE 6.2
Stepwise Configuration and Operating Bands of ANT#1 to ANT#3

Step	Sub-6 GHz 5G NR	IEEE 802.11ac	IEEE 802.11ax	C-band	X-band	Impedance Matching
ANT#1	–	Yes	Yes	–	–	Good
ANT#2	–	Yes	–	Yes	–	Good
ANT#3	Yes	Yes	Yes	Yes	Yes	Good

6.3.4　Geometry, Design, and Analysis of the Proposed Two-Antenna MIMO

Figure 6.16 depicts the geometry of the proposed two-antenna MIMO with and without loading the decoupling structure. As seen in Figure 6.16A, two identical monopole antenna elements (ANT#3) are closely deployed side-by-side of each other, with a gap distance of only 0.12λ (λ is the free space wavelength at 5 GHz), and it is much narrower than the one reported in Ref. [28] with a gap distance of $>1/4\lambda$. Nevertheless, this gap distance can still provide enough space for the deployment of decoupling structure (five concentric ring elements), as shown in Figure 6.16B. Furthermore, this narrow gap distance can also ensure that the desired antenna and MIMO diversity performances remain unaffected in a rich multipath fading environment.

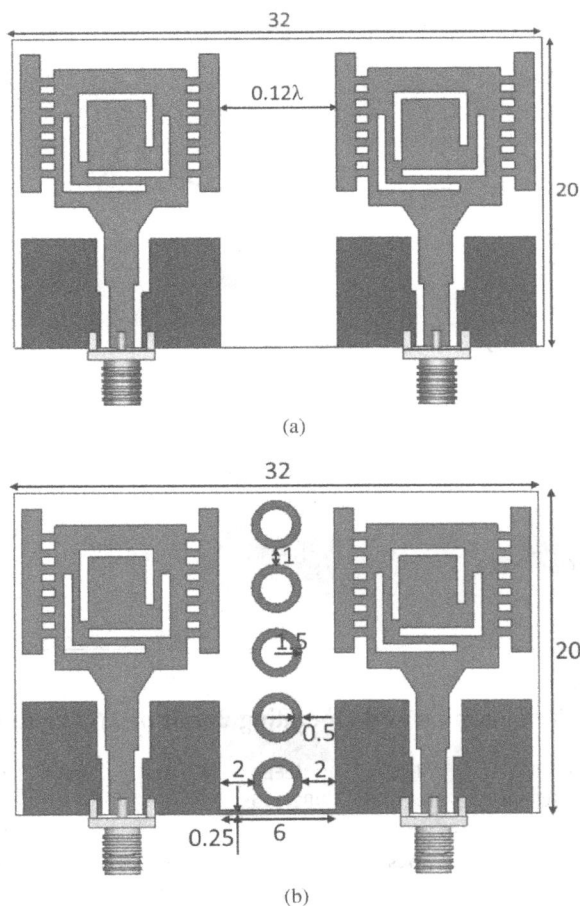

FIGURE 6.16　Geometry of the two-antenna MIMO, (a) without decoupling structure and (b) with decoupling structure.

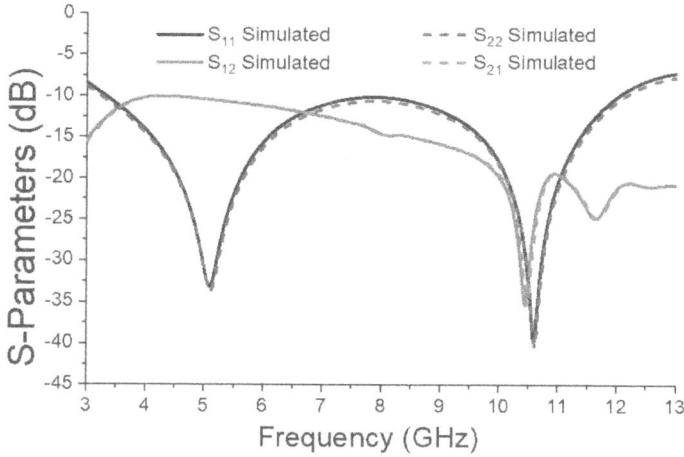

FIGURE 6.17 Simulated S-parameters of the two-antenna MIMO without loading the decoupling structure.

6.3.4.1 Analysis of the Two-Antenna MIMO without Decoupling Structure

Figure 6.17 plots the simulated reflection coefficient (S_{11} and S_{22}) and isolation (S_{12} and S_{21}) curves of the two-antenna MIMO without loading the decoupling structure. Here, both the S_{11} and S_{22} are almost identical, and they have shown wide 10 dB impedance bandwidths of 81.08% (3.3–7.8 GHz) and 40% (8.0–12.0 GHz). However, the isolation level (S_{12} or S_{21}) between the two antenna elements at around 4 GHz is approximately 10 dB, which is undesirable, because as per the requirement of industry and IEEE standards, the minimum isolation between two adjacent antenna elements should be greater than 15 dB, so that each antenna element will produce independent communication paths, resulting in a higher data rate as well as uninterrupted internet access, wireless, and satellite services.

Figure 6.18 shows the electric field intensity (V/m) distribution across the two-antenna MIMO at a resonant frequency 5GHz. As shown in Figure 6.18a, when

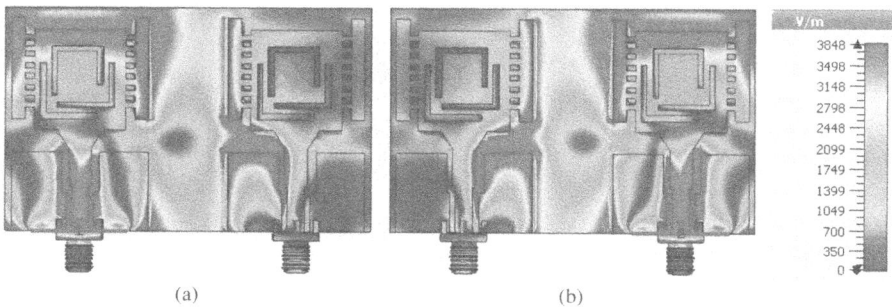

FIGURE 6.18 Electric field intensity (V/m) of the two-antenna MIMO without decoupling structure: (a) antenna element#1 excited and (b) antenna element#2 excited.

antenna element#1 is excited (while antenna element#2 is terminated by 50 Ω load impedance), one can see that antenna element#2 is highly influenced (coupled) by the strong electric field generated from the antenna element#1. It is also validated when antenna element#2 is excited, while antenna element#1 is terminated with 50 Ω load impedance, as seen in Figure 6.18b. This clearly indicates that both antenna elements are strongly mutually coupled with each other.

6.3.4.2   Analysis of the Two-Antenna MIMO with Decoupling Structure

To reduce the mutual coupling between the two antenna elements, a decoupling structure is loaded between the antenna elements (within the 0.12λ gap distance), as shown in Figure 6.16b, in which the decoupling structure comprised five concentric ring elements. In this case, each ring has an inner and outer radius of 1 mm and 1.5 mm, respectively, and they are printed vertically with a gap distance of 1 mm from each other. Furthermore, a very thin strip of 0.25 mm in height is added in between the antenna elements in order to connect the ground planes. By further observing the reflection coefficient curves in Figure 6.19, it is apparent that the deployment of the decoupling structure, as well as connecting the ground plane, does not affect the impedance bandwidth (as seen from the S_{11}/S_{22} curves) of both antenna elements. Notably, the decoupling structure has significantly enhanced the isolation level between the two antenna elements, and a level of >20 dB (seen from S_{12}/S_{21} curves) is observed, which was verified by applying two plane waves across the length of the decoupling structure [41]. The isolation below −20 dB is achieved as the equidistance spaced circles act as reflectors and absorb the surface wave which, therefore, serves as a wide stop band filter and nullifies the surface wave propagation between Ant.1 and Ant.2 [42].

To further comprehend the validation of isolating structure, electric field distribution (V/m) analysis at a resonant frequency of 5GHz is shown in Figure 6.20, where it can be easily seen that antenna element#2 is being prevented from the strong electric

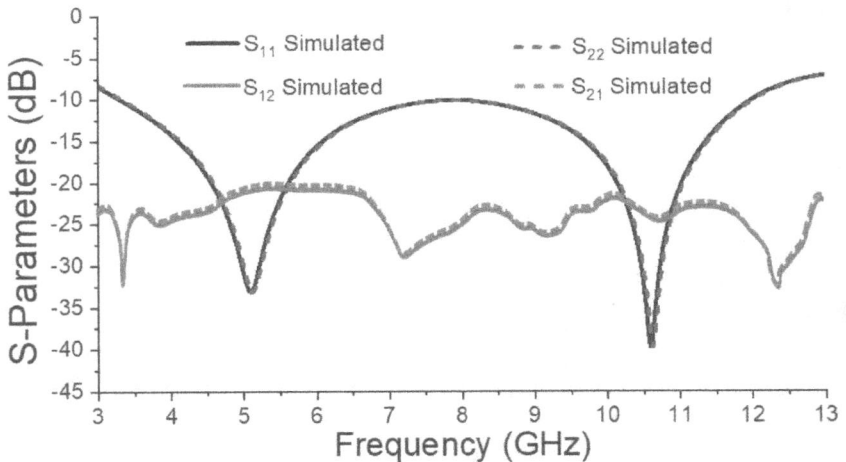

FIGURE 6.19   S-parameters of two-port MIMO antenna with decoupling structure.

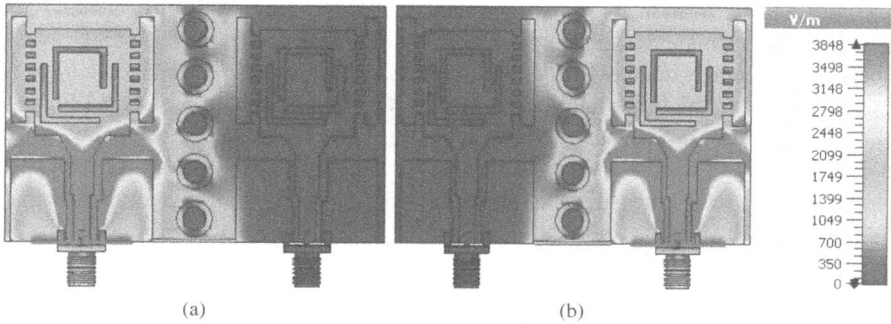

(a) (b)

FIGURE 6.20 Electric field intensity (V/m) of the proposed two-antenna MIMO with decoupling structure: (a) antenna element#1 excited and (b) antenna element#2 excited.

field of antenna element#1, which is due to the five concentric ring elements that act as a band stop filter to attenuate the correlated signals coming from antenna element#1. The same phenomenon is also shown when antenna element#2 is excited, while antenna element#1 is terminated with a 50 Ω load impedance.

6.3.5 RESULTS AND DISCUSSION OF THE PROPOSED TWO-ANTENNA MIMO

To implement the proposed two-antenna MIMO for wireless and satellite applications functioning in sub-6 GHz 5G NR, 802.11ac/ax, C-band, and X-band, a prototype was manufactured and depicted in Figure 6.21. The scattering, radiating,

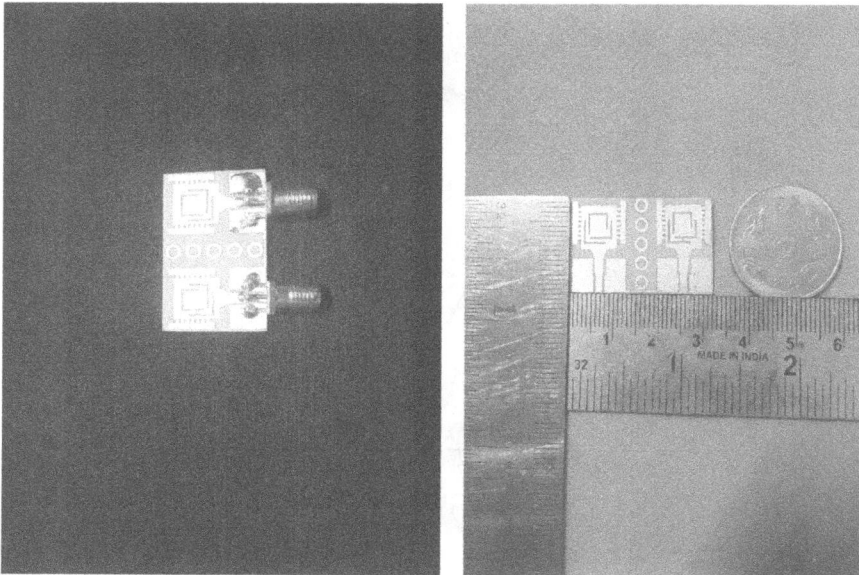

FIGURE 6.21 Fabricated prototype of the two-antenna MIMO.

and diversity performances of the fabricated proposed prototype were verified with simulated performances, and they are discussed later in the chapter. Notably, when analyzing the performances of the two-antenna MIMO, only one antenna element is excited, whereas the other antenna element is terminated with a 50 Ω load imped-ance. The Rohde & Schwarz ZNH18 network analyzer, having a frequency range of 30 KHz to 18 GHz, was used for measuring S-parameters, whereas the anechoic chamber was used to measure the radiation characteristics of the proposed antenna.

6.3.5.1 Simulated and Measured Reflection Coefficient and Isolation

As the reflection coefficient (S_{11}) and isolation (S_{12}) are analogous to S_{22} and S_{21}, respectively, Figure 6.22 only shows the simulated and measured S_{11} and S_{12} characteristics of the proposed two-antenna MIMO. In this figure, both simu-lated and measured results are well-validated with each other, and the observed

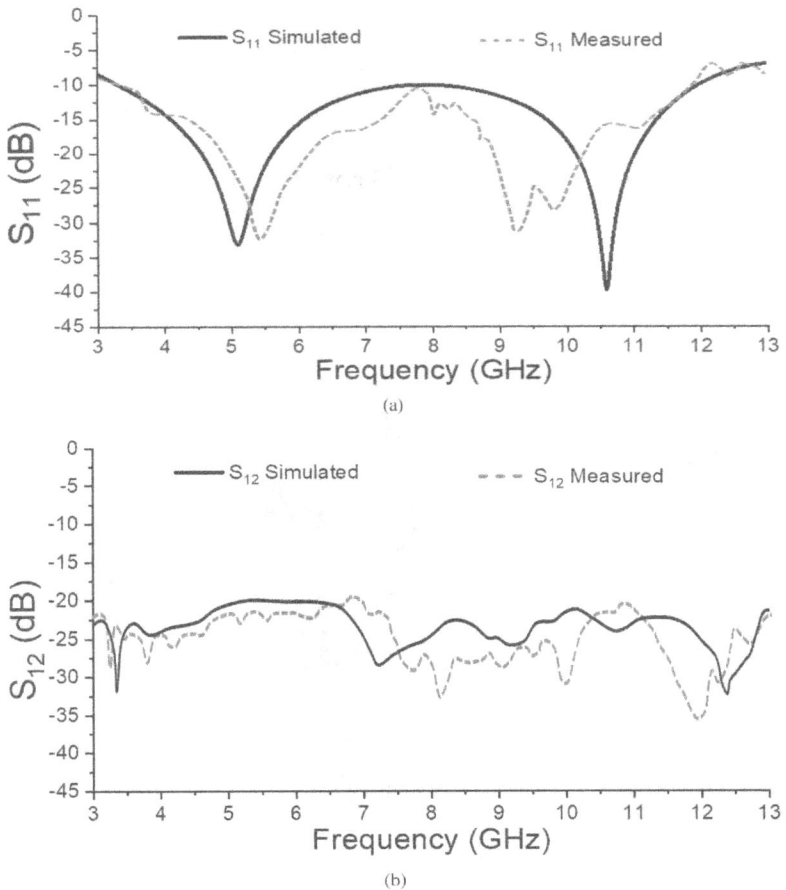

FIGURE 6.22 Simulated and measured S-parameters of the proposed two-antenna MIMO: (a) reflection coefficient S_{11} and (b) isolation, S_{12}.

deviation (especially in the high band) may be due to manufacturing tolerances and minor fabricating errors. Nevertheless, a dual-band operation is clearly shown in Figure 6.22a, in which the measured low-band and high-band operation have exhibited wide 10 dB impedance bandwidth of 80.54% (3.3–7.75 GHz) and 41.2% (7.9–12.0 GHz), respectively, while isolation of larger than 20 dB was achieved across the two operating bands of interest, as seen in Figure 6.22b.

6.3.5.2 Simulated and Measured Far Field Radiation Patterns

The radiation patterns of the proposed two-antennas MIMO across the E-plane and H-plane are depicted in Figure 6.23a–d. As shown in Figures 6.23a and b, across the two bands of interest of 5.0 GHz and 10.5 GHz, the two antenna elements at E-plane are exhibiting near-omnidirectional patterns and eight-shaped patterns for the copolarization (copolar) and cross-polarization (cross-pol) radiation, respectively. Furthermore, it is also noteworthy that at both frequencies, the radiation patterns of antenna element#1 are exact mirror images of those shown in antenna element#2. As for its H-plane counterparts, as shown in Figure 6.23c and d, across the two bands of interest of 5.0 GHz and 10.5 GHz, respectively, the two antenna elements at H-plane

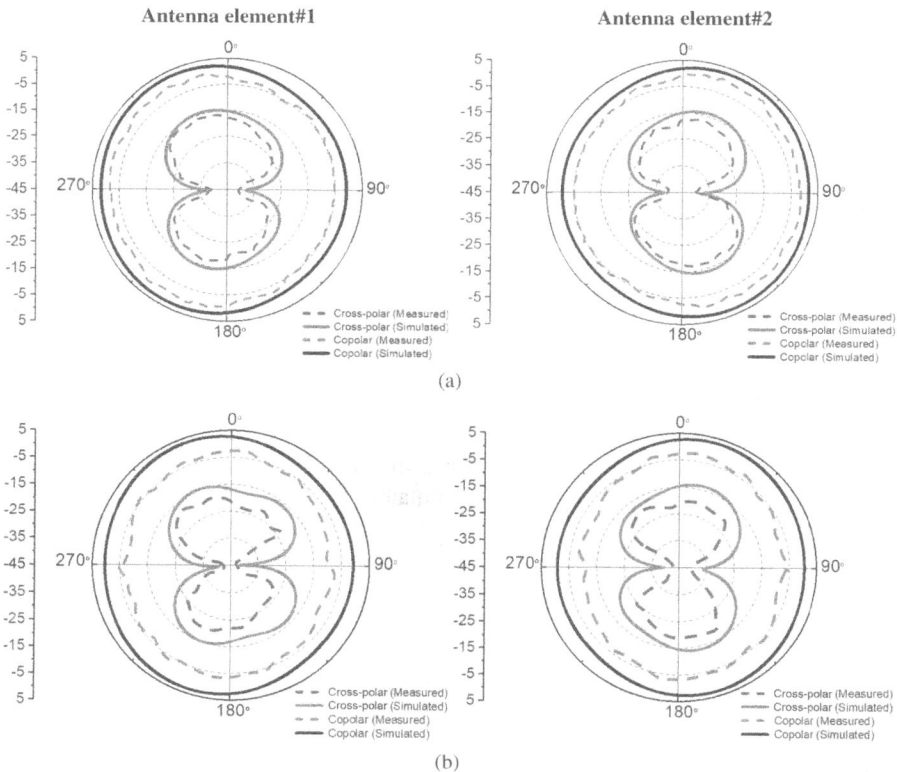

FIGURE 6.23 Radiation patterns of antenna element#1 and antenna element#2: (a) 5.0 GHz, E-plane, (b) 10.5 GHz, E-plane, (c) 5.0 GHz, H-plane, and (d) 10.5 GHz, H-plane. (*Continued*)

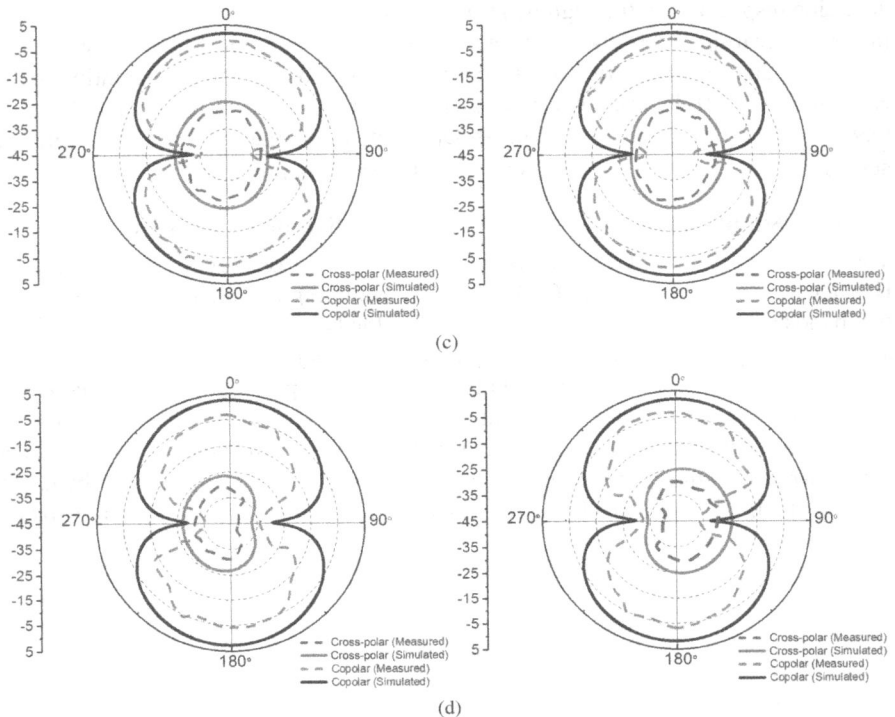

FIGURE 6.23   (Continued)

are exhibiting bi-directional patterns (copolar) and broadside patterns (cross-pol). These results demonstrated that the proposed two-antenna MIMO has offered acceptable radiation characteristics to meet the desire MIMO diversity performances.

6.3.5.3 Simulated and Measured Realized Gain and Radiation Efficiency

As the two antenna elements are identical to each other, we only plot the gain and efficiency of antenna element#1, which are illustrated in Figure 6.24. In this figure, between 3.5 GHz and 12 GHz, the simulated gain was 3.0–4.66 dBi, while the measured gain was 2.65–4.00 dBi. As for its corresponding radiation efficiency, the simulated gain ranged from 72.5% to 82%, while the measured gain was between 69.64% and 76.8%. Thus, the two-antenna MIMO has exhibited stable gain and efficiency throughout the operating bands of interest. The simulated and measured peak gain and peak efficiency are compared and presented in Table 6.3, and one can see that these results can ensure good quality of communication.

6.3.6 Diversity Performance Analysis

To prove the potency of the two-antenna MIMO, the diversity performances matrix including envelope correlation coefficient (ECC), diversity gain (DG), mean effective

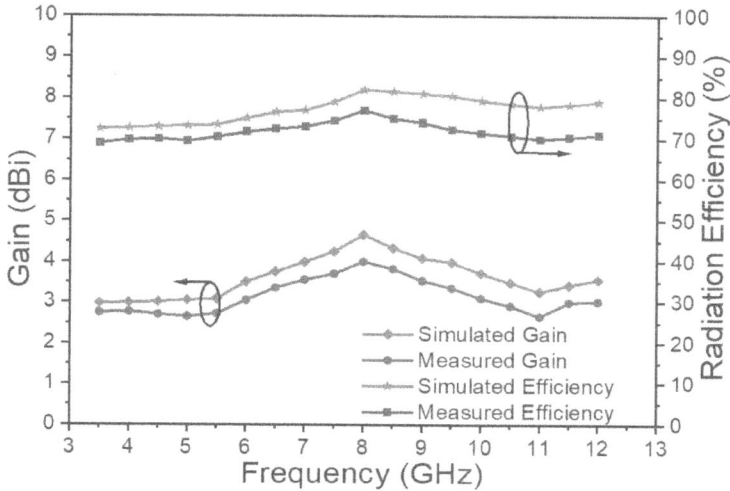

FIGURE 6.24 Gain and radiation efficiency of the proposed two-port MIMO antenna.

gain (MEG), channel capacity loss (CCL), channel capacity, and total active reflection coefficient (TARC), is essential, and thus is further verified through simulation as well as measurement.

6.3.6.1 ECC

To know how independent antenna elements radiate throughout the operating band, the ECC is an essential parameter to be investigated. Ideally, the ECC value should be equal to zero, which indicates that antenna elements radiate independently by producing uncorrelated radiations. However, in a rich fading environment, the value of ECC is not equivalent to zero. Notably, the ECC value of the proposed two-antenna MIMO can be calculated via the S-parameters, as denoted in Equation (6.4), however, one condition to be noted is that the two antenna elements must have an efficiency of near 100% so that an accurate ECC can be determined. Thus, it is always

---

TABLE 6.3

Comparison of Simulated and Measured Peak Gain and Peak Efficiency

Functioning Band	Simulated Peak Gain (dBi)	Measured Peak Gain (dBi)	Simulated Peak Efficiency (η)%	Measured Peak Efficiency (η)%
Sub-6 GHz 5G NR & C-band uplink	3.00	2.77	72.99	70.10
IEEE 802.11ac	3.08	2.72	73.50	70.50
IEEE 802.11ax & C-band downlink	3.75	3.35	76.50	72.49
X-band	3.48	2.91	78.48	70.80

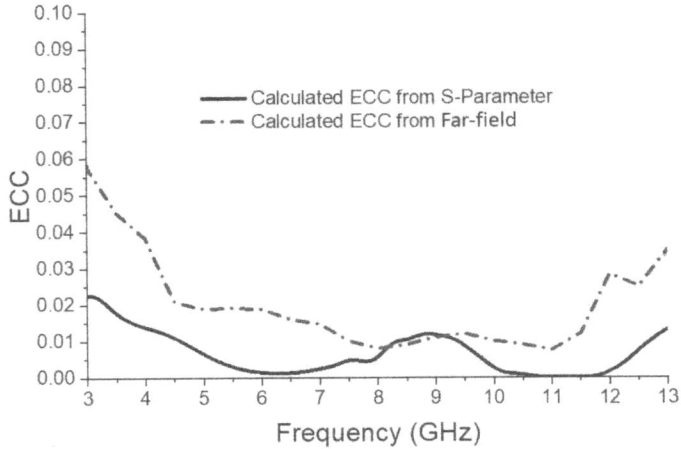

FIGURE 6.25 ECC of the proposed two-antenna MIMO.

better to apply Equation (6.5) via the far-field radiation patterns to determine the ECC, as the calculation of ECC via this equation is more accurate.

$$\rho_e = \frac{\left|S_{11}^*S_{12} + S_{21}^*S_{22}\right|^2}{\left(1 - |S_{11}|^2 - |S_{21}|^2\right)\left(1 - |S_{22}|^2 - |S_{12}|^2\right)} \qquad (6.4)$$

$$\rho_e = \frac{\left|\iint_{4\pi}\left[\overline{F_1}(\theta,\varnothing) * \overline{F_2}(\theta,\varnothing)\right]d\Omega\right|}{\iint_{4\pi}\left|\overline{F_1}(\theta,\varnothing)\right|^2 d\Omega \iint_{4\pi}\left|\overline{F_2}(\theta,\varnothing)\right|^2 d\Omega} \qquad (6.5)$$

where $\overline{F_i}(\theta,\varnothing)$ is the 3D field pattern of the antenna when the ith port is excited and Ω is a solid angle.

Figure 6.25 shows the ECC values for various frequency ranges. The ECC values obtained from Equations (6.4) and ((6.5) are well below 0.05, which are close to zero throughout the bands of interest of the two-antenna MIMO. This confirms that both the antenna elements are uncorrelated and, hence, contribute to increasing the data rate of the system.

6.3.6.2   DG dB

The DG is another metric of interest as it helps the MIMO antenna to resolve multipath signals, thus improving the receiver's ability to recover intelligent data from multipath signals at a fixed rate of transmission. The DG is used to determine the increment in signal-to-noise ratio (SNR) magnitude of each path due to the introduction of the spatial diversity scheme, calculated using the Equation (6.6).

$$DG = 10\sqrt{1 - |\rho_e|^2} \qquad (6.6)$$

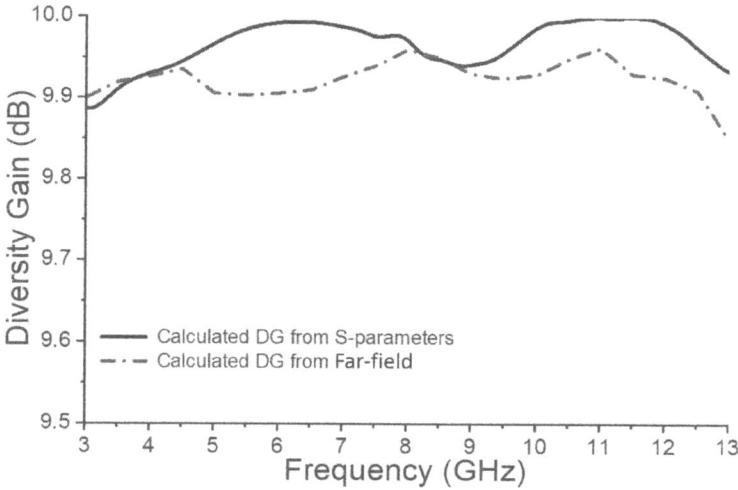

FIGURE 6.26   Diversity gain of the proposed two-antenna MIMO.

Figure 6.26 shows the DG of the proposed two-antenna MIMO. The calculated DG values obtained from far-field and S-parameters are close to 10 dB. This confirms that both the antenna elements are strongly uncorrelated and are good candidates for MIMO applications.

6.3.6.3   MEG

The MEG is a parameter used to determine the performance of antenna elements in a rich multipath fading practical environment. MEG is the ratio of mean received power to the mean incident power at the antenna element, and it is obtained using the efficiency method denoted in Equation (6.7).

$$MEG_i = 0.5\mu_{irad} = 0.5\left(1 - \sum_{j=1}^{K}|S_{ij}|^2\right) \qquad (6.7)$$

where K is the number of antenna elements, i is the excited antenna, and η_{irad} is the radiation efficiency of the ith antenna.

By expanding Equation (6.7), the MEG of each antenna element can be calculated by using Equations (6.8) and (6.9):

$$MEG_1 = 0.5\left(1 - |S_{11}|^2 - |S_{12}|^2\right) \qquad (6.8)$$

$$MEG_2 = 0.5\left(1 - |S_{21}|^2 - |S_{22}|^2\right) \qquad (6.9)$$

where, MEG_1 and MEG_2 are the MEG of antenna element#1 and antenna element#2, respectively. As shown in Figure 6.27, the calculated MEG of the two antenna

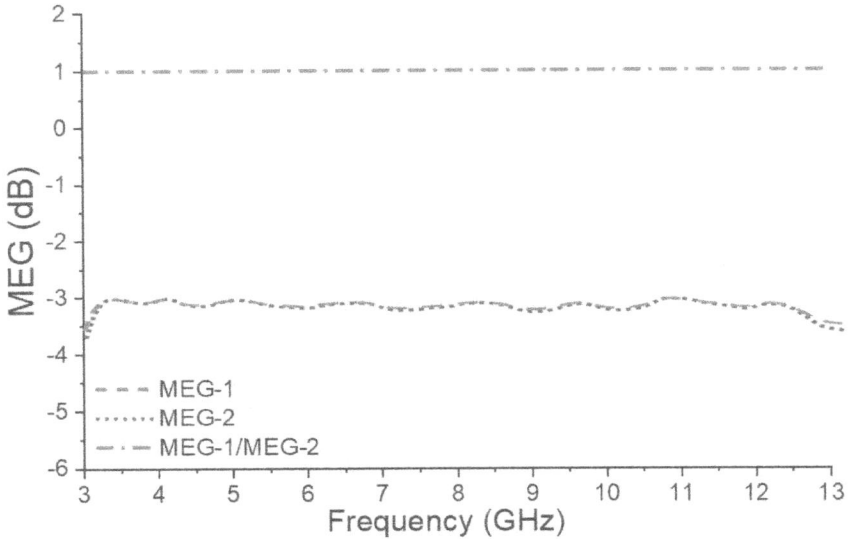

FIGURE 6.27   Mean effective gain of the proposed two-antenna MIMO.

elements is identical (approximately -3 dB) across the bands of interest, thus, their corresponding ratio (MEG_1/MEG_2) is 1. Table 6.4 concludes the MEG values of the proposed two-antenna MIMO. In this table, the two antenna elements have acquired very good values, which shows that the proposed two-antenna MIMO has maximum DG, better isolation across functioning bands, and smaller losses in diversity performances. Moreover, because the MEG ratios of the two antenna elements are closer to 1, it also validates better diversity performance from the proposed two-antenna MIMO under a very rich multipath fading environment of wireless channels.

TABLE 6.4

Mean Effective Gains (MEG) of the Proposed Two-Antenna MIMO

Frequency (GHz)	MEG (-dB) of Antenna Elements			
	Antenna Element #1	Antenna Element #2	Ratio of antenna Element #1/Antenna Element #2	Ratio of Antenna Element #2/Antenna Element #1
Sub-6 GHz 5G and C-band uplink	−3.10	−3.11	0.99	1.00
IEEE 802.11ac	−3.21	−3.22	0.99	1.00
IEEE 802.11ax 5G and C-band downlink	−3.15	−3.14	1.00	0.99
X-band	−3.05	−3.04	1.00	0.99

6.3.6.4   TARC

The TARC validates the diversity performance of a MIMO antenna and is calculated by using Equation (6.10).

$$\Gamma = \frac{\sqrt{\left(\left|S_{ii} + S_{ij}e^{j\theta}\right|^2\right) + \left(\left|S_{ji} + S_{jj}e^{j\theta}\right|^2\right)}}{\sqrt{2}} \tag{6.10}$$

where θ is the input phase angle, which is varied from $0°$ to $180°$ at intervals of $30°$ and S_{ii} and S_{jj} are the reflection coefficients (dB) of antenna element#1 and antenna element#2, respectively.

Figure 6.28 illustrates the measured TARC values, which are almost stable and below -10 dB in the entire operating bands. This validates that the proposed two-antenna MIMO has obtained good isolation and is a good candidate for integration with the phase shifter system.

6.3.6.5   CCL

The CCL is an essential metric to characterize the diversity performance of a MIMO antenna, as it confirms the higher bound on rate of transmission. Thus, to ensure a high data rate, the CCL values must be below 0.5 bits/s/Hz across the operational band, and is estimated using Equations (6.11) through (6.16).

$$C_{loss} = -log_2 det\left(\mu^R\right) \tag{6.11}$$

where, Ψ^R is the correlation matrix and is described as:

$$\mu^R = \begin{bmatrix} \mu_{11} & \mu_{12} \\ \mu_{21} & \mu_{22} \end{bmatrix} \tag{6.12}$$

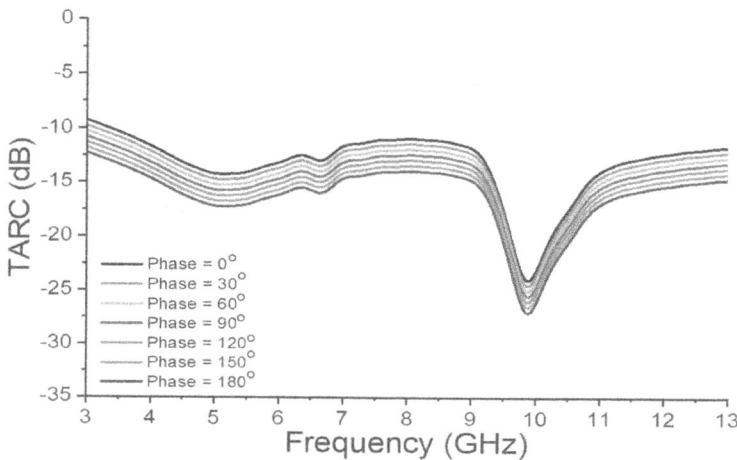

FIGURE 6.28   Total active reflection coefficient of the two-antenna MIMO.

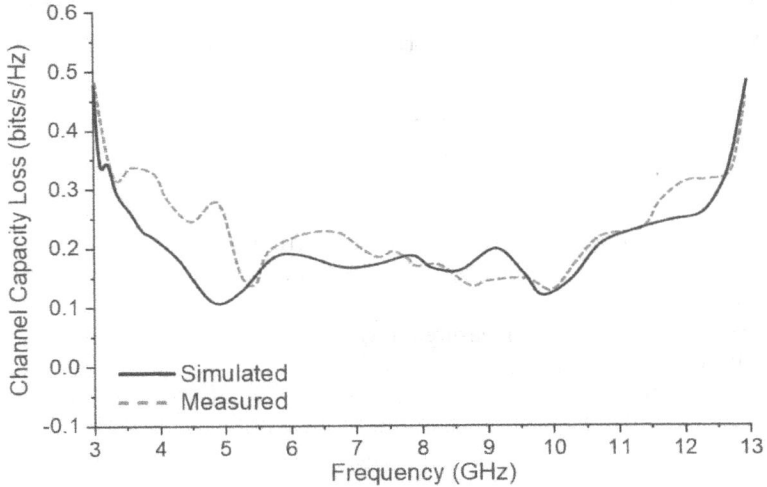

FIGURE 6.29   Channel capacity loss of the proposed two-antenna MIMO.

where,

$$\mu_{11} = 1 - \left(|S_{11}|^2 + |S_{12}|^2 \right) \tag{6.13}$$

$$\mu_{12} = -\left(S_{11}^* S_{12} + S_{21}^* S_{22} \right) \tag{6.14}$$

$$\mu_{21} = -\left(S_{22}^* S_{21} + S_{12}^* S_{11} \right) \tag{6.15}$$

$$\mu_{22} = 1 - \left(|S_{22}|^2 + |S_{21}|^2 \right) \tag{6.16}$$

The simulated and measured CCL values of the proposed two-antenna MIMO using the S-parameters are demonstrated in Figure 6.29. Here, average values of CCL less than 0.35 bits/s/Hz are observed in the entire band, which ensures better performance of the MIMO antenna by fulfilling the limits defined by the industry standards. Furthermore, the small discrepancy observed between the simulated and measured CCL values may be due to fabrication tolerances.

6.3.6.6   Channel Capacity (bits/s/Hz)

The channel capacity is vital parameter used to compute the multiplexing performance of the MIMO antenna. The channel capacity of the proposed two-antenna MIMO is computed using the equation in Ref. [39]. Figure 6.30 depicts the comparison graph of the channel capacity for the single input single output (SISO) and MIMO antenna. From Figure 6.30 it is easily observed that the channel capacity of the proposed two-antenna MIMO is greater than 10.00 bits/s/Hz, in the entire band

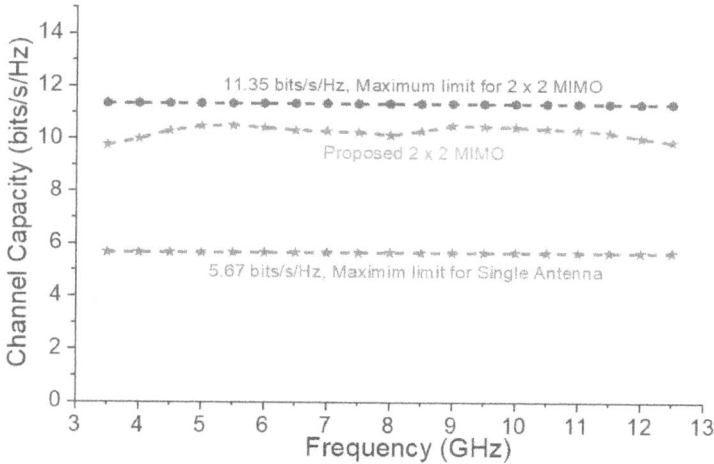

FIGURE 6.30 Channel capacity of the proposed two-antenna MIMO.

of interest, that is, approximately 1.78 times higher as compared to the maximum limit of an ideal SISO antenna (about 5.65 bps/Hz). Further, it is also noticed that the channel capacity values are very near to the maximum limit for an ideal 2×2 MIMO system (11.35 bps/Hz).

6.3.7 PERFORMANCE COMPARISON OF THE TWO-ANTENNA MIMO WITH EXISTING STATE-OF-THE-ART TECHNOLOGY

The performance comparison of the proposed two-antenna MIMO with other existing two-antenna MIMO types, including bandwidth, dimensions, decoupling structure, gain, efficiency, ECC, and DG are mentioned in Table 6.5. Here, it is observed that the proposed two-antenna MIMO has exhibited wide dual impedance bandwidths with compact dimensions, and good ECC and DG values.

From Table 6.5, the following features of the proposed two-antenna MIMO are observed:

- It has the smallest physical dimensions (volume size) as compared to all the antennas mentioned in Table 6.1.
- Unlike Refs. [29, 32, 35–38], the proposed antenna has exhibited a very wide dual bandwidth operation.
- It has a higher gain as compared to Refs. [21, 28, 29, 32, 34–36, 38].
- Unlike Refs. [21 and 34], the reported antenna has higher ECC and DG.
- The decoupling structure applied in this research is very simple and easy to design, and it can yield high isolation of >20 dB as compared to antennas reported in Refs. [20, 21, 23, 28].

TABLE 6.5
Performance Comparison of Proposed Two-Antenna MIMO with Existing State-of-the-Art Technology

Ref.	Dimension (mm³)	Sub	BW (GHz)	Gain (dBi)	ECC	DG	Isol (dB)	Decoupling Structure
[20]	32 × 98 × 1	FR-4	0.6–0.7 1.7–1.9 2.4–2.7 3.2–4.1 5.1–5.9	5.14	0.04	9.8	>15	Not used
[21]	56 × 37 × 1.6	FR-4	2.24–2.50 3.60–3.99 4.40–4.60 5.71–5.90	2	0.08	9.5	>15	Not used
[23]	70 × 52 × 1.6	FR-4	3.10–3.21 6.20–6.33 7.60–7.90	5.84	0.025	9.5	>31	Not used
[28]	90 × 21 × 1.6	FR-4	2.22–2.54 3.14–3.90 5.30–5.90	3.22	0.01	10	>20	Not used
[29]	34 × 34 × 1.44	FR-4	3.50–3.60 5.00–5.40	4.7	0.01	–	>19	U-shape slot in ground plane
[32]	20 × 35 × 0.8	FR-4 epoxy	3.34–3.87	2.5	0.01	–	>20	T-shape ground stub
[34]	59 × 55 × 8.1	FR-4	3.00–7.00	4	0.2	8.94	>20	DGS
[37]	38.2 × 95.94 × 1.6	FR-4	2.43–2.50	4.25	0.008	9.99	>24	Fractal EBG
[38]	55 × 28 × 1.6	FR-4	2.01–3.92	2	0.01	9.8	>15	EBG
[39]	100 × 60 × 1	Jeans	1.34–3.92 4.34–6.34	5	0.04	9.0	>18	Meta-inspired
[40]	47.5 × 40 × 1.6	FR-4	3.35–3.78	3.5	0.05	–	>15	Metamaterial
Reported Work	32 × 20 × 0.8	FR-4	3.3–7.8 8.0–12.0	4.0	0.05	9.9	>20	Concentric rings

6.3.8 Concluding Remarks

A two-antenna MIMO functioning in the sub-6 GHz 5G NR, IEEE 802.11ac/ax, C-band, and X-band has been successfully studied. Besides showing very wide dual 10 dB impedance bandwidths of 81.08% (3.3–7.8 GHz) and 40% (8.0–12.0 GHz), the reported two-antenna MIMO has also exhibited a gain of >3 dBi and efficiency greater than 69% throughout the two bands of interest. By loading a decoupling structure (five concentric ring elements) between the two adjacent antenna elements, a very desirable isolation above >20 dB is obtained. The MIMO performance metrics such as ECC (<0.05), DG (>9.9 dB), CCL (<0.35/bits/s/Hz), TARC (<–10 dB), and MEG1/MEG2 ratio (approximately equal to 1) are investigated, and their corresponding values are well within the acceptable practical values. Furthermore, the calculated channel capacity is larger than 10.00 bits/s/Hz. Therefore, because of its compact size, good scattering and radiation characteristics, better diversity performance, the proposed two-antenna MIMO is a potential in futuristic devices for above aforementioned wireless and satellite applications.

REFERENCES

1. Kulkarni, J. & Sim, C.-Y.-D. "Wideband CPW-fed oval-shaped monopole antenna for wi-Fi5 and wi-Fi6 applications." *Progress in Electromagnetics Research C*, vol. 107, pp. 173–182, 2021.
2. Guo, Q., Zhang, J., Zhu, J. & Yan, D. "A compact multiband dielectric resonator antenna for wireless communications." *Microwave and Optical Technology Letters*, vol. 62, pp. 2945–2952, 2020.
3. Gong, Y., Yang, S., Li, B., Chen, Y., Tong, F. & Yu, C. "Multi-band and high gain antenna using AMC ground characterized with four zero-phases of reflection coefficient." *IEEE Access*, vol. 8, pp. 171457–171468, 2020.
4. Rajalakshmi, P. & Gunavathi, N. "Compact modified hexagonal spiral resonator-based tri-band patch antenna with octagonal slot for Wi-Fi/WLAN applications." *Progress in Electromagnetics Research C*, vol. 106, pp. 77–87, 2020.
5. Yang, Y., Zhang, F., Zhang, Y. & Li, X. "Design and analysis of a novel miniaturized dual-band omnidirectional antenna for Wi-Fi applications." *Progress in Electromagnetics Research M*, vol. 94, pp. 95–103, 2020.
6. Aziz, A., Motagaly, A., Ibrahim, A., Rouby, W. & Abdalla, M. "A printed expanded graphite paper based dual band antenna for conformal wireless applications." *AEU — International Journal of Electronics and Communications*, vol. 110, pp. 1–7, 2019.
7. Kulkarni, J. & Sim, C. Y. D., "Low-profile, compact multi-band monopole antenna for futuristic wireless applications." *2020 IEEE International Conference on Electronics, Computing and Communication Technologies (CONECCT)*, Bangalore, India, 2020, pp. 1–5.
8. Kumar, A., Althuwayb, A. A. & Al-Hasan, M. J. "Wideband triple resonance patch antenna for 5G Wi-Fi spectrum." *Progress in Electromagnetics Research Letters*, vol. 93, pp. 89–97, 2020.
9. Abbasi, N., Langley, R. & Bashir, S. "Multiband shorted monopole antenna." *Journal of Electromagnetic Waves and Applications*, vol. 28, pp. 618–633, 2014.
10. Saraswat, R. & Kumar, M. "A vertex-fed hexa-band frequency reconfigurable antenna for wireless applications." *International Journal of RF and Microwave Computer-Aided Engineering*, vol. 29, pp. 1–13, 2019.

11. Jing, J., Pang, J., Lin, H., Qui, Z. & Liu, C. "A multiband compact low-profile planar antenna based on multiple resonator stub." *Progress in Electromagnetics Research Letters*, vol. 94, pp. 1–7, 2020.
12. Kumar, Y., Gangwar, R. & Kanaujia, B. "Asymmetrical mirror imaged monopole antenna with modified ground structure for DBDP radiations." *International Journal of Electronics*, vol. 107, pp. 1–24, 2020.
13. Kulkarni, J., Kulkarni, N. & Desai, A. "Development of H-shaped monopole antenna for IEEE 802.11a and HIPERLAN 2 applications in the laptop computer." *International Journal of RF and Microwave Computer-Aided Engineering*, vol. 30, no. 7, pp. 1–14, 2020.
14. Sim, C. Y. D., Chen, C. C., Zhang, X. Y. & Lee, Y. L. "Very small-size uniplanar printed monopole antenna for dual-band WLAN laptop computer applications." *IEEE Transactions on Antennas and Propagation*, vol. 65, pp. 2916–2922, 2017.
15. Kulkarni, J. "Multi-band printed monopole antenna conforming bandwidth requirement of GSM/WLAN/WiMAX standards." *Progress in Electromagnetics Research Letters*, vol. 91, pp. 59–66, 2020.
16. Kulkarni, J. "An ultra-thin, dual band, sub 6 GHz, 5G and WLAN antenna for next generation laptop computers." *Circuit World*, vol. 45, pp. 363–370, 2020.
17. Sim, C., Liu, H. & Huang, C. "Wideband MIMO antenna array design for future Mobile devices operating in the 5G NR frequency bands n77/n78/n79 and LTE band 46." *IEEE Antennas and Wireless Propagation Letters*, vol. 19, pp. 74–78, 2020.
18. Kulkarni, J., Desai, A. & Sim, C. Y. D. "Wideband four-Port MIMO antenna array with high isolation for future wireless systems." *AEU — International Journal of Electronics and Communications*, vol. 128, 2020, doi: 10.1016/j.aeue.2020.153507.
19. Alharbi, A. G., Kulkarni, J., Desai, A., Sim, C.-Y.-D. & Poddar, A. "A multi-slot two-antenna MIMO with high isolation for sub-6 GHz 5G/IEEE802.11ac/ax/C-band/X-band wireless and satellite applications." *Electronics*, vol. 11, p. 473, 2022.
20. Hassan, M. M., Zahid, Z., Khan, A. A., Rashid, I., Rauf, A., Maqsood, M. & Bhatti, F. A. "Two element MIMO antenna with frequency reconfigurable characteristics utilizing RF MEMS for 5G applications." *Journal of Electromagnetic Waves and Applications*, vol. 34, no. 9, pp. 1210–1224, 2020.
21. Chouhan, S., Panda, D. K., Kushwah, V. S. & Singhal, S. "Spider-shaped fractal MIMO antenna for WLAN/WiMAX/Wi-Fi/Bluetooth/C-band applications." *AEU — International Journal of Electronics and Communications*, vol. 110, pp. 1–8, 2019.
22. Soltani, S., Lotfi, P. & Murch, R. D. "A dual-band multiport MIMO slot antenna for WLAN applications." *IEEE Antennas and Wireless Propagation Letters*, vol. 16, pp. 529–532, 2017.
23. Babu, K. V. & Anuradha, B. "Design and performance analysis of tri-band Wang shaped MIMO antenna." *International Journal of Information Technology*, vol. 12, pp. 559–566, 2020.
24. Ouahabil, M. E., Zakriti, A., Essaaidi, M., Dkiouak, A. & Elftouh, H. "A miniaturized dual-band MIMO antenna with low mutual coupling for wireless applications." *Progress in Electromagnetics Research C*, vol. 93, pp. 93–101, 2019.
25. Kulkarni, J., Sim, C. Y. D. & Deshpande, V. "Low-profile, compact, two port MIMO antenna conforming Wi-Fi-5/Wi-Fi-6/V2X/DSRC/INSAT-C for wireless industrial applications." *2020 IEEE 17th India Council International Conference (INDICON)*, New Delhi, India, 2020, pp. 1–5.
26. Dkiouak, A., Zakriti, A. & Ouahabi, M. E. "Design of a compact dual-band MIMO antenna with high isolation for WLAN and X-band satellite by using orthogonal polarization." *Journal of Electromagnetic Waves and Applications*, vol. 34, no. 9, pp. 1254–1267, 2019.

27. Kumar, M. & Nath, V. "Design and development of triple-band compact ACS-fed MIMO antenna for 2.4/3.5/5 GHz WLAN/WiMAX applications." *Analog Integrated Circuits and Signal Processing*, vol. 103, pp. 461–470, 2020.
28. Ekrami, H. & Jam, S. "A compact triple-band dual-element MIMO antenna with high port-to-port isolation for wireless applications." *International Journal of Electronics and Communications*, vol. 96, pp. 219–227, 2018.
29. Islam, S. N. & Das, S. "Dual band CPW fed MIMO antenna with polarization diversity and improved gain." *International Journal of RF and Microwave Computer-Aided Engineering*, vol. 30, no. 4, e22128, 2020.
30. Tiwari, R. N., Singh, P., Kanaujia, B. K., Kumar, S. & Gupta, S. K. "A low profile dual band MIMO antenna for LTE/Bluetooth/Wi-Fi/WLAN applications." *Journal of Electromagnetic Waves and Applications*, vol. 34, no. 9, pp. 1239–1253, 2020.
31. Kulkarni, J., Desai, A. & Sim, C. Y. D. "Two port CPW-fed MIMO antenna with wide bandwidth and high isolation for future wireless applications." *International Journal of RF and Microwave Computer-Aided Engineering*, vol. 31, e22700, 2021.
32. Saurabh, A. K. & Meshram, M. K. "Compact sub-6 GHz 5G-multiple-input-multiple-output antenna system with enhanced isolation." *International Journal of RF and Microwave Computer-Aided Engineering*, vol. 30, no. 8, e22246, 2020.
33. Khan, A. A., Jamaluddin, M. H., Aqeel, S., Nasir, J., Kazim, J. R. & Owais, O. "Dual-band MIMO dielectric resonator antenna for WiMAX/WLAN applications." *IET Microwaves, Antennas & Propagation*, vol. 11, no. 1, pp. 113–120, 2017.
34. Kumari, T., Das, G., Sharma, A. & Gangwar, R. K. "Design approach for dual element hybrid MIMO antenna arrangement for wideband applications." *International Journal of RF and Microwave Computer-Aided Engineering*, vol. 29, no. 1, e21486, 2019.
35. Kumar, M. & Nath, V. "Analysis of low mutual coupling compact multi-band microstrip patch antenna and its array using defected ground structure." *Engineering Science and Technology, an International Journal*, vol. 19, pp. 866–874, 2016.
36. Pandhare, R. A., Zade, P. L. & Abegaonkar, M. P. "Miniaturized microstrip antenna array using defected ground structure with enhanced performance." *Engineering Science and Technology, an International Journal*, vol. 19, pp. 1360–1367, 2016.
37. Sharma, K. & Pandey, G. P. "Two port compact MIMO antenna for ISM band applications." *Progress in Electromagnetics Research C*, vol. 100, pp. 173–185, 2020.
38. Biswas, A. K. & Chakraborty, U. "Reduced mutual coupling of compact MIMO antenna designed for WLAN and WiMAX applications." *International Journal of RF and Microwave Computer-Aided Engineering*, vol. 29, no. 3, e21629, 2018.
39. Roy, S. & Chakraborty, U. "Mutual coupling reduction in a multi-band MIMO antenna using meta-inspired decoupling network." *Wireless Personal Communications*, vol. 114, pp. 3231–3246, 2020.
40. Wang, C., Yang, X. S. & Wang, B. Z. "A metamaterial-based compact broadband planar monopole MIMO antenna with high isolation." *Microwave and Optical Technology Letters*, vol. 62, no. 9, pp. 1–6, 2020.
41. Kulkarni, J., Desai, A. Sim, Chow-Yen Desmond. "Wideband four-port MIMO antenna array with high isolation for future wireless systems." *AEU — International Journal of Electronics and Communications*, vol. 128, 2021. https://doi.org/10.1016/j.aeue.2020.153507
42. Pirasteh, A., Roshani, S., & Roshani, S. "Compact microstrip lowpass filter with ultra-sharp response using a square-loaded modified T-shaped resonator." *Turkish Journal of Electrical Engineering & Computer Sciences*, vol. 26, pp. 1736–1746, 2018.

7 Cutting-Edge Design Approaches for Circularly Polarized Antenna

7.1 INTRODUCTION

This chapter discusses a case study on broadband circularly polarized (CP) antennas and their design techniques in detail. The CP antennas play a very important role in modern global navigation satellite system (GNSS) systems, satellite communication, global positioning system (GPS) receivers, and wireless power transmission. Moreover, the CP antennas are critical for modern wireless communication systems due to the growing demands for large capacity and higher data rate.

7.2 CASE STUDY: CP ROTATED L-SHAPED ANTENNA WITH J-SHAPED DEFECTED GROUND STRUCTURE FOR WIRELESS LOCAL AREA NETWORKS AND VEHICLE TO EVERYTHING APPLICATIONS

The authors in Ref. [1] have reported a novel and compact microstrip-line-fed printed antenna for wideband CP radiation. The designed antenna utilizes a crescent-shaped substrate, rotated L-shaped monopole, and defected ground structure (DGS) to achieve a wide 3 dB axial ratio (AR) bandwidth and 10 dB impedance bandwidth (ZBW) across the entire 5 GHz wireless local area network (WLAN) and Vehicle to Everything (V2X) operational bands. As the substrate of the proposed antenna is only 0.8 mm thick, it has a very low profile of 0.014 λ in terms of free space wavelength (λ) at 5.5 GHz. The proposed CP antenna exhibits overlapping 10 dB ZBW and 3 dB axial ratio bandwidth (ARBW) of 22.05% (4.80–5.99 GHz), along with broadside far-field patterns, gain greater than 2.5 dBi, and efficiency above 85% throughout the desired operating band. Therefore, it is a good candidate for WLAN and V2X communication applications.

7.2.1 OVERVIEW OF CP DESIGN IN THE LITERATURE

The CP radiation and high radiating performance antennas with their favorable features, such as being low profile, lightweight, and a planner structure, are essential for modern wireless communication including Industrial Internet of Things (IIoT), machine-to-machine communication in aerospace, security, commercial, agricultural as well as V2X applications [2, 3]. In this context, several antenna designers and researchers have reported outstanding CP/dual-CP antennas to meet the growing demand for broad bandwidth and multimedia applications in the near future [4–19]. Among them, slots with various geometries of the radiating element are gaining

 DOI: 10.1201/9781003331018-7

more attention [4–19] as they can be utilized to generate CP radiations easily for desired wireless bands.

The antenna reported in Ref. [4] is designed using an asymmetrical Y-shaped dielectric resonator (DR) material for WLAN and worldwide interoperability for microwave access (WiMAX) operations with a designed footprint of 120×70 mm^2, whereas a CP dual-band modified Ψ-shaped antenna that has a dimension of 130×85 mm^2 is discussed in Ref. [5]. However, both antennas have very large dimensions. The antennas with reduced footprint are discussed in Ref. [6–8]. The annular ring slot antenna in Ref. [6] is designed on a 50.3×51.5 mm^2 Teflon substrate, whereas the perturbed Ψ-shaped antenna in Ref. [7] and fractal inspired slot antenna in Ref. [8] are printed on 58.2×47.7 mm^2 RT-Duroid and 39×46 mm^2 TLY-5 substrate, respectively. Even though the CP antennas in Ref. [6–8] have exhibited smaller dimensions and dual-band operations, they are designed on expensive substrates.

To reduce the cost, the antennas in Ref. [9–12] are designed on cost-effective and easily available FR-4 substrate. Here, a fan-shaped CP antenna operating in the 2.4 GHz WLAN band is reported in Ref. [9], while in Ref. [10] has reported a fractal defected ground patch antenna operating in the 1.55 GHz band. In Ref. [11], a truncated square aperture patch antenna operating in the 2.4/5 GHz WLAN band has been investigated, whereas a gap-coupled patch antenna operating in the 3.6/5.3 WLAN/WiMAX band is discussed in Ref. [12]. However, the antennas in Ref. [9–12] have demonstrated a very narrow ZBW and do not cover the entire band of operation. Wideband antennas designed on cost-effective FR-4 substrate have been reported in Ref. [13–17]. The square-slot CP antenna in Ref. [13] has a ZBW of 36.41% (4.65–6.72 GHz), and the perturbation patch antenna with an open slot [14] has shown a very wide ZBW of 103.5% (2.40–7.55 GHz). Notably, the inverted L-shaped CP antenna design in Ref. [15] has demonstrated a wide ZBW of 50.9% (3.48–5.86 GHz) for C-band application, and the CP microstrip dual-band antenna (with L-shaped radiator and two circular strips) studied in Ref. [16] has also shown a wide ZBWs of 46.4% (1.82–2.92) and 40.5% (3.15–4.75 GHz). Nonetheless, to cover both sub-6 GHz and X-band applications, Ref. [17] has reported a uniplanar slot antenna with a loaded metallic reflector at the bottom, and dual broadband ZBWs of 21.27% (5.46–6.76 GHz) and 24.65% (8.18–10.48 GHz) were achieved. Even though the antennas in Ref. [13–17] have also achieved wide ZBWs, they do not have matching overlapping ARBW. A frequency tunable dual-folded inverted-L CP antenna [18] and CP V-shaped slot antenna [19] with overlapping 3 dB ARBW and 10 dB ZBW have been investigated recently, however, the use of diodes and vias in Refs. [18] and [19], respectively, increased their manufacturing complexity.

Compared with the aforementioned various CP antenna techniques, the proposed CP antenna design is a simple patch design that is composed of a microstrip-line-fed rotated L-shaped antenna with J-shaped DGS, and it has a very wide overlapping 3 dB ARBW and 10 dB ZBW. In this design, a non-centered rotated L-shaped strip and J-shaped DGS (narrow slit and slots) are etched on the upper and bottom layer of the crescent-shaped FR-4 substrates, respectively, to generate very wide impedance matching and CP characteristics across the entire 5 GHz WLAN and V2X band. By utilizing this structure, near overlapping 3 dB ARBW and 10 dB ZBW of 31.5% (4.80–5.99 GHz) is achieved. As the proposed CP antenna has exhibited antenna gain of greater than 2.5 dBi, efficiency above 85%, and desirable CP radiations across the entire 5 GHz WLAN and V2X band, it is a very suitable CP antenna design for modern wireless communications.

7.2.2 GEOMETRY AND DESIGN OF PROPOSED ROTATED L-SHAPED ANTENNA

Figure 7.1a and b show the geometry of the proposed CP antenna that is printed on a lossy 0.8 mm thick crescent-shaped FR-4 dielectric with a dielectric constant (ϵ_r) of 4.3 and loss tangent (tan δ) of 0.025. Its corresponding layered view is depicted in Figure 7.1c. The optimized dimensions of the proposed CP antenna are articulated in Table 7.1, and it has a volume size of $30 \times 30 \times 0.8$ mm^3 ($0.55 \times 0.55 \times 0.014$ λ^3). The upper surface layer of the crescent-shaped FR-4 substrate is comprised of a non-centered rotated L-shaped monopole excited by a 50 Ω microstrip transmission feeding line, while the bottom surface layer is mainly a J-shaped DGS, as shown in Figure 7.1a and b, respectively. Here, the 50 Ω microstrip feeding line has a size of ($L_F \times W_F$) and it is deployed at a distance of W_a, located on the left edge of the substrate. As for the rotated L-shaped monopole that is connected to the microstrip feeding line, it has vertical and horizontal sections of size ($L_m \times W_m$) and ($L_n \times W_n$), respectively, as seen in Figure 7.1a. As depicted in Figure 7.1b, the left opening stub section of the J-shaped DGS is composed of two rectangular stubs of sizes ($L_1 \times W_1$) and ($L_2 \times W_2$), and a very narrow slit of size ($L_s \times W_s$) is loaded into this opening stub section. Here, a vertical rectangular stub of size ($L_5 \times W_3$) is linked to the bottom of this left opening stub section, and it is connected to a horizontal bottom stub of size ($W \times L_4$). Notably, the right section of the J-shaped DGS comprises two vertical stubs in which the narrow opening vertical stub has a size of ($L_7 \times W_5$) and the bottom right-side stub section can be considered as a protruded rectangular stub section of size ($L_6 \times W_4$). Four identical circular slots (with radius R_2 spaced at a distance L9 from each other) are vertically loaded into this protruded rectangular section to achieve good impedance matching and CP radiations across the 5 GHz WLAN and V2X bands. As for the crescent cut (with radius R_1) at the top-right corner of the FR-4 substrate, it is to further achieve wide ZBW and ARBW across the desired operating bands. The proposed CP antenna is simulated and optimized using CST MWS software® [20].

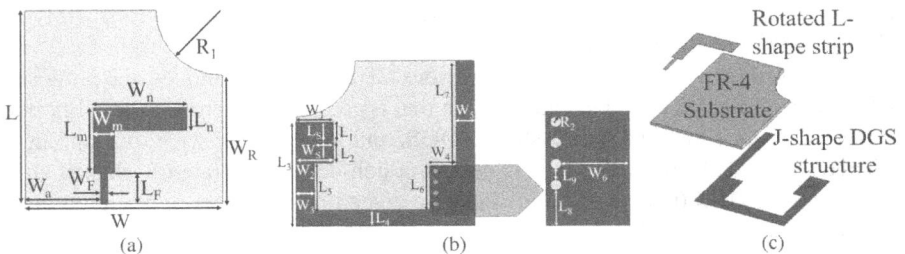

FIGURE 7.1 Geometry of the proposed CP antenna (all dimensions are in millimeters): (a) front view, (b) back view, and (c) layered view.

TABLE 7.1
Optimized Dimensions of the Proposed CP Antenna

Parameters	Values (mm)	Parameters	Values (mm)	Parameters	Values (mm)
L	30.0	L_7	19.0	W_1	6.2
L_F	4.8	L_8	3.5	W_2	6.2
L_m	10.2	L_9	1.0	W_3	3.0
L_n	3.5	L_S	0.2	W_4	4.4
L_1	3.4	W	30.0	W_5	3.0
L_2	3.8	W_F	1.0	W_6	6.0
L_3	19.0	W_m	3.2	W_S	2.8
L_4	3.0	W_n	14.2	R_1	10
L_5	8.6	W_a	11	R_2	0.5
L_6	8.0	W_R	20.0		

7.2.3 STEP-WISE EVOLUTION OF THE PROPOSED CP ANTENNA

The step-wise design evolution to analyze the working mechanism of the proposed CP antenna is depicted in Figures 7.2 and 7.3, showing the reflection coefficient (S_{11}) and 3 dB AR curves, respectively. Initially, as shown in Figure 7.2, in Step 1, a vertical non-centered rectangular strip along with a rotated L-shaped monopole fed by a microstrip line with partial J-shaped DGS is designed. In Step 2, a narrow slit ($Ls \times Ws$) is further loaded into the left opening section of the J-shaped structure. In Step 3, the right vertical section is reduced to two non-equal width vertical stubs, and it can be regarded that the bottom-right corner has a protruded rectangular stub section ($L6 \times W_4$). Next, in Step 4, the top-right corner of the FR-4 substrate is removed with a size of a quarter-circle that has a radius of R_1. Finally, in Step 5, four vertical circular slots (each with radius R_2) are loaded into the protruded rectangular stub section.

The S_{11} and AR curves based on the step-wise evolution from Step 1 to Step 5 of the proposed CP antenna are analyzed in Figure 7.3a and b, respectively. It is apparent that the antenna in Step 1 induces a linear polarization (LP) and has a very narrow bandwidth. In Step 2 and Step 3, the deployment of the two rectangular stubs (separated by the narrow slit), as well as the protruded rectangular stub section

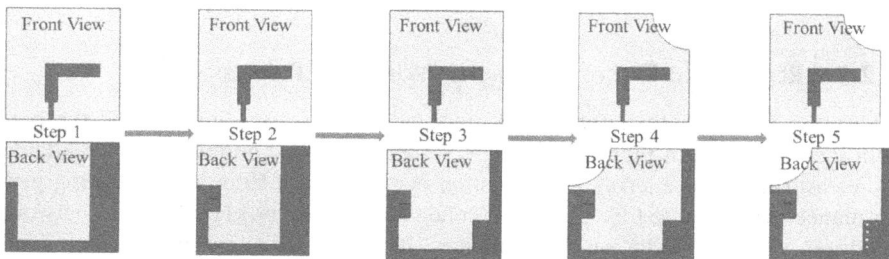

FIGURE 7.2   Step-wise design evaluation of the proposed CP antenna.

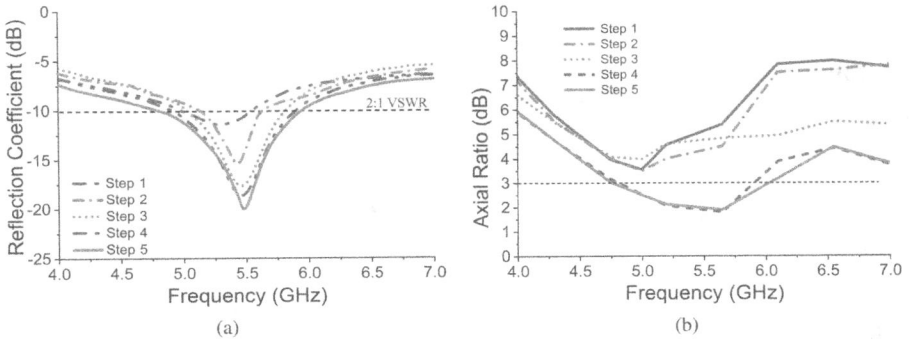

FIGURE 7.3 Step-wise design of the proposed CP antenna, (a) S_{11} (dB), (b) axial ratio.

placed at the bottom-right corner of the J-shaped structure, perturbs the current paths and haves in generating more electric field current that results in enhancing the 10 dB ZBW. Here, even though a significant increase in the bandwidth is noticed, the antenna is linearly polarized, and the desired 5 GHz WLAN band is not covered. It is further noticed from Figure 7.3b that desirable CP bandwidth performances (with AR <3 dB) are achieved by removing the top-right corner of the substrate with a quarter-circle of radius R_1. Finally, the four vertical circular slots can aid in achieving the desired 10 dB ZBW and 3 dB ARBW of 22.55% (4.76–5.97 GHz).

7.2.4 SURFACE CURRENT DISTRIBUTION FOR UNDERSTANDING THE CP MECHANISM

To know how the CP radiation is excited, it can be easily observed (at the resonating frequency of 5.5 GHz) by determining the rotation of the resultant surface current distribution (J) A/m at phase angle 0° and 90°, as shown in Figure 7.4a and b, respectively. The surface current is observed on the rotated L-shaped monopole on the front side as well as on the J-shaped strip on the backside. Here, it is apparent that the resultant current J lies in the second and third quadrant at a phase angle of 0° and 90°, respectively. It can be noticed that the resultant current J rotates in a counter-clockwise direction and are orthogonal to each other as the phase changes from 0° to 90°. Therefore, as the resultant current rotates in a counter-clockwise direction, a right-hand CP (RHCP) is achieved.

7.2.5 RESULTS AND DISCUSSION OF THE PROPOSED CP ANTENNA

The proposed CP antenna was fabricated and its corresponding top and the bottom view is shown in Figure 7.5a and b, respectively. The prototype was validated by measuring the scattering and radiation performances. Here, the scattering performance was validated by using the Rohde & Schwarz (9 kHz–16 GHz) network analyzer, while the radiation performances (including radiation patterns, gain, and radiation efficiency) were achieved in a calibrated anechoic chamber, as shown in Figure 7.5c.

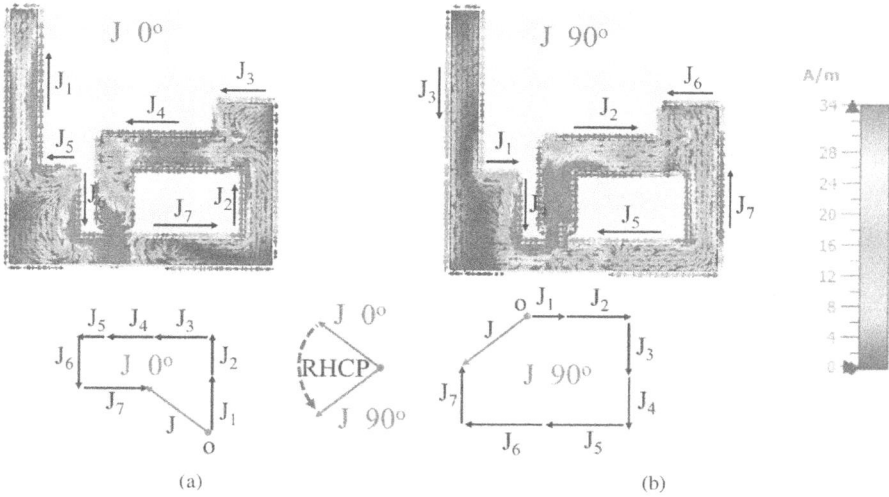

FIGURE 7.4 Simulated surface current (A/m) distribution of the proposed CP antenna at 5.5 GHz at 0° (a) and at 90° (b).

7.2.5.1 Reflection Coefficient (S_{11}) dB and 3 dB ARBW

Figure 7.6a shows the simulated and measured S_{11} of the proposed CP antenna. The simulated and measured 10 dB ZBW are 22.55% (4.76–5.97 GHz) and 22.05% (4.80–5.99 GHz). Compared with Figure 7.6b, a wide CP bandwidth is achieved across the desired band of interest. The simulated and measured 3 dB ARBW of the proposed CP antenna is 22.55% (4.76–5.97 GHz) and 22.05% (4.80–5.99 GHz). The 10 dB ZBW and 3 dB ARBW completely overlap. Notably, a very slight deviation between the simulated and measured results is seen, which may be due to a slight fabrication tolerance of the proposed CP antenna prototype.

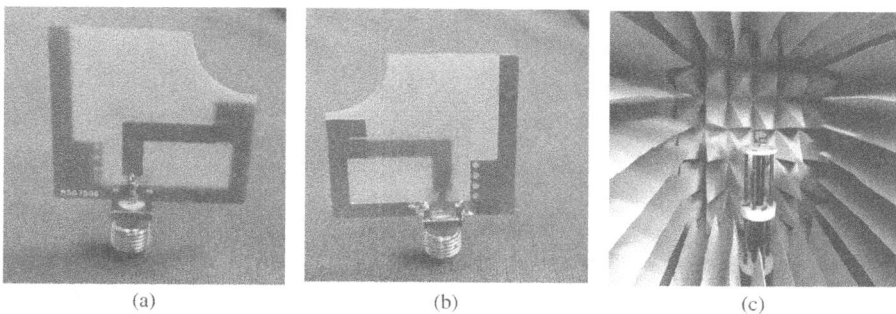

FIGURE 7.5 Fabricated prototype and measurement setup of the proposed CP antenna: (a) front view, (b) back view, and (c) antenna mounted inside an anechoic chamber.

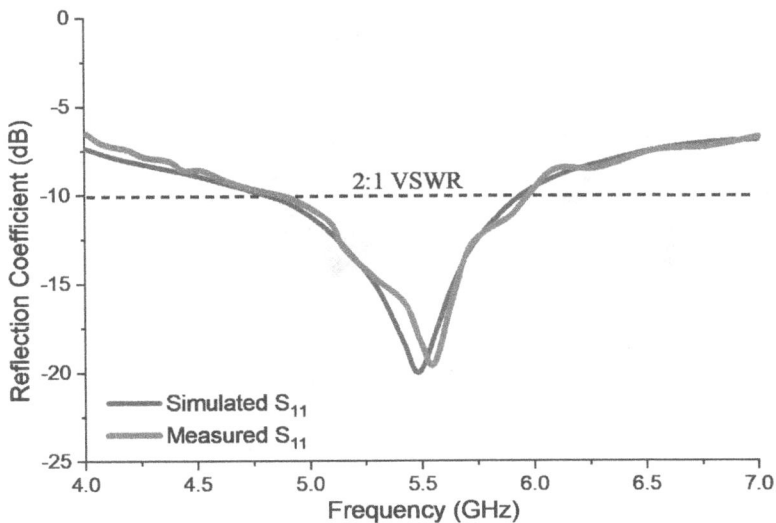

(a)

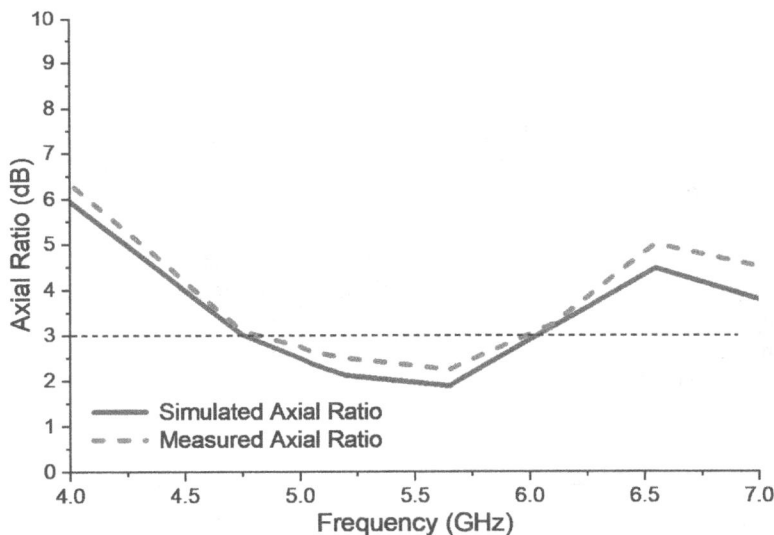

(b)

FIGURE 7.6   Measured and simulated results of the proposed CP antenna: (a) S_{11} (dB) and (b) axial ratio (dB).

7.2.5.2   Radiation Patterns of the Proposed CP Antenna

Figure 7.7 shows the comparison of two dimensional (2D) simulated and measured radiation pattern of the proposed CP antenna (at 5.5 GHz) in the x-z plane (E-plane) and y-z plane (H-plane). Here, the left-hand circularly polarized (LHCP) waves excited by the proposed CP antenna are tilted at a ±30° broadside

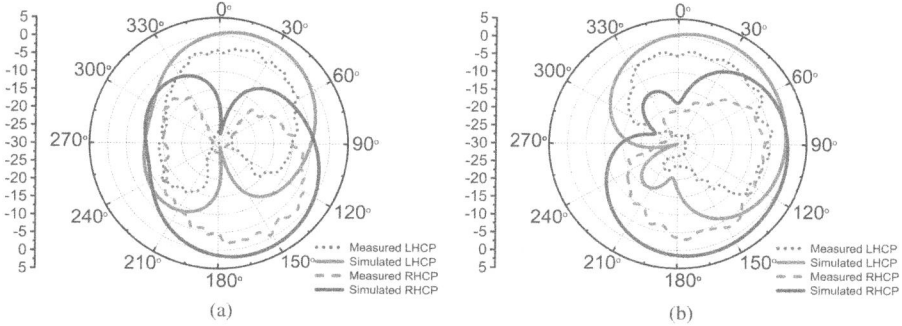

FIGURE 7.7   The measured and simulated 2D radiation pattern of the proposed CP antenna at 5.5 GHz: (a) E-plane and (b) H-plane.

direction, whereas the right-hand circularly polarized (RHCP) waves are excited at ±120° of LHCP. There is a slight deviation in the measured and simulated results, which may be due to fabrication tolerances and the use of coaxial cable inside the anechoic chamber.

7.2.5.3   Measured and Simulated Gain and Radiation Efficiency

Figure 7.8 shows the gain and efficiency curves of the proposed CP antenna. The simulated gain was calculated using Friis formula given in Ref. [21]. As shown in Figure 7.8, a measured gain ranging between 2.5 dBi and 3.5 dBi is observed, whereas the simulated gain is greater than 3 dBi throughout the operating band. Likewise, the measured and simulated efficiency was above 85% and 88%, respectively, across the band of interest.

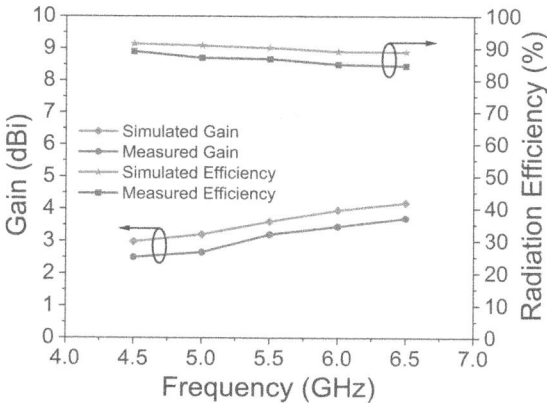

FIGURE 7.8   Gain and efficiency plots of the proposed CP antenna.

7.2.6 Performance Comparison of the Proposed CP Antenna

Table 7.2 shows the performance comparison of the proposed CP antenna with other existing reported state-of-the-art technology, which include size 10 dB ZBW, 3 dB ARBW, gain, and antenna design technique.

As depicted in Table 7.2, the salient features of the proposed CP antenna are listed as follows:

- As compared to all the reported state-of-the-art technology, the proposed CP antenna is compact in size (occupies the smallest area), unlike in Refs. [2–12] and [14–19].
- Unlike Refs. [4, 5, and 17], it does not require any additional ground plane, reactive component, or a three-dimensional (3D) structure for the excitation of bands.
- It does not use expensive substrate as compared to Refs. [4, 6–8].
- The proposed CP antenna is very compact and has very wide ZBW and ARBW, unlike Refs. [9–11].
- The 3 dB ARBW overlaps with the 10 dB ZBW, meaning that the operating CP bandwidth is very wide and comparable with the excited ZBW, which is a unique feature. In comparison, the 3 dB ARBW in Refs. [2–17] is much narrower than the 10 dB ZBW.

7.2.7 Concluding Remarks

A novel CP antenna for 5 GHz WLAN and V2X communication with CP radiation has been studied successfully. The proposed CP antenna has a simple geometry and a very small designed footprint of 30×30 mm^2. It is also easy to fabricate, less costly, and can be easily integrated into any wireless device with restricted space. The proposed CP antenna exhibits a measured 3 dB ARBW of (4.8–5.99 GHz) with overlapping measured 10 dB ZBW. The antenna also exhibits broadside radiation patterns, gain above 2 dBi, and efficiency greater than 85% throughout the operating band. This confirms the applicability of the proposed CP antenna design for 5 GHz WLAN and V2X communication.

7.3 CASE STUDY: BROADBAND AND COMPACT CP MIMO ANTENNA WITH CONCENTRIC RINGS AND OVAL SLOTS FOR 5G APPLICATIONS

The authors in Ref. [22] have investigated a broadband and compact size CP two-port MIMO antenna with a footprint of 25 mm $\times$ 20 mm. The designed oval-shaped MIMO antenna employs concentric rings with oval slots (CROS), along with a circular radiator and two open-ended parallel protruded stubs. The CP radiation is achieved by embedding three oval slots, in which two of them are deployed at the left and right side of the concentric rings and the third is deployed at the top section of the concentric rings. The measured 10 dB impedance bandwidth of the proposed MIMO antenna was 46.30% (3.12–5.00 GHz), and its corresponding 3 dB ARBW

TABLE 7.2
Performance Comparison of the Proposed CP Antenna

Ref	Size (mm²)	Thickness (mm)	ZBW (GHz)	ARBW (GHz)	Substrate	Gain	Antenna Design
[2]	63 × 58.4	1.5	1.5–3.3	1.98–3.02	FR-4	2.5	Chifre-shaped monopole along with asymmetric fed
[3]	60 × 50	0.8	2.21–2.86, 5.05–6.54	2.24–2.56, 5.01–5.94	FR-4	2.33	Annular-slot antenna loaded with a lightning-shaped slot
[4]	120 × 70	20.5	2.20–4.18	2.38–2.5	Taconic RF-35	4.11	Y-shaped DR antenna
[5]	130 × 85	6.6	3.4–4.8	3.48–3.92, 4.54–4.77	FR-4	7	Modified Ψ Shape
[6]	50.3 × 51.5	0.5	5.4–7.0	5.67–5.97	Teflon	7.6	Annular Ring with U-shaped slot antenna
[7]	58.2 × 47.7	0.254	5.0–7.0	4.9–5.7	RT-Duroid	7	Perturbed Ψ-shaped antenna
[8]	39 × 46	1.6	3.43–4.64, 5.08–9.00	3.26–3.91, 6.33–8.49	TLY-5	2	CPW-fed fractal-inspired wide slot antenna
[9]	38 × 42	1.6	2.4–2.484	2.43–2.46	FR-4	–	Circular disc sector patch is truncated to obtain fan shape
[10]	45 × 45	3.18	1.55–1.58	1.572–1.578	FR-4	1.7	Etched fractal defected ground
[11]	40 × 54	1.6	2.40–2.48, 5.72–5.87	2.40–2.49, 5.58–5.93	FR-4	3	Truncated square aperture slot
[13]	21.8 × 20	0.74	4.65–6.72	4.85–5.37	FR-4	3	Square slot with modified corner at the left-bottom of the square slot and two orthogonal equal size patches
[15]	32 × 32	1.6	3.48–5.86	4.71–5.54	FR-4	5	Inverted L-shape antenna with slits, notch, square, strips, and stubs on the ground plane
[16]	45 × 45	1.0	1.82–2.92, 3.15–4.75	1.88–2.42, 3.20–4.83	FR-4	4.8	Upper L-shaped radiator and two circular strips on the ground
[17]	60 × 60	9.66	5.46–6.76, 8.18–10.48	5.7–6.25, 8.55–10	FR-4	7	Metallic reflector and RF pin diode
[18]	66 × 10	1	3.28–3.70, 6.42–6.75, 13.75–14.2	3.28–3.70, 6.42–6.75, 13.75–14.2	FR-4	2.68	Frequency tunable dual folded inverted-L antenna with varactor loaded split-ring resonator structures
[19]	42 × 44	0.8	2.19–4.6	2.19–4.6	FR-4	–	Z-shaped feedline and two right-angled V-slots
Reported	30 × 30	0.8	4.80–5.99	4.80–5.99	FR-4	2.5	Rotated L-shaped monopole and J-shaped defected ground structure

was 41.34% (3.30–5.02 GHz). Furthermore, very wide beamwidths of $137° \pm 0.2$ and $154° \pm 0.2$ were measured in the right-hand circular polarization (RHCP) and left-hand circular polarization (LHCP) radiation patterns, respectively. The minimum achieved isolation between the antenna elements is 18.50 dB without using any additional decoupling structure, and the calculated ECC less than 0.03.

7.3.1 EXISTING CP MIMO ANTENNAS

CP antennas are vital to wireless communication, satellite communication, WLAN, and 5G communication systems, as it offers several advantages compared to linearly polarized antennas [23–26], such as preventing multipath interference and fading. Therefore, in view of implementing the concept of "Work from Home" and "Learn from Home," the number of wireless devices becoming connected is increasing day by day. To fulfill the demands of wireless users and to achieve quality of Service (QoS) and smooth wireless communication, CP antennas play very important roles in upcoming wireless devices. Apart from this, the CP antennas also helps in overcoming the orientation problem between transmitter and receiver, the multipath fading problem, and, at the same time, provide higher performance, better mobility, suppressed multipath interference, and better weather penetration as compared to LP antennas [27]. Furthermore, MIMO antennas are desired as they offer larger capacity with higher data rates, low latency, faster mobility, high data throughput, and reliable communication [28, 29]. Therefore, MIMO antenna designs with CP senses have been reported in recent years [30–38].

Stacked patch MIMO antennas have the features of wideband/multiband and performance enhancement. Therefore, the work in Refs. [30–32] has applied the stacked patch technique to the CP MIMO antennas. In Ref. [30], corner-truncated square slot and patch configuration are applied to the MIMO antenna to yield CP radiations in 2.45 GHz and 5.8 GHz bands. In Ref. [31], the CP is achieved by protruding a stub from the right end of the ground plane, whereas the method applied by Ref. [32] is to chamfer the two opposite corners of the ring slot as well as the resonant patch. However, the MIMO antennas in Ref. [30, 32] have exhibited larger dimensions, and the antennas in Ref. [30, 31] have used expensive substrates. Notably, applying the stacked patch technique will inevitably increase the overall profile of the designated antenna, which makes it difficult to integrate into a slim wireless device. Even though Ref. [33] has introduced a compact CP MIMO antenna designed using two truncated corner square patches with parasitic periodic metallic plates [33], it requires expensive Taconic substrate (same as [30, 31]) that results in higher manufacturing expenses.

To minimize the manufacturing expenses, the works reported in Ref. [34–38] have introduced MIMO antenna designs using low-cost FR-4 substrate. In Ref. [34], an eyebrow-shaped strip is applied to enhance the CP performance of the MIMO antenna, but it has a very large dimension of 85 mm × 73 mm. To reduce the overall MIMO antenna size to 21 mm × 46 mm, the split-ring resonator was introduced to the MIMO antenna and good CP radiation is achieved by using an offset feeding technique and defective ground structure (DGS) [13]. To further decrease the dimensions to 27 mm × 27 mm, the MIMO antenna design in Ref. [14] employs two orthogonally designed T-shaped patches as well as three L-shaped parasitic patches in the ground plane to yield good CP radiation. However, Refs. [37] and [38] have shown poor isolation between their respective adjacent antenna elements. To increase the isolation between adjacent antenna elements, Ref. [37] has applied the spatial diversity method (adjacent elements separated by a distance of 13.75 mm), and good CP radiation is achieved by applying 90° phased slots at the center of the truncated patch. As for Ref. [38], the designed microstrip-fed MIMO antenna is composed of a patch containing two L-shaped radiators along with a wide hexagonal slot radiator on the ground plane, and good 3 dB ARBW of 34.38% and isolation of greater than 17 dB is achieved.

In this chapter, a very low-profile ($0.27 \lambda \times 0.22 \lambda \times 0.01 \lambda$) (where λ is calculated at the lowest frequency of 3.30 GHz), oval-shaped CP two-port MIMO antenna for 5G applications is presented. Each CP antenna element is fed by a 50 Ω microstrip line and employs CROS, along with a circular radiator and two open-ended parallel protruded stubs. The oval-shaped slots loaded on the left as well as right side of the concentric rings are for achieving CP radiation, whereas the one that is loaded at the top section is to enhance the ARBW. The proposed CP MIMO antenna can yield RHCP and LHCP radiation when Ant. 1 and Ant. 2 are excited, respectively. Furthermore, wide 10 dB impedance bandwidth of 46.3% and 3 dB ARBW of 41.34% are also achieved. Notably, the separation distance (center-to-center) of the two antenna elements is only 0.16λ and isolation of up to 18.50 dB can be achieved without applying any additional decoupling structure.

7.3.2 GEOMETRY AND DESIGN ANALYSIS OF PROPOSED MIMO ANTENNA

The design layout and fabricated prototype of the proposed CP MIMO/diversity antenna are illustrated in Figure 7.9a and b, respectively and parameters with their optimized values are articulated in Table 7.3. The CP MIMO antenna is composed

TABLE 7.3
Dimensions of Proposed MIMO Antenna

Parameter	Value (mm)
L	20
L_1	2
L_2	6
L_3	0.5
L_4	0.5
L_5	7
W	25
W_1	2.4
W_2	0.4
W_3	15
W_4	0.7
r_1	2.5
r_{2r}	3.5
r_3	4.5

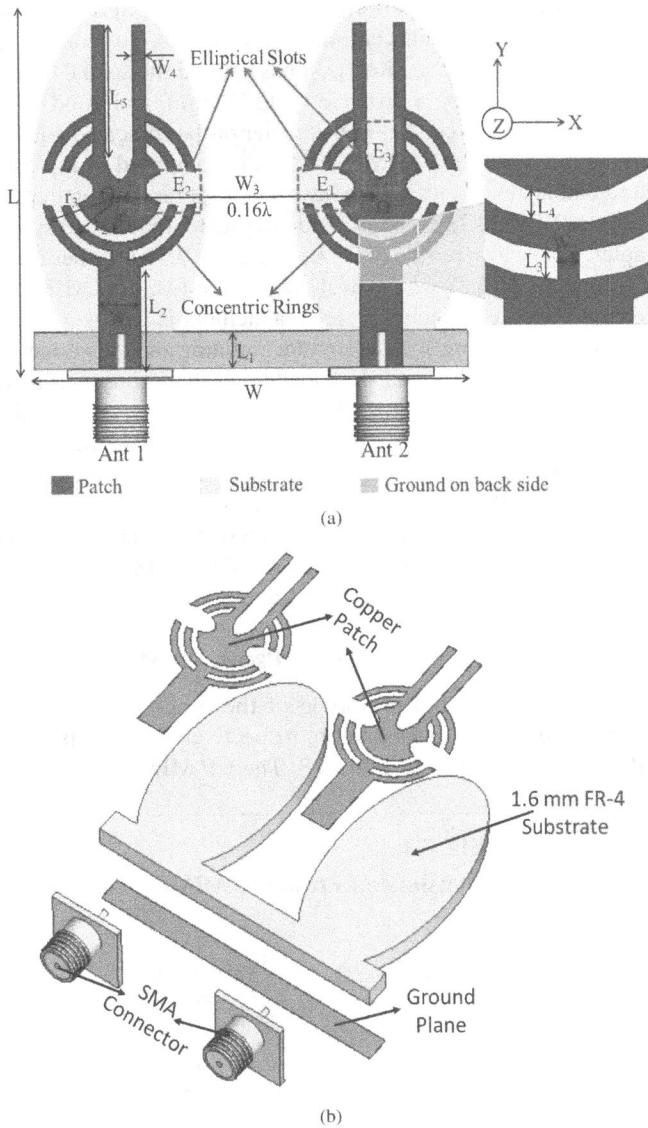

FIGURE 7.9 The geometry of the proposed antenna: (a) front view and (b) breakdown view.

of two identical antenna elements, namely, Ant. 1 and Ant. 2 (each has a partial ground of size W × L$_1$), which are printed on an economical FR-4 substrate (ε_r = 4.3, δ = 0.025) measuring 25 mm × 20 mm × 1.6 mm. As Ant. 1 is the mirror image of Ant. 2, they can support polarization diversity, meaning that Ant. 1 and Ant. 2 radiate RHCP and LHCP waves, respectively. Furthermore, to achieve good isolation, the center-to-center spacing between the two antenna elements (including a partial air gap) is set to be 0.16 λ.

In the proposed structure, each antenna element (Ant. 1 and Ant. 2) is composed of two concentric rings (radius r_2 and r_3) and one circular patch (radius r_1), which shares the same center location "O." Here, two elliptical slots having a major and minor axis of 2 mm and 1 mm, respectively, are loaded on the left and right side of the antenna element, and an elliptical slot having a major and minor axis of 3 mm and 1 mm, respectively, is loaded on the top section of the antenna element between two open-ended protruded stubs (each with a dimension of $L_5 \times W_4$). The two antenna elements shared the same partial ground plane of dimension $L_1 \times W$, which is designed on the back of the substrate. The proposed CP MIMO antenna is simulated (including the SMA connectors) using the 3D EM simulator CST MWS®.

7.3.2.1 Step-Wise Design of Single Antenna Element

The step-wise design of the single antenna element is presented in Figure 7.10. Here, five major steps are analyzed to comprehend the design mechanism, and their corresponding 10 dB impedance bandwidth and 3 dB ARBW are illustrated in Figure 7.11a and b, respectively.

As depicted in Figure 7.10, the proposed antenna element design is stemmed from (Step 1) a circular ring-slot monopole antenna that is loaded by a concentric ring at center location "O" with radius r_2 and r_3. Notably, this concentric ring is turned into a "split-ring" type by introducing a small, shorted section of width 0.5 mm, and resonance at 4.10 GHz is induced with a broad bandwidth of 23.5% (3.60–4.60 GHz), as visualized in Figure 7.11a. To further increase the bandwidth, in Step 2, a circular radiator of radius r_1 is loaded into the center ring-slot position, and by further observing Figure 7.11a, even though this circular radiator can improve the bandwidth to 38.5% (4.00–5.55 GHz), the resonance mode is shifted toward the higher frequency spectrum (at 4.6 GHz). The broadband behavior in Step 2 is obtained due to the unpunctured and longer path of currents flowing through the circular radiator.

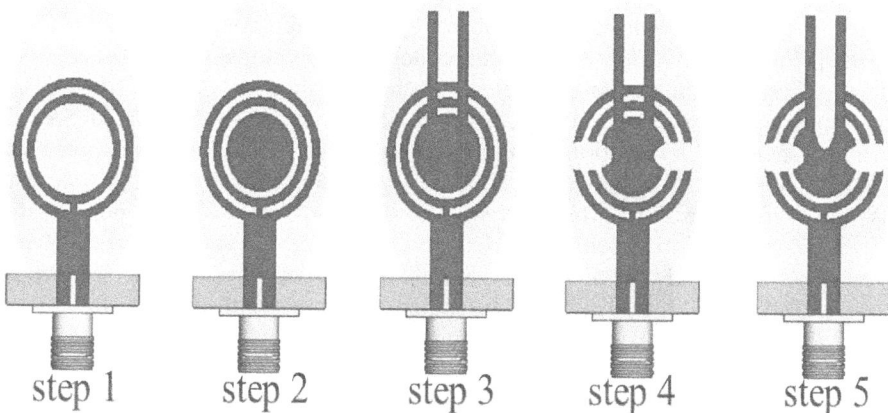

FIGURE 7.10 Step-wise design of proposed antenna element.

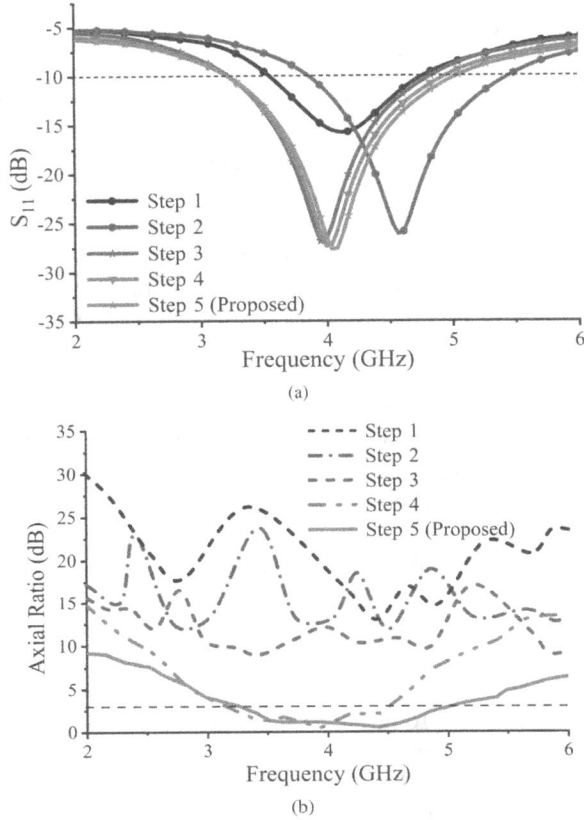

FIGURE 7.11   The (a) S_{11} (dB) and (b) axial ratio (dB).

To shift the resonance back to approximately 4 GHz, as well as further improve the operational bandwidth to occupy the complete 5G new radio (NR) band n77/n78/n79, in Step 3, two open-ended parallel stubs are protruded from the top section of the antenna element, and a broad bandwidth of (3.18–4.82 GHz) is achieved. However, from Figure 7.11b, it is noted that Step 1 to Step 3 can only exhibit LP radiations as their corresponding AR values are larger than 10 dB. Therefore, to achieve CP radiation across the desired bands of interest, as shown in Figure 7.10 (Step 4), two elliptical slots are loaded to perturb the surface current distribution on the left and right side of the radiator. As illustrated in Figure 7.11, the antenna in Step 4 has shown a broad 3 dB ARBW of 3.20–4.50 GHz with desirable 10 dB impedance bandwidth of 3.18–4.90 GHz. Nevertheless, the ARBW of this antenna (Step 4) can only cover the 5G NR band n77/n78 (3.30–4.20 GHz). Therefore, to further increase the ARBW, a top-loaded elliptical slot that acts as a perturbation element is introduced, and the antenna is now denoted as Step 5 (or proposed antenna element) in Figure 7.10. As shown in Figure 7.11, the proposed antenna element can yield a wide 3 dB

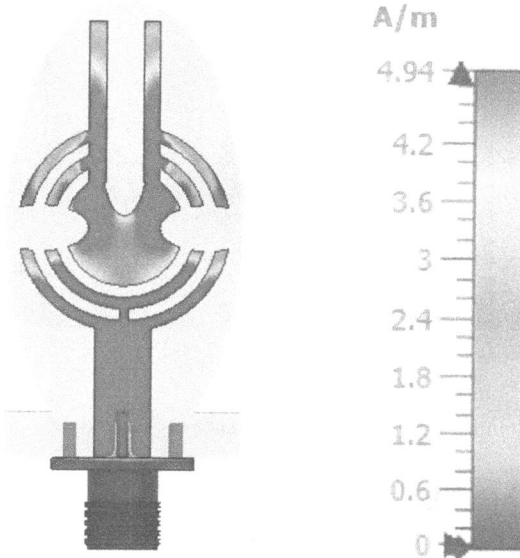

FIGURE 7.12 Surface current distribution of the proposed single antenna element.

ARBW of 42.42% (3.25–5.00 GHz) with desirable 10 dB impedance bandwidth of 44.50% (3.18–5.00 GHz), and it can fully cover the entire 5G NR band n77/n78/n79 (3.30–5.00 GHz).

7.3.2.2 Surface Current Distribution of Proposed Antenna

To analyze the working of the proposed antenna, its corresponding simulated surface current (A/m) distribution is analyzed in Figure 7.12. From Figure 7.12, it is observed that at 4 GHz, a maximum current is flowing through the circular radiator and at the same time, an equal amount of current flowing through the concentric rings, which proves that it helps in generating the resonance as well as widening the bandwidth at 4 GHz. Further, equal amount of current is also observed in the two open-ended parallel stub, which slightly helps in bandwidth enhancement and tuning the resonance in frequency range of (3.18–5.00 GHz).

7.3.2.3 Amplitude Ratio and Phase Difference of Proposed Antenna

To generate a good CP radiation for an antenna, the horizontal electric field (E_x) and the vertical electric field (E_y) must have an equal amplitude of approximately 1 dB (or 0 dB) with a phase difference (PD) of 90° throughout the bands of interest. In this regard, to fully comprehend the CP excitation of the proposed antenna element (Step 5), Figure 7.13 depicts its corresponding calculated amplitude ratio (E_x/E_y) and PD of the two orthogonal electric field components. Here, the amplitude ratio is closer to 0 dB with PD of near 90° between them. Therefore, the proposed antenna element is a good CP antenna that can be further applied as a MIMO configuration.

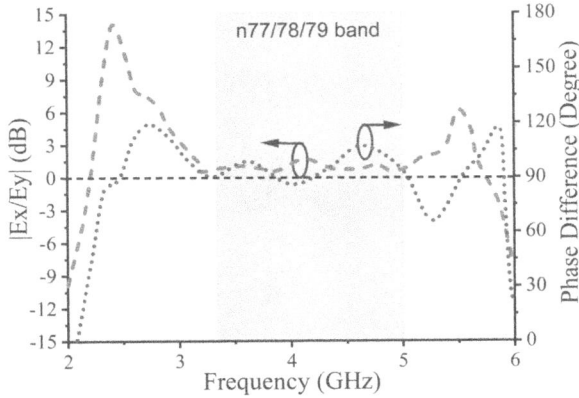

FIGURE 7.13 Amplitude ratio and phase difference of the proposed antenna.

7.3.2.4 MIMO Analysis

The simulated S_{11}, S_{22} parameters for the two-port MIMO antenna geometry shown in Figure 7.9a is obtained by activating Ant. 1 and terminating Ant. 2 with 50 Ω impedance load and vice versa, whereas the transmitting coefficient S_{12}, S_{21} are obtained by activating both the Ant. 1 and Ant. 2, simultaneously. The curves of S_{22} and S_{21} are not shown for brevity. The simulated S_{11} in Figure 7.14 illustrates that the bandwidth remains almost the same when the antenna is transformed from single antenna to two antenna elements. The bandwidth obtained for the two-port MIMO antenna array is 44.50% (3.18–5.00 GHz). Similarly, from the transmission coefficient curve S_{12} shown in Figure 7.14, the isolation between Ant. 1 and Ant. 2 is larger than 18.5 dB throughout the operating band.

To validate the isolation between Ant. 1 and Ant. 2, the electric field intensity (V/m) distribution on the surface of the two-port MIMO antenna array is illustrated in

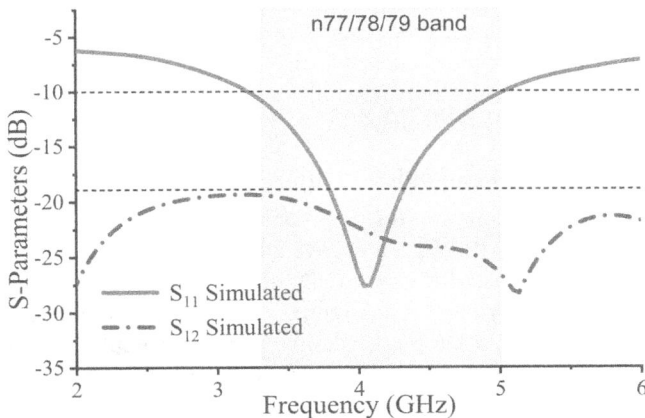

FIGURE 7.14 Amplitude ratio and phase difference of the proposed antenna.

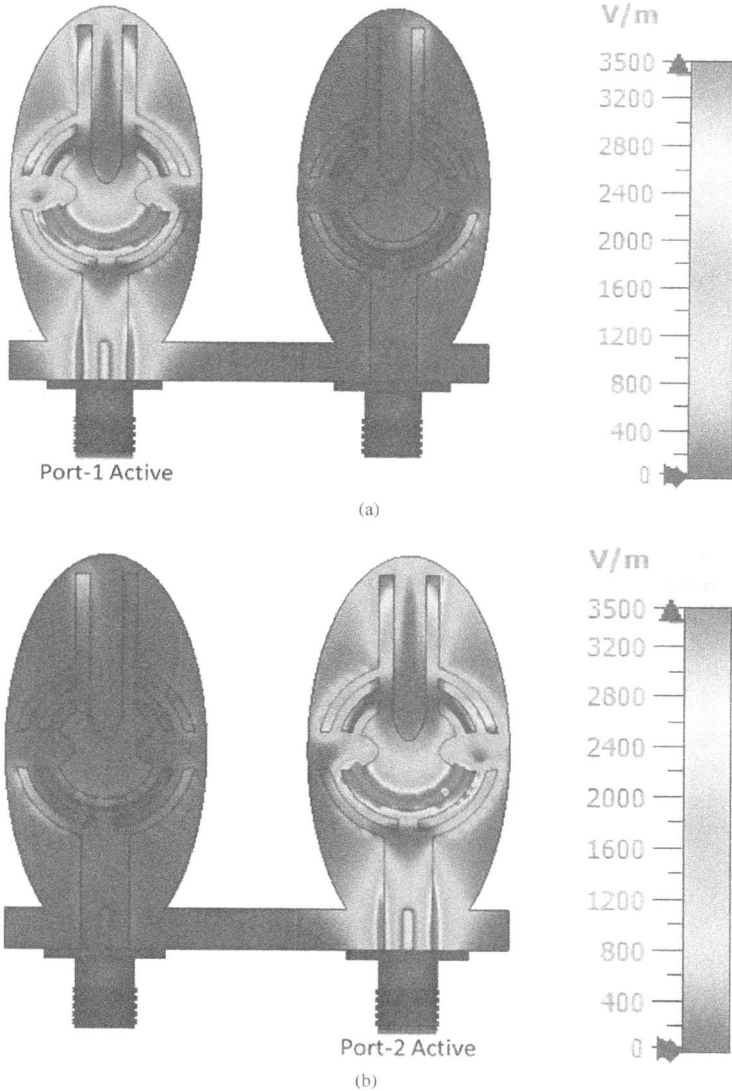

FIGURE 7.15 Electric field intensity of the proposed MIMO antenna (a) when Ant. 1 is excited and (b) when Ant. 2 is excited.

Figures 7.15a and b, respectively. For analyzing the effect, Ant. 1 is excited, while Ant. 2 is kept terminated with 50 Ω matched impedance load. Under this scenario, from Figure 7.15a, it is evident that the deployment of antennas at 0.16 λ center-to-center distance protects the current flowing out from Ant. 1 and also suppresses the propagation of surface wave through the ground plane without affecting the impedance matching and radiation performance. Notably, the same is also verified when Ant. 2 is excited while the Ant. 1 is kept properly terminated with 50 Ω matched load.

7.3.3 Parametric Analysis

To validate the geometry of the proposed MIMO antenna, the RF performance of the proposed MIMO antenna in terms of S-parameters is analyzed. The parametric analysis is carried out by varying antenna parameters like radius r_1, open-ended stubs L_5 and center-to-center distance w_3 between the antenna elements, keeping all the other parameters constant.

7.3.3.1   Effect of Varying Radius r_1

The inner solid circular radius r_1 is varied from 2 mm to 2.75 mm as shown in Figure 7.16. It is analyzed that the -10 dB impedance bandwidth increases with an increase in radius up to 2.5 mm. However, a further increase in the radius results in bandwidth reduction (as seen from the green line curve) which may be due to the strong coupling between solid radius r_1 and concentric rings when the radius is increased beyond 2.5 mm. Therefore, the optimum value of radius r_1 is selected as 2.5 mm, which covers the desired band of operation.

7.3.3.2   Effect of Varying Open-Ended Stubs L_5

Figure 7.17 illustrates the variation of open-ended stubs L_5 from 6 mm to 7.5 mm. It can be observed that increasing the length of the stubs shifts the resonance toward a lower frequency, whereas a small effect is observed on the bandwidth. The value of L_5 is considered as 7 mm, considering the best RF performance.

7.3.3.3   Effect of Varying Center-to-Center Distance W_3

The effect of varying center-to-center distance W_3 from 11 mm to 17 mm on S_{11} and S_{12} is analyzed in Figure 7.18. The curves for S_{11} remain almost the same for all the values of W_3, whereas better S_{12} curves are observed when the values of W_3 are increased. For W_3 values of 11 mm and 13 mm, the isolation is less than 17 dB,

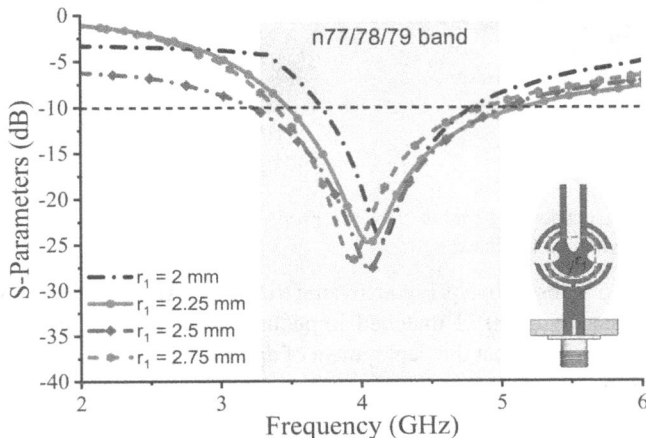

FIGURE 7.16   Parametric variation of the inner circular radius r_1.

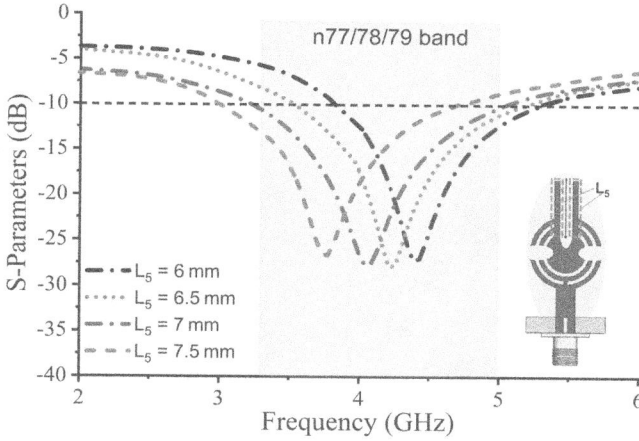

FIGURE 7.17   Parametric variation of open-ended stubs L_5.

whereas for W_3 of 15 mm and 17 mm, the isolation is greater than 18.5 dB. Even though the isolation value is greater at 17 mm, to restrict the area of the proposed MIMO antenna, the optimum value of 15 mm is considered.

7.3.4 RESULTS AND DISCUSSION

The proposed two-port CP MIMO antenna was fabricated, as illustrated in Figure 7.19, and its typical results, such as reflection coefficients (S_{11} and S_{22}), transmission coefficients (S_{12} and S_{21}), AR, 2D and 3D radiation patterns, gain, and efficiency, were measured and compared with the simulated results.

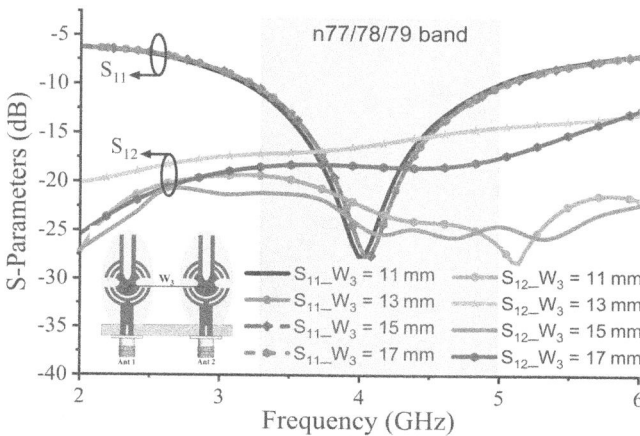

FIGURE 7.18   Amplitude ratio and phase difference of the proposed antenna.

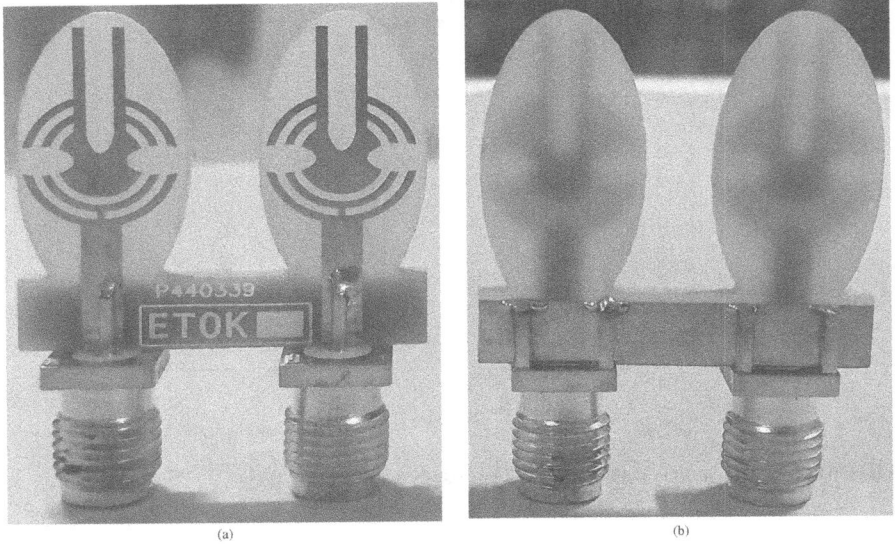

FIGURE 7.19   Fabricated prototype of the proposed MIMO antenna: (a) front view and (b) back view.

7.3.4.1   Measured and Simulated S-Parameters

The measured and simulated S-parameters (S_{11} and S_{12}) are plotted in Figure 7.20. While measuring S_{11} and S_{22}, Ant. 1 was activated, whereas Ant. 2 was terminated with 50 Ω impedance and vice versa. While measuring S_{12} and S_{21}, both Ant. 1 and Ant. 2 were activated simultaneously. Due to analogy, S_{22} and S_{12} are not shown for brevity. Here, the measured 10 dB impedance bandwidth was approximately 46.30% (3.12–5.00 GHz) with a resonance frequency of 4 GHz, and the measured isolation level between antenna elements was larger than 18.5 dB across the bands of interest.

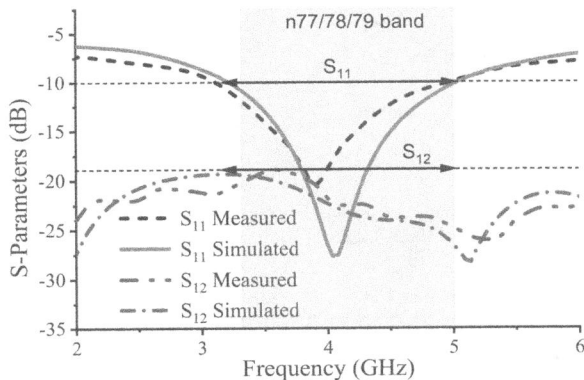

FIGURE 7.20   Measured and simulated S-parameters of proposed antenna.

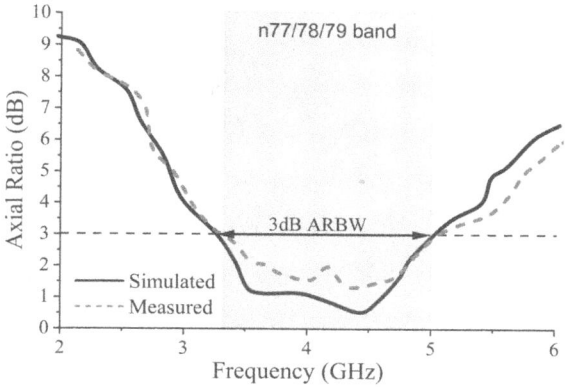

FIGURE 7.21   Measured and simulated axial ratio of the proposed antenna.

7.3.4.2   Measured and Simulated AR

The measured and simulated AR diagram of the proposed CP MIMO antenna (Ant. 1) is plotted in Figure 7.21, and a very wide 3 dB ARBW of 41.34% (3.30–5.02 GHz) was measured. By comparing Figures 7.20 and 7.21, the measured 3 dB ARBW has overlapped with the impedance bandwidth across the desired bands of interest. The simulated and measured results are well validated, and the slight deviation could be due to soldering and fabrication tolerances.

7.3.4.3   Vector Current Distribution of Proposed Antenna

The simulated vector current distributions of the proposed CP MIMO antenna at 4 GHz (with different phases 0°, 90°, 180°, and 270°) are studied in Figure 7.22a–d, when both the antenna elements are excited simultaneously. In Figure 7.22a, at 0°

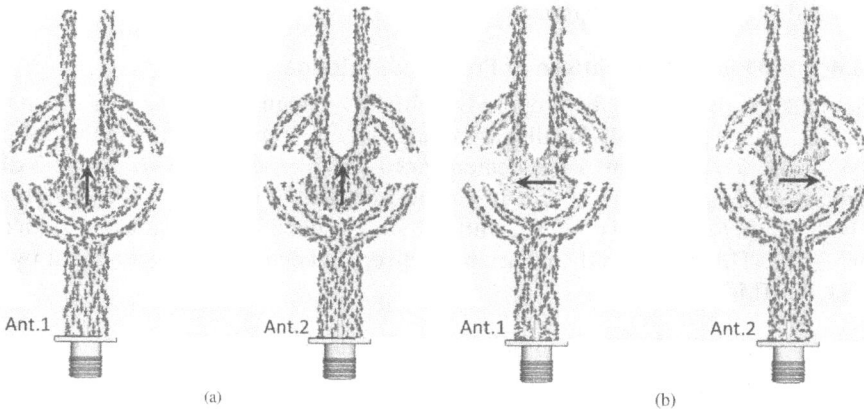

FIGURE 7.22   Vector current distributions of Ant. 1 and Ant. 2 at (a) 0°, (b) 90°, (c) 180°, and (d) 270°. (*Continued*)

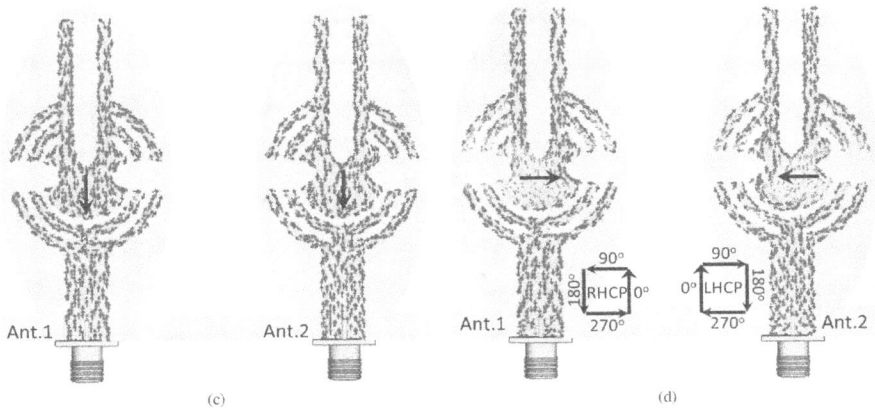

FIGURE 7.22   (*Continued*)

phase, the maximum current flows in the +y of both Ant. 1 and Ant. 2, whereas at 180° phase, the current at both Ant. 1 and Ant. 2 are flowing in the -y direction, as illustrated in Figure 7.22c. At 90° and 270° phases (see Figure 7.22b and d), the vector current distributions at Ant. 1 and Ant. 2 are the same in magnitude but flow in opposite phases. Thus, the current vectors are changing with time and rotate in an counterclockwise manner for Ant. 1 (exciting RHCP wave), whereas the current vectors at Ant. 2 are rotating in a clockwise direction (exciting LHCP wave).

7.3.4.4   Radiation Characteristic Measurement Setup of Proposed Antenna

Figure 7.23a depicts the block diagram, whereas Figure 7.23b illustrates the actual setup of radiation characteristic measurement of the proposed MIMO antenna inside the anechoic chamber. This setup is used to measure 2D and 3D radiation patterns, gain, and MIMO diversity parameters.

7.3.4.5   2D Radiation Pattern of Proposed Antenna

The radiation patterns of the proposed CP MIMO antenna at 4 GHz (when Ant. 1 and Ant. 2 are excited individually) are visualized in Figure 7.24. In both the planes ($\phi = 0°$ and $\phi = 90°$), Ant. 1 has demonstrated RHCP operation with a wide 3 dB angular beamwidth of 137° ± 0.2. In comparison, Ant. 2 exhibits LHCP operation in both planes ($\phi = 0°$ and $\phi = 90°$), with a slightly wider 3 dB angular beamwidth of 154° ± 0.2. This validates the polarization diversity behavior of the proposed two-port CP MIMO antenna.

7.3.4.6   3D Radiation Pattern of Proposed Antenna

The simulated 3D radiation patterns at 4 GHz are illustrated in Figure 7.25. It is visualized that 3D radiation patterns of Ant. 1 and Ant. 2 are oblique dipole patterns that form mirror images to each other, which further validates that the proposed MIMO antenna has good radiation performance.

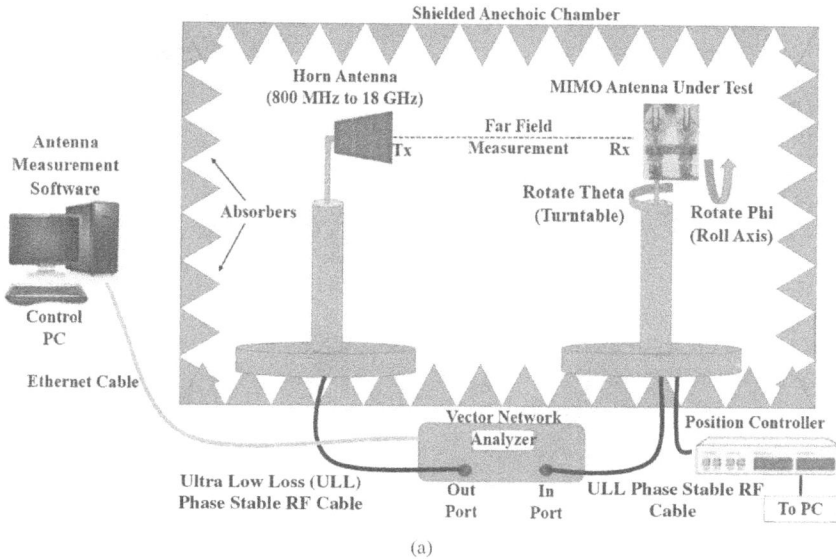

(a)

(b)

FIGURE 7.23 Radiation characteristic measurement setup: (a) block diagram and (b) actual setup in an anechoic chamber.

7.3.4.7   Gain and Efficiency of Proposed Antenna

The measured and simulated gain and efficiency of the proposed CP MIMO antenna are shown in Figure 7.26. The measured gain and efficiency were approximately 2.0–2.5 dBi and 70–75%, respectively, with minimal deviation across 3.12–5.00 GHz.

7.3.5   Mimo Diversity Parameters

7.3.5.1   ECC of Proposed Antenna

The ECC is used to verify the performance of the proposed CP MIMO antenna for diversity applications. It is calculated from far-field patterns using formula

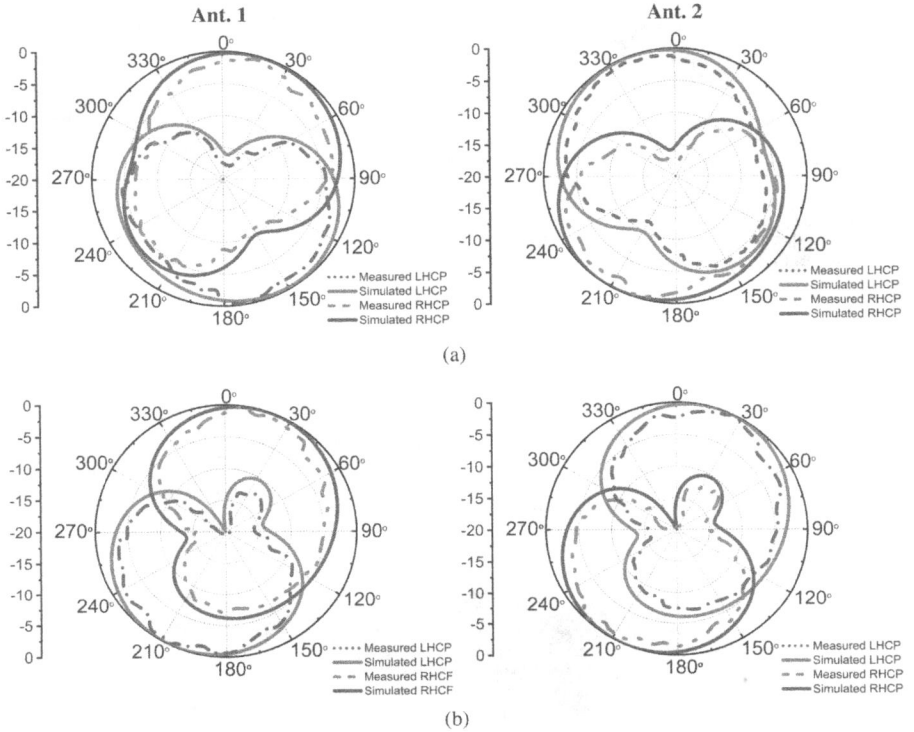

FIGURE 7.24 Two-dimensional radiation pattern of the proposed antenna: (a) E-plane and (b) H-plane.

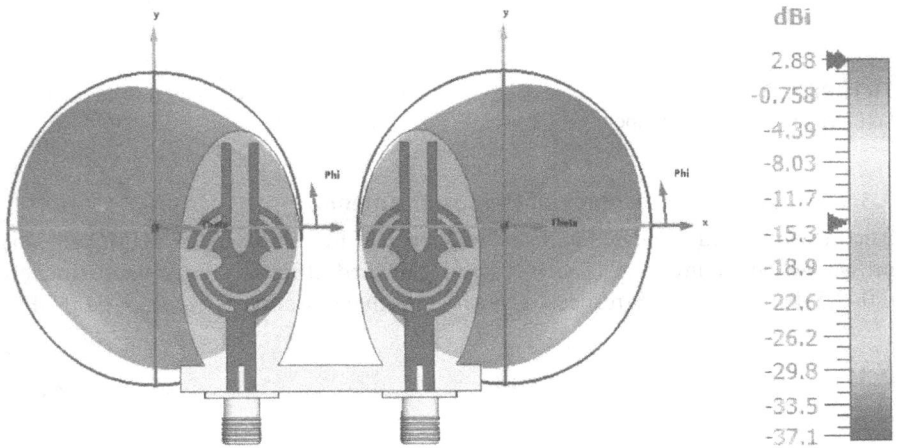

FIGURE 7.25 Three-dimensional radiation pattern of proposed antenna.

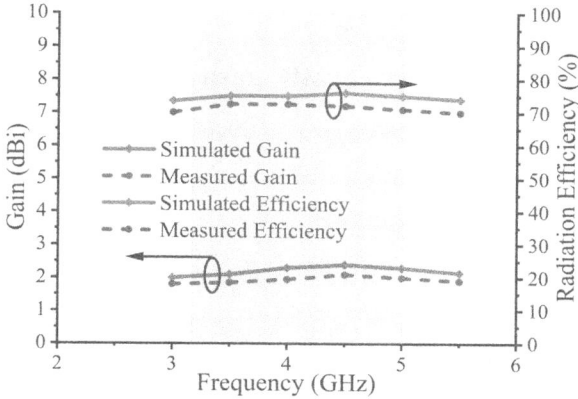

FIGURE 7.26   Gain and efficiency of proposed antenna.

mentioned in Ref. [39]. As shown in Figure 7.27, the ECC values are less than 0.03 across the operating bands, which confirms that the designed CP MIMO antenna has good isolation and offers the best performance under the influence of multipath fading environments.

7.3.5.2   Mean Effective Gain of Proposed Antenna

Mean effective gain (MEG) is a crucial parameter for the diversity performance analysis of any MIMO system. Therefore, the MEG is defined as the ratio of power received by the MIMO antenna to the power received by isotropic antenna. In a MIMO system, the MEG is calculated using Equations (7.1–7.3).

$$MEG_i = 0.5\left[1 - \sum_{j=1}^{M}\left|S_{ij}\right|^2\right] \qquad (7.1)$$

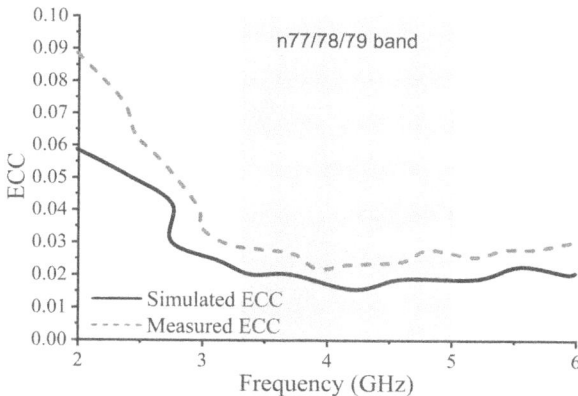

FIGURE 7.27   ECC of the proposed antenna.

TABLE 7.4
MEG Values of Proposed MIMO Antenna

Frequency (GHz)	MEG-1	MEG-2	MEG-1/MEG-2
3.5	-3.82	-3.83	0.997
4	-3.82	-3.85	0.992
4.5	-3.83	-3.84	0.997
5	-3.85	-3.82	1.007

Where, M is the number of antenna elements and "i" is the port number. By expanding the above equation, the MEG-1 for port-1 is found as:

$$MEG_1 = 0.5\left[1 - |S_{11}|^2 - |S_{12}|^2\right] \tag{7.2}$$

and the MEG-2 for port-2 is found as:

$$MEG_2 = 0.5\left[1 - |S_{21}|^2 - |S_{22}|^2\right] \tag{7.3}$$

For better diversity metrics, the MEG-1 and MEG-2 values, as per industry standard, should be -3 $\leq$ MEG (dB) $\leq$ -12, whereas the ratio of MEG-1/MEG-2 should be approximately equal to 1. From Table 7.4, it is validated that the MEG values as well as the ratio are within the well-defined limit.

7.3.5.3 Channel Capacity of Proposed Antenna

The ergodic channel capacity of the proposed MIMO antenna across the operating bands is analyzed in Figure 7.28 using Equation (7.4).

$$C = k\left\{log_2 \det\left[I + \eta\frac{SNR}{k}HH^*\right]\right\} \tag{7.4}$$

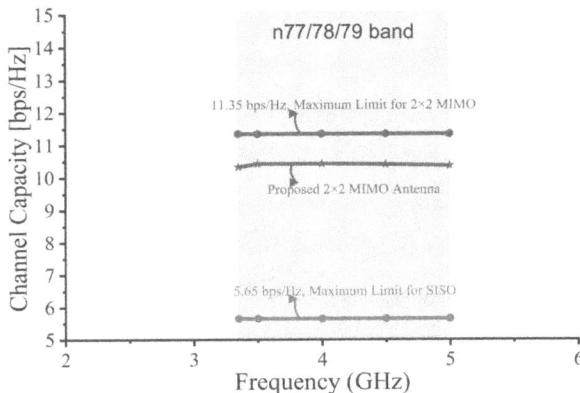

FIGURE 7.28 Channel capacity of the proposed antenna.

TABLE 7.5
Channel Capacity Values of
Proposed MIMO Antenna

Frequency (GHz)	Channel Capacity (bps/Hz)
3.5	10.43
4	10.43
4.5	10.41
5	10.37

where in Equation (7.4), k is the number of antenna elements, SNR defines the mean signal to noise ratio, [I] denotes an identity matrix, η indicates the efficiency, [H] is the normalized channel matrix considered as frequency independent over the operating bands, and H^* denotes the transpose conjugate matrix of H.

The MIMO system channel model should be selected first while calculating the channel capacity — the model for ray tracing or the associated statistical model is mostly used. Based on the correlation matrix method, the MIMO antenna's channel capacity is calculated. The calculated channel capacity of the proposed array is indicated in Table 7.5 as well as in Figure 7.28, which is between 10 bps/Hz and 10.5 bps/Hz, considering simulated efficiencies (η) and averaging 10,000 Rayleigh fading realizations with 20 dB SNR in the identically and independently distributed propagation condition [29]. For maximum channel capacity, all channels have no correlation, and fading matrix [H][H*] is converted into an identity matrix.

The channel capacity of the proposed two-port MIMO antenna is above 10.3 bps/Hz throughout the operating band, which is about 1.82 times greater than the maximum limit of an ideal single antenna (about 5.65 bps/Hz). Compared with the maximum limit for an ideal two-port MIMO antenna, the proposed MIMO antenna have exhibited good channel capacities.

7.3.5.4   Total Active Reflection Coefficient of Proposed Antenna

The total active reflection coefficient (TARC) between Ant. 1 and Ant. 2 is calculated using Equation (7.5).

$$\Gamma = \frac{\sqrt{\left(\left|S_{ii} + S_{ij}e^{j\theta}\right|^2\right) + \left(\left|S_{ji} + S_{jj}e^{j\theta}\right|^2\right)}}{\sqrt{2}} \tag{7.5}$$

where θ is the input phase angle varied from 0° to 120° at an interval of 30° and S_{ii} and S_{ij} are the reflection coefficients of the port i and port j, respectively. Figure 7.29 depicts the TARC values for antenna elements, where it is visualized that under the variation of phase angle, the performances of the proposed MIMO antenna remain unaltered in the scattering environment.

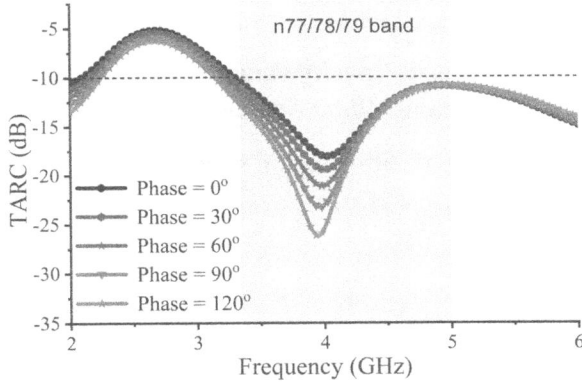

FIGURE 7.29   Total active reflection coefficient of proposed antenna.

7.3.5.5   Multiplexing Efficiency of Proposed Antenna

Multiplexing efficiency is the signal-to-noise ratio between the imperfect MIMO antenna and the ideal antenna and can be calculated using Formula (7.6) [40]:

$$\eta_{mux} = \sqrt{\eta_i \eta_j \left(1 - |\rho_c|\right)} \tag{7.6}$$

As illustrated in Figure 7.30, both the simulated as well as calculated multiplexing efficiency is higher than −3 dB throughout the operating band, which satisfies the requirements of MIMO wireless communication systems.

7.3.6   PERFORMANCE COMPARISON OF PROPOSED ANTENNA WITH EXISTING STATE-OF-THE-ART TECHNOLOGY

The performance comparison of designed CP MIMO antenna with recent literature is compared in Table 7.6. In Table 7.6, it can be noted that the proposed MIMO

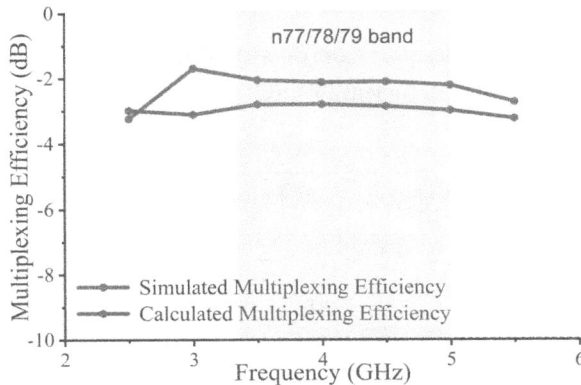

FIGURE 7.30   Simulated and calculated the multiplexing efficiency of the proposed antenna.

TABLE 7.6
Comparison of Proposed Antenna with Recent CP MIMO Antennas

Ref.	Size (mm²)	Bands (GHz)	ARBW (GHz)	Isol. (dB)	Gain (dBi)	ECC
[29]	119 × 119	2.4–2.48 5.72–5.87	2.41–2.49 5.62–5.96	30	10	<0.03
[30]	13.7 × 36.2	5.20–6.30	5.20–6.30	20	6	<0.004
[31]	96 × 96	2.36–2.53	2.30–2.50	25	8	<0.004
[32]	45 × 32	5.10–5.85	5.10–5.85	20	5	<0.05
[33]	85 × 73	3.50–8.20	4.60–7.60	20	10	<0.4
[34]	21 × 46	3.18–3.90 4.94–7.73	6.65–6.92	15	3.4	<0.1
[35]	27 × 27	2.20–6.80	4.00–4.47 5.10–5.22	14	5.5	–
[36]	27.6 × 97	5.49–6.02	5.74–5.83	34	5.34	<0.1
[37]	25 × 25	3.00–11.0	4.00–5.50	17	4.9	<0.15
Reported	25 × 20	3.12–5.00	3.30–5.02	18	2.5	<0.01

antenna has the smallest dimension as well as covers the complete 5G sub-6 GHz band with overlapping 3 dB ARBW in the desired band compared to the reported CP MIMO antennas.

7.3.7 CONCLUDING REMARKS

A compact size dual CP MIMO antenna with broad ARBW has been successfully investigated. The proposed CP MIMO geometry is composed of two antenna elements arranged in a mirrored pattern to each other in which Ant. 1 and Ant. 2 can induce RHCP and LHCP radiation, respectively. The proposed CP MIMO antenna has demonstrated a broad impedance bandwidth of 46.30% (3.12–5.00 GHz) with wide ARBW of 41.34% (3.30–5.02 GHz). Furthermore, peak gain of up to 2.5 dBi and efficiency of greater than 70% were measured. Therefore, the proposed dual CP MIMO antenna is a good candidate for 5G NR band n77/n78/n79 communication applications.

REFERENCES

1. Kulkarni, J., Sim, C.-Y.-D., Poddar, A. K., Rohde, U. L. & Alharbi, A. G. "A compact circularly polarized rotated L-shaped antenna with J-shaped defected ground structure for WLAN and V2X applications." *Progress In Electromagnetics Research Letters*, vol. 102, pp. 135–143, 2022.
2. Han, R. C. & Zhong, S. S. "Broadband circularly-polarised chifre-shaped monopole antenna with asymmetric feed." *Electronics letters*, vol. 52, no. 4, pp. 256–258, 2016.
3. Ge, L., Sim, C. Y. D., Su, H. L., Lu, J. Y. & Ku, C. "Single-layer dual-broadband circularly polarised annular-slot antenna for WLAN applications." *IET Microwaves, Antennas & Propagation*, vol. 12, no. 1, pp. 99–107, 2018.

4. Altaf, A. & Seo, M. "Dual-band circularly polarized dielectric resonator antenna for WLAN and WiMAX applications." *Sensors*, vol. 20, no. 4, p. 1137, 2020.
5. Deshmukh, A. A. & Odhekar, A. A. "Dual band circularly polarized modified Ψ-shape microstrip antenna." *Progress In Electromagnetics Research C*, vol. 115, pp. 161–174, 2021.
6. Rasool, N., Kama, H., Abdul Basit, M. & Abdullah, M. "A low profile high gain ultra lightweight circularly polarized annular ring slot antenna for airborne and airship applications." *IEEE Access*, vol. 7, pp. 155048–155056, 2019.
7. Mondal, T., Maity, S., Ghatak, R. & Chaudhuri, S. R. B. "Design and analysis of a wideband circularly polarised perturbed psi-shaped antenna." *IET Microwaves, Antennas & Propagation*, vol. 12, no. 9, pp. 1582–1586, 2018.
8. Saraswat, K. & Harish, A. R., "A dual band circularly polarized 45° rotated rectangular slot antenna with parasitic patch." *International Journal of Electronics and Communications*, vol. 123, 2020. https://doi.org/10.1016/j.aeue.2020.153260
9. Mathew, S., Anitha, R., Vinesh, P. V. & Vasudevan, K. "Circularly polarized sector-shaped patch antenna for WLAN applications." *Procedia Computer Science*, vol. *46*, pp. 1274–1277, 2015.
10. Wei, K., Li, J. Y., Wang, L., Xu, R. & Xing, Z. J. "A new technique to design circularly polarized microstrip antenna by fractal defected ground structure." *IEEE Transactions on Antennas and Propagation*, vol. 65, no. 7, pp. 3721–3725, 2017.
11. Niyamanon, S., Senathong, R. & Phongcharoenpanich, C. "Dual-frequency circularly polarized truncated square aperture patch antenna with slant strip and L-shaped slot for WLAN applications." *International Journal of Antennas and Propagation*, vol. 2018, pp. 1–11, 2018.
12. Singh, V., Mishra, B. & Singh, R. "Anchor shape gap coupled patch antenna for WiMAX and WLAN applications." *COMPEL - The International Journal for Computation and Mathematics in Electrical and Electronic Engineering*, vol. 38, pp. 263–286, 2019.
13. Midya, M., Bhattacharjee, S. & Mitra, M. "Compact CPW-fed circularly polarized antenna for WLAN application." *Progress In Electromagnetics Research M*, vol. 67, pp. 65–73, 2018.
14. Fu, Q., Feng, Q. & Chen, H. "Design and optimization of CPW-fed broadband circularly polarized antenna for multiple communication systems." *Progress In Electromagnetics Research Letters*, vol. 99, pp. 65–74, 2021.
15. Srivastava, K., Mishra, B. & Singh, R., "Microstrip-line-fed inverted L-shaped circularly polarized antenna for C-band applications." *International Journal of Microwave and Wireless Technologies*, pp. 1–9, 2021. https://doi.org/10.1017/S1759078721000635
16. Yu, Z., Huang, L., Gao, Q. & He, B. "A compact dual-band wideband circularly polarized microstrip antenna for sub-6G application." *Progress In Electromagnetics Research Letters*, vol. 100, pp. 99–107, 2021.
17. Mohanty, A. & Behera, B. R. "Dynamically switched dual-band dual-polarized dual-sense low-profile compact slot circularly polarized antenna assisted with high gain reflector for sub-6GHz and X-band applications." *Progress In Electromagnetics Research C*, vol. 110, pp. 197–212, 2021.
18. Praveen Kumar, P. C. & Rao, P. T. "Frequency-tunable circularly polarized twin dual folded inverted-L antenna with varactor-loaded split-ring resonator structures." *International Journal of Communication Systems*, e4715, 2021. https://doi.org/10.1002/dac.4715
19. Huang, H. F. & Wang, B. "A circularly polarized slot antenna with enhanced axial ratio bandwidth." *2017 Sixth Asia-Pacific Conference on Antennas and Propagation (APCAP)*, pp. 1–3, 2017.
20. CST-Computer imulation Technology Studio Suite, 2017, https://www.cst.com.

21. Kulkarni, J. "Multi-band printed monopole antenna conforming bandwidth requirement of GSM/WLAN/WiMAX standards." *Progress In Electromagnetics Research Letters*, vol. 91, pp. 59–66, 2020.

22. Kulkarni, J., Sim, C.-Y.-D., Gangwar, R. K. & Anguera, J. "Broadband and compact circularly polarized MIMO antenna with concentric rings and oval slots for 5G application." *IEEE Access*, vol. 10, pp. 29925–29936, 2022. https://doi.org/10.1109/ACCESS.2022.3157914

23. Gao, S., Luo, Q. & Zhu, F. *Circularly Polarized Antennas*, John Wiley & Sons, Ltd. West Sussex, UK, 2014, pp. 1–26.

24. Banerjee, U., Karmakar, A. & Saha, A. "A review on circularly polarized antennas, trends and advances." *International Journal of Microwave and Wireless Technologies*, vol. 12, pp. 1–22, 2020.

25. Sharma, S. & Tripathi, C. C. "A comprehensive study on circularly polarized antenna." *2016 Second International Innovative Applications of Computational Intelligence on Power, Energy and Controls with their Impact on Humanity (CIPECH)*, pp. 234–239, 2016.

26. Samsuzzaman, M. & Islam, M. "Circularly polarized broadband printed antenna for wireless applications." *Sensors*, vol. 18, no. 12, p. 4261, 2018.

27. Parchin, N. O., Basherlou, H. J. & Abd-Alhameed, R. A. "Dual circularly polarized crescent-shaped slot antenna for 5G front-end systems." *Progress in Electromagnetics Research Letters*, vol. 91, pp. 41–48, 2020.

28. Kulkarni, J., Desai, A. & Sim, C. Y. D. "Wideband four-port MIMO antenna array with high isolation for future wireless systems." *AEU – International Journal of Electronics and Communications*, vol. 128, pp. 1–14, 2021.

29. Kulkarni, J., Desai, A. & Sim, C. Y. D. "Two port CPW-fed MIMO antenna with wide bandwidth and high isolation for future wireless applications." *International Journal of RF and Microwave Computer-Aided Engineering*, vol. 31, pp. 1–16, 2021.

30. Zhang, E., Michel, A., Nepa, P. & Qui, J. "Compact dual-band circularly polarized stacked patch antenna for microwave-radio-frequency identification multiple-input-multiple-output application." *International Journal of Antennas and Propagation*, vol. 2021, pp. 1–13, 2021.

31. Ullah, U., Mabrouk, I. B. & Koziel, S. "Enhanced-performance circularly polarized MIMO antenna with Polarization/Pattern diversity." *IEEE Access*, vol. 8, pp. 11887–11895, 2020.

32. Zhang, E., Michel, A., Pino, M. R., Nepa, P. & Qiu, J. "A dual circularly polarized patch antenna with high isolation for MIMO WLAN applications." *IEEE Access*, vol. 8, pp. 117833–117840, 2020.

33. Tran, H., Hussain, N. & Le, T. "Low-profile wideband circularly polarized MIMO antenna with polarization diversity for WLAN applications." *AEU – International Journal of Electronics and Communications*, vol. 108, pp. 172–180, 2019.

34. Jalali, M., Naser-Moghadasi, M. & Sadeghzadeh, R. A. "Dual circularly polarized multilayer MIMO antenna array with an enhanced SR-feeding network for C-band application." *International Journal of Microwave and Wireless Technologies*, vol. 9, no. 8, pp. 1741–1748, 2017.

35. Ameen, M., Ahmad, O. & Chaudhary, R. "Single split-ring resonator loaded self-decoupled dual-polarized MIMO antenna for mid-band 5G and C-band applications." *AEU – International Journal of Electronics and Communications*, vol. 124, pp. 1–31, 2020.

36. Xiong, X., Ling, B. W.-K., Zhang, H. & Zhang, G. "Coplanar waveguide fed multiple input multiple output antenna with higher isolation and multi-sense circular polarization." *Journal of Electromagnetic Waves and Applications*, vol. 32, pp. 685–694, 2018.

37. Malviya, L., Panigrahi, R. K. & Kartikeyan, M. V. "Circularly polarized 2×2 MIMO antenna for WLAN applications." *Progress in Electromagnetics Research C*, vol. 66, pp. 97–107, 2016.
38. Saxena, S., Kanaujia, B. K., Dwari, S., Kumar, S. & Tiwari, R. "A compact dual-polarized MIMO antenna with distinct diversity performance for UWB applications." *IEEE Antennas and Wireless Propagation Letters*, vol. 16, pp. 3096–3099, 2017.
39. Kumar, S., Lee, G. H., Kim, D. H., Choi, H. C. & Kim, K. W. "Dual circularly polarized planar four-Port MIMO antenna with wide axial-ratio bandwidth." *Sensors*, vol. 20, no. 19, p. 5610, 2020.
40. Yin, W., Chen, S., Chang, J., Li, C. & Khamas, S. K. "CPW fed compact UWB 4-element MIMO antenna with high isolation." *Sensors*, vol. 21, no. 8, p. 2688, 2021.

8 Maximizing Wireless Performance
Advancement in Wideband Four-Port MIMO Antennas

8.1 INTRODUCTION TO MIMO ANTENNAS

The importance of a four-port multiple input multiple output (MIMO) antenna for wireless local area network (WLAN) and Wi-Fi 6 applications lies in its ability to enhance wireless communication performance and enable higher data rates.

Here are some key reasons why a four-port MIMO antenna is important for WLAN and Wi-Fi 6 applications:

- **Increased data throughput**: MIMO technology utilizes multiple antennas to transmit and receive data simultaneously, thereby increasing the overall data throughput. With a four-port MIMO antenna, multiple data streams can be transmitted and received, leading to higher data rates and improved network capacity. This is particularly beneficial in Wi-Fi 6, which supports higher bandwidth and more concurrent connections.
- **Improved signal quality and coverage:** By employing multiple antennas, a four-port MIMO antenna can leverage spatial diversity and beamforming techniques to enhance signal quality and coverage. It can mitigate multipath interference and improve the reliability of wireless connections, even in challenging environments with obstacles and reflections.
- **Increased network capacity:** With more antenna ports, a four-port MIMO antenna can support a higher number of concurrent connections. This is crucial in WLAN and Wi-Fi 6 applications, where multiple devices connect to the network simultaneously. By efficiently handling multiple data streams, a four-port MIMO antenna helps to alleviate network congestion and maintain smooth and stable connections for all connected devices.
- **Enhanced spectral efficiency:** MIMO technology allows for the efficient utilization of available radio spectrum by transmitting multiple data streams in parallel. This increases the spectral efficiency, enabling more data to be transmitted within a given frequency band. A four-port MIMO antenna contributes to improved spectral efficiency, making better use of the available frequency resources, and optimizing the overall network performance.

DOI: 10.1201/9781003331018-8

258 Advanced Planar Antennas for Wireless Technology

In summary, a four-port MIMO antenna plays a vital role in WLAN and Wi-Fi 6 applications by enabling higher data rates, improving signal quality and coverage, and increasing network capacity.

8.2 CASE STUDY: FOUR-PORT GROUND COUPLED COPLANAR WAVEGUIDE GROUND-FED MIMO ANTENNA FOR WIRELESS APPLICATIONS

The authors in Ref. [1] have developed a coplanar waveguide ground (CPWG)-fed four-port MIMO antenna having a designed footprint of 1.65 λ × 1.65 λ (at 5.5 GHz) for Wi-Fi 5 (5.15–5.85 GHz) and Wi-Fi 6 (5.925–7.125 GHz) access point (AP) and repeater applications. The single antenna element consists of a Y-shaped monopole top-loaded by an inverted-C structure (ICS), in which the two open ends of the ICS are coupled to the two-sided ground, forming a symmetrical ground-coupled loop antenna structure. Further, four identical antenna elements are arranged in a sequential clockwise manner forming a four-port MIMO antenna structure. To enhance the isolation between the adjacent antenna elements, the partial ground planes of all four elements are meticulously connected by a whirligig isolating structure. Besides yielding a very wide 10 dB impedance bandwidth of 35.45% (5.15–7.37 GHz) with isolation better than 25 dB, the designed four-port MIMO antenna has also demonstrated acceptable gain and radiation efficiency of 4 dBi and 74%, respectively. Furthermore, desirable MIMO performances like envelope correlation coefficient (ECC) less than 0.008 and channel capacity of about 21 bps/Hz are also calculated across the bands of interest.

8.2.1 EXISTING STATE-OF-THE-ART TECHNOLOGY

The tremendous growth in users across homes, factories, offices, and public places demands well-established WLANs that can provide seamless internet access. As the wireless networking standard (IEEE 802.11ac) released in 2014 cannot keep up with the increasing demand, it does urgently require an upgrade, and presently, a much newer version known as the IEEE 802.11ax in the form of IEEE 802.11ac (Wi-Fi 5) and IEEE 802.11ax (Wi-Fi 6) is already replacing the older version. The useful techniques of Wi-Fi 5 combined with the increased capacity of Wi-Fi 6 will help in achieving the growing user needs [2]. Due to the numerous wireless devices in dense environments for multiple users, it is crucial to design appropriate antennas with multiuser multiple input multiple output (MU-MIMO) configurations used by Wi-Fi 5 but with extended capabilities [3] required by Wi-Fi 6. At the moment, most of the routers using Wi-Fi 5 technology are capable of handling four data streams of users simultaneously, whereas the MU-MIMO antennas of Wi-Fi 6 are capable of supporting at least eight spatially divided data streams simultaneously for eight users without any delays in buffering along the down/up links between the wireless devices and APs. Thus, the wireless networks with Wi-Fi 6 are capable of handling big data transmissions without a data buffer delay.

The merging of MIMO technology and the latest IEEE standards can fulfill the user demands to an even greater extent as a larger number of users per AP can enjoy data-intensive video-streaming applications simultaneously. Thus, there is an urgent demand to design a MIMO antenna with operational bandwidths that can cover the operational bandwidths of Wi-Fi 5 (5.15–5.85 GHz) and Wi-Fi 6 (5.925–7.125 GHz). From the literature, many two-port or four-port MIMO antennas have been proposed [4–21], in which the MIMO antenna structure design usually applies either the separated or connected ground methods. According to authors in Ref. [22], the connected ground MIMO antennas are of great value as it is easier to interpret the signals correctly concerning the reference level. In Refs. [4] and [5], a two-port connected ground MIMO antenna with an operating bandwidth of 1.48–3.8 GHz and 27.5–28.35 GHz has been studied for mobile terminals, respectively. From these two designs working in the sub-6 GHz and millimeter wave (mmWave) bands, one can see that the two-port antenna designs with the connected ground can achieve acceptable isolation with MIMO diversity performances, especially the isolation of the mmWave [5] is >40 dB, but the planar size of the antenna [4] is rather large at 100×65 mm^2. Nevertheless, the two-port MIMO antenna design configuration is not desirable for accommodating a much greater number of users and, thus, the development of a four-port MIMO antenna type is presently the research hotspot in the area of Wi-Fi MIMO antenna design.

Four-port MIMO antennas with separate ground [6–11] and connected ground [12–15] have been extensively investigated using microstrip and probe feeding techniques. Even though the MIMO antennas with the separated ground can be designed to cover the operational bands for WLAN, worldwide interoperability for microwave access (WiMAX), and ultra-wide band (UWB) applications, the disconnected ground makes the antenna less practical for its use in commercial applications. In Ref. [12], a dual-band MIMO antenna of size 46×21 mm^2 is proposed to operate at 2.4–2.5 GHz and 4.9–5.72 GHz in the two-port and four-port configurations, respectively. However, as the MIMO antenna has different operating frequency modes in the two-port and four-port configurations, it is unfeasible for practical applications if all the four ports are needed to be used for both the WLAN bands. Furthermore, the ECC and isolation values of Ref. [12] are not desirable. Therefore, to have an antenna with all four elements working in the same frequency band, an inverted L-shaped monopole antenna is arranged symmetrically to form a MIMO structure (of size 40×40 mm^2) with a wide bandwidth of 2.70–4.94 GHz [13], but the MIMO diversity performances in terms of ECC and isolation need further improvement. In Ref. [14], a four-port MIMO antenna with UWB operation (3.1–10.6 GHz) is investigated, and it achieves ECC <0.001 and isolation >20 dB. However, author in Ref. [14] has only experimentally validated the two-port MIMO type, and its corresponding four-port MIMO type is not tested experimentally for isolation and MIMO diversity performances. In Ref. [15], a four-port MIMO antenna (of size 24×20 mm^2) with symmetrically oriented elements radiating at the mmWave frequency band is studied. Even though good isolation can be yielded due to spatial diversity, it cannot cover the operational bandwidths for Wi-Fi 5/6 applications.

To achieve larger operational bandwidths, the CPW-fed technique is usually applied as the feeding method in the antenna design because it can offer more advantages over the microstrip feeding type, such as providing easier means of series and parallel connection with passive and active elements, better impedance matching, and improving the antenna gain [16]. Therefore, recently, the two-port [17–19] and four-port [20, 21, 23] MIMO antenna designs using CPW feed have been studied with separate [17, 20, 21] and connected ground [18, 19, 23] configurations. Here, even though the aforementioned two-port CPW-fed MIMO antennas have attained good performance in terms of scattering parameters (S-parameters) and MIMO diversity, only authors in Refs. [17] and [19] can yield very good dual-band and UWB operation, respectively, but as mentioned earlier, the two-port structure is insufficient for accommodating a greater number of users. As reported in Refs. [20, 21], the four-port MIMO antennas can well cover the Wi-Fi 5 operational bandwidths, but due to its separated ground configuration, it limited their practical usage. In Ref. [23], a low-profile four-port UWB MIMO antenna for portable wireless applications is studied, and the four monopole elements are perpendicularly arranged for achieving higher polarization diversity. In this case, a gap is optimized between the antenna elements and CPW feed for reducing the mutual coupling, and an UWB (3–20 GHz) with isolation >17 dB, ECC <0.1, and an efficiency of 75% are achieved. By further observing the works reported in Refs. [16–21, 23], one can see that the MIMO antenna with CPW feed can yield higher isolation and lower ECC values with connected ground working in the Wi-Fi 5/6 operational bands.

In this chapter, a four-port MIMO structure with a CPWG feeding method is proposed for Wi-Fi 5 and Wi-Fi 6 applications. The CPWG feeding method is chosen in this work because it provides lower losses enabling effective use of low-cost FR-4 substrate, provides high isolation, flexibility in geometry, wider bandwidth, better impedance matching, and, more importantly, enhances the antenna effectiveness. Furthermore, the additional partial ground loaded behind the CPW feeding line in this work is also used to interconnect the isolation structure that can eventually yield very good isolation level >25 dB across the bands of interest. Here, the main radiator of each antenna element is a symmetrical ground-coupled loop antenna type, which is mainly composed of a Y-shaped monopole, an ICS that has its two open-end coupled to the two coplanar grounds printed on the top side. Notably, the proposed four-port MIMO antenna is realized by symmetrically arranging the four identical antenna elements in a sequential rotational manner so that good polarization diversity can be achieved. To further achieve high isolation of >25 dB across the operational bandwidths (36.14%, 5.10–7.35 GHz), here, a novel method of meticulously connecting the partial ground planes of all four antenna elements by a whirligig isolating structure is proposed. Other typical MIMO antenna performances such as ECC, and average efficiency will also be explicitly shown and validated by measuring the experimental values over the fabricated prototype.

8.2.2 DESIGN AND ANALYSIS OF THE PROPOSED CPWG-FED ANTENNA ELEMENT

Figure 8.1 depicts the geometry and layout of the proposed CPWG-fed antenna element. As depicted in Figure 8.1, the proposed antenna element is printed on a double-sided, single-layer FR-4 substrate (dielectric constant 4.4 and tan δ 0.025) that has an overall dimension of $40 \times 40 \times 0.8$ mm³.

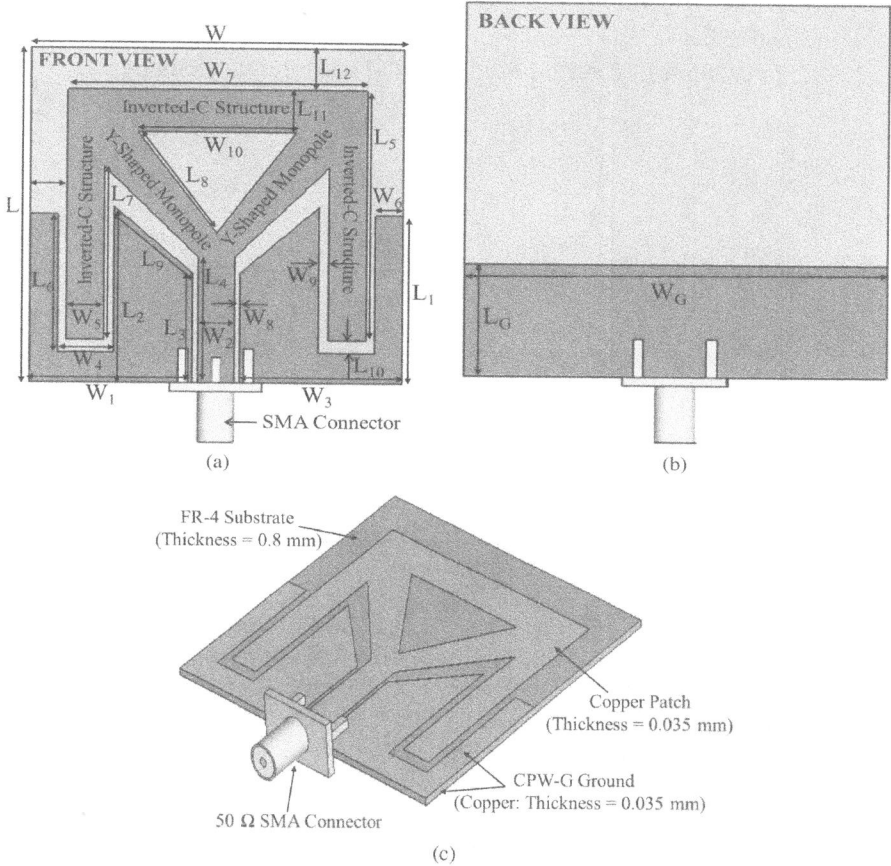

FRONT VIEW

W

W_7

L_{12}

Inverted-C Structure L_{11}

W_{10}

L_8

Inverted-C Structure

L_7

L_9

L_4

W_5

L_2

L_3

W_2

W_8

W_4

W_1

W_3

Y-Shaped Monopole

Y-Shaped Monopole

Inverted-C Structure

L_5

W_6

W_9

L_1

L_{10}

L

L

SMA Connector

(a)

BACK VIEW

W_G

L_G

(b)

FR-4 Substrate
(Thickness = 0.8 mm)

Copper Patch
(Thickness = 0.035 mm)

CPW-G Ground
(Copper: Thickness = 0.035 mm)

50 Ω SMA Connector

(c)

FIGURE 8.1 Structural layout of the proposed CPWG-fed antenna: (a) front view, (b) back view, and (c) perspective view.

The top side (top view) of the antenna element is depicted in Figure 8.1a, in which the main radiator (symmetrical ground-coupled loop antenna structure) is excited by a 50 Ω transmission line of size 15 mm (L_4) × 4 mm (W_2), with two narrow feeding gaps of width $W_8 = 0.5$ mm that keeps the two "tapered and slotted" ground planes on the left and right side of the 50 Ω transmission line. Here, the two "tapered and slotted" ground planes have an initial size of 21 mm (L_1) × 17.5 mm (W_1), and each tapered corner section has a height and base of 8 mm and 8.5 mm, respectively, for achieving good impedance matching. Notably, the two slots that are loaded on the two ground planes each have a slotted size of 17.5 mm (L_6) × 6 mm (W_4). As for the main radiator, it is composed of a Y-shaped monopole top-loaded by an ICS, in which the horizontal section of the ICS has a width of 24 mm ($W_7 - 2 \times W_5$) and a length of 4 mm, while its two corresponding vertical sections (with its two open ends coupled to the two ground planes) each has a length and width of 30 mm (L_5) and

TABLE 8.1

Dimensions of the Proposed CPWG-Fed Antenna Element

Parameter	Value (mm)	Parameter	Value (mm)
L	40	W	40
L_1	21	W1	17.5
L_2	21	W2	4
L_3	13	W3	17.5
L_4	15	W4	6
L_5	30	W5	4
L_6	17.5	W6	3
L_7	21.5	W7	32
L_8	15.65	W8	0.5
L_9	11.67	W9	1.0
L_{10}	1.5	W10	18.26
L_{11}	5	LG	12
L_{12}	5	WG	40

4 mm (W_5), respectively. To realize the Y-shaped monopole, the top section of the Y-shaped monopole can be seen as an isosceles triangle slot of two equal side lengths ($L_8 = 15.65$ mm) with a base width of $W_{10} = 18.26$ mm. As shown in Figure 8.1b, a partial ground of size 12 mm (L_G) × 40 mm (W_G) is printed on the bottom side (back view) of the FR-4 substrate, which forms the CPWG feeding structure. The perspective view and the detailed dimensions of the proposed antenna element are presented in Figure 8.1c and Table 8.1, respectively.

To further comprehend the design mechanism of the proposed CPWG-fed antenna element, the step-wise evolution of the antenna element along with its corresponding reflection coefficient diagram is explained in Figure 8.2. The electromagnetic computer simulation tool Microwave Studio Suit (CST Microwave Studio®) is used to develop and simulate the proposed antenna element. The initial step (Step 1) of this work stems from a CPWG-fed "Y-shaped" monopole with two truncated CPW ground planes, and this structure can excite a good lower resonant frequency (f_L) at 5.2 GHz with a frequency range of 4.81–5.65 GHz (low band), but this low band is not wide enough to cover the entire bands of interest for Wi-Fi 5 (5.15–5.85 GHz) and Wi-Fi 6 (5.925–7.125 GHz). Nevertheless, one can also see from the Step 1 structure that another adjacent higher resonant frequency at approximately 6.3 GHz (f_H) has also been induced but it is not fully impedance matched. Therefore, our next step (Step 2) is to further provide good matching to the impedance of this adjacent f_H. As shown in Figure 8.2, if a horizontal radiating strip of size 32 × 5 mm² is top-loaded to the "Y-shaped" monopole, even though the f_L is shifted from 5.2 GHz to 5.5 GHz with a low band of (5.23–6.26 GHz) and the f_H is successfully matched, the combination of the low band and high band (excited by f_L and f_H, respectively) still fails to yield the required bands of interest. Hence, the next step is to further match the impedance of f_H and shift it to the higher spectrum. In Step 3, two vertical radiators

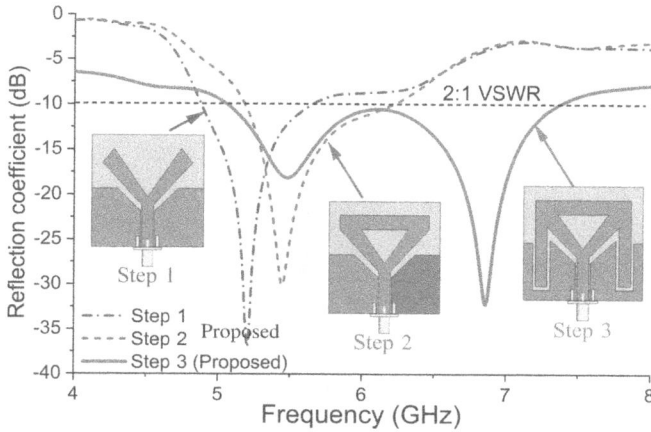

FIGURE 8.2   Step-wise design of the proposed coplanar waveguide ground-fed antenna element (ground-coupled loop antenna).

are extended from the two upper-edge corners of the Step 2 structure, and a portion of their open ends are coupled to the two CPW grounds (each has an open-slotted size of 17.5 mm × 6 mm), which is analogous to an ICS that coupled to the two CPW grounds, forming a ground-coupled loop antenna type (now denoted as the proposed antenna element). By further observing the two resonant frequencies of this proposed antenna element, the f_L remains at 5.5 GHz, whereas the f_H is now shifted to the higher spectrum at 6.7 GHz with a very good impedance match. Therefore, due to the merging of the low band and high band, a wide 10 dB impedance bandwidth of 36.14% (5.10–7.35 GHz) is achieved, and it can meet the bands of interest for Wi-Fi 5 (5.15–5.85 GHz) and Wi-Fi 6 (5.925–7.125 GHz).

The performance analysis of the proposed antenna in terms of reflection coefficient (S_{11}) is further carried out by varying several parameters one-by-one, and keeping other parameters constant. At first, the length of strip L_1 is varied from 18 mm to 22 mm in step increments of 2 mm, as depicted in Figure 8.3. At $L_1 = 18$mm and 22 mm, the antenna fails to cover the entire (5.15–7.37 GHz) band. Therefore, the optimum value of $L_1 = 20$mm, is considered as it covers the entire desired band.

The second parameter considered for parametric variation is the air gap L_{10} between the inverted U element and CPWG plane as shown in Figure 8.4. Here, the parameter is varied from 0.5 mm to 2.5 mm in three steps. When $L_{10} = 0.5$ mm, the S_{11} curve lies from 6 GHz to 6.8 GHz only, whereas at $L_{10} = 2.5$ mm, the S_{11} varies from (5.45–6.00 GHz) and from (6.20–7.00 GHz) only. Here, $L_{10} = 1.5$, covers the entire desired band from (5.15–7.37 GHz) and is therefore considered as an optimized value of L_{10}.

Further, the length of horizontal strip of inverted U shape is varied from 4 mm to 6 mm, as illustrated in Figure 8.5. As shown in Figure 8.5, the desired band is obtained for $L_{11} = 5$ mm, whereas the values of $L_{11} = 4$ mm and 6 mm, produces almost similar graphs which fail to cover the entire desired band. Hence, the value of $L_{11} = 5$ mm was selected considering the best radio frequency (RF) performance.

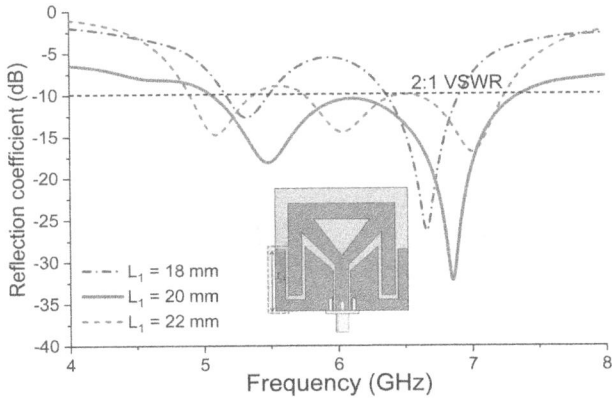

FIGURE 8.3 Parametric variation of L_1.

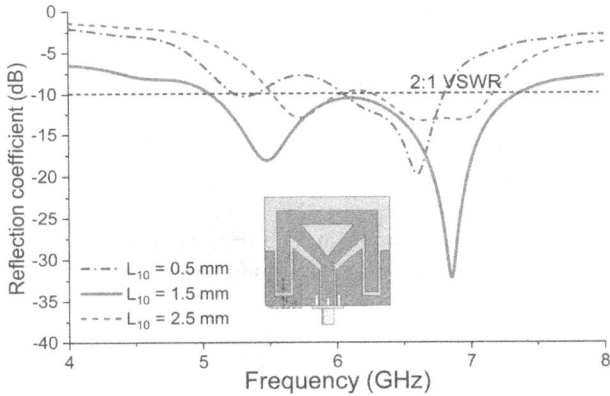

FIGURE 8.4 Parametric variation of L_{10}.

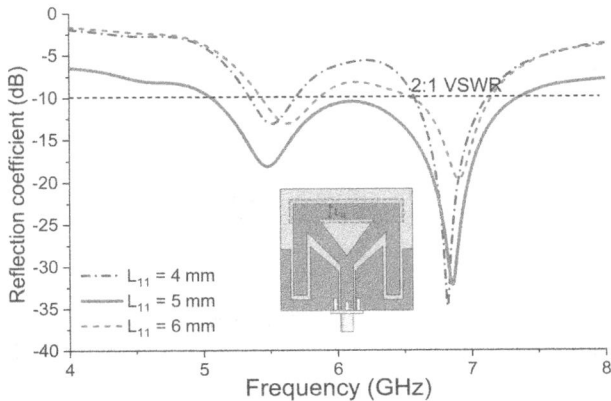

FIGURE 8.5 Parametric variation of L_{11}.

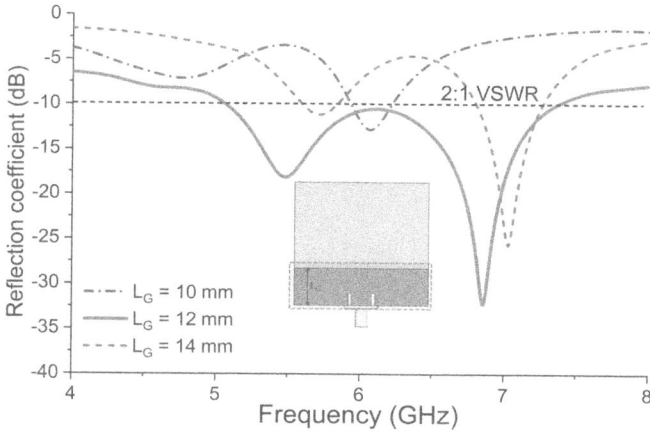

FIGURE 8.6   Parametric variation of L_G.

Finally, the height L_G of the ground plane was varied from 10 mm to 14 mm in steps of 2 mm as seen from Figure 8.6. Here, it can be easily visualized that L_G = 10 mm fails to produce any resonance whereas L_G = 14 mm produces a single resonance of (6.6–7.10 GHz). Therefore, the optimum value of L_G = 12 mm is considered as it covers the desired operating band.

Figure 8.7 analyzes the surface current distribution of the proposed antenna element at 5.5 GHz and 6.8 GHz. At both resonances (5.5 GHz and 6.8 GHz), the maximum amount of current flows throughout the radiating structure, including the "Y-shaped" monopole strip and ICS. This clearly shows that the deployment of these structures allows the current in various directions for a longer time, which will assist to widen the band in the frequency range of (5.10–7.35 GHz). As seen in Figure 8.7A,

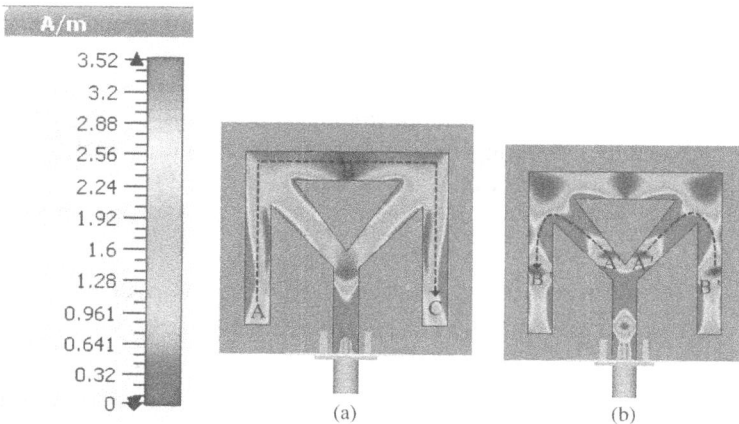

FIGURE 8.7   Surface current distribution of the proposed coplanar waveguide ground-fed antenna element at (a) f_L = 5.5 GHz and (b) f_H = 6.8 GHz.

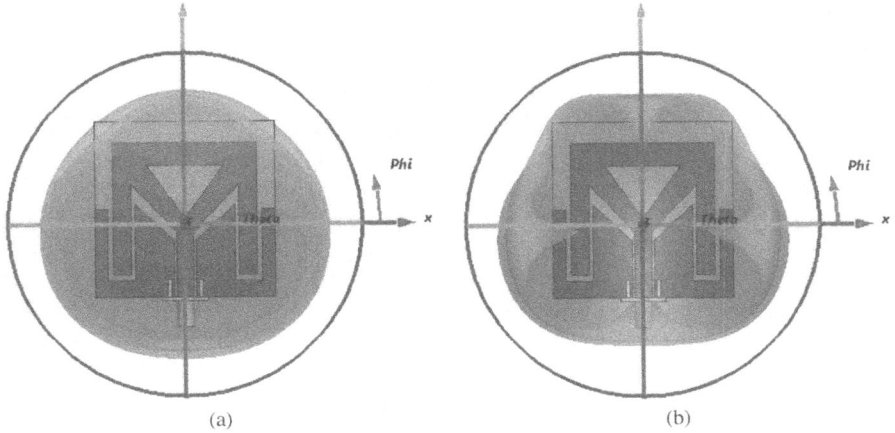

(a)   (b)

FIGURE 8.8   Three-dimensional radiation patterns of the proposed antenna element at (a) 5.5 GHz (A) and (b) 6.8 GHz.

the electrical path of the f_L (5.5 GHz mode) is from point A (current maximum) to point B (current null) and to point C (current maximum), thus, f_L is analogous to a loop mode with a near half-wavelength current distribution along the ICS. As for the electrical path for f_H (6.8 GHz mode), as depicted in Figure 8.7B, it is from point A′ (current null) to point B′ (current null) but has a current maximum between them. Therefore, f_H is a near half-wavelength current distribution along the Y-shaped monopole and the vertical section of the ICS.

The simulated three-dimensional (3D) patterns of the proposed antenna element at f_L (5.5 GHz) and f_H (6.8 GHz) are depicted in Figure 8.8a and b, respectively. From Figure 8.8a, the radiation patterns are nearly omnidirectional, whereas from Figure 8.8b, the radiation patterns exhibit nearly omnidirectional radiation patterns with few null points. However, these null points are not deep enough and will have minimum effect.

The realized gain and radiation efficiency of the proposed antenna element are depicted in Figure 8.9. It is easily seen that the proposed antenna element offers a stable gain and radiation efficiency of approximately 3.5–4.0 dBi and 70–74%, respectively, throughout the operating bands of interest. This confirms that the proposed antenna element is a potential candidate for Wi-Fi 5 and Wi-Fi 6 access applications where stable gain and efficiency are essential to ensure continuous internet access.

8.2.3 Design Analysis of the Proposed Four-Port MIMO Antenna without Whirligig Isolating Structure

The optimized antenna element as described in Section 8.2.2 is further reconfigured into a four-antenna element MIMO type, and these four identical antenna elements are arranged in a sequential rotational manner so that good polarization diversity can be achieved between any two adjacent antenna elements, as shown in Figure 8.10a. Here, the proposed four-port MIMO antenna is also built on an FR-4 substrate of a

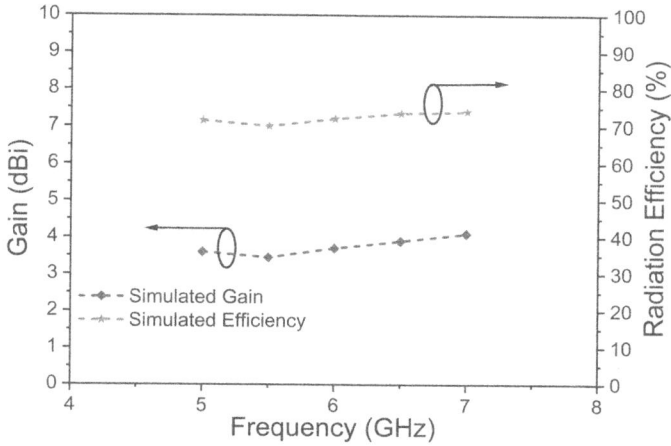

FIGURE 8.9 Gain and radiation efficiency of the proposed antenna element.

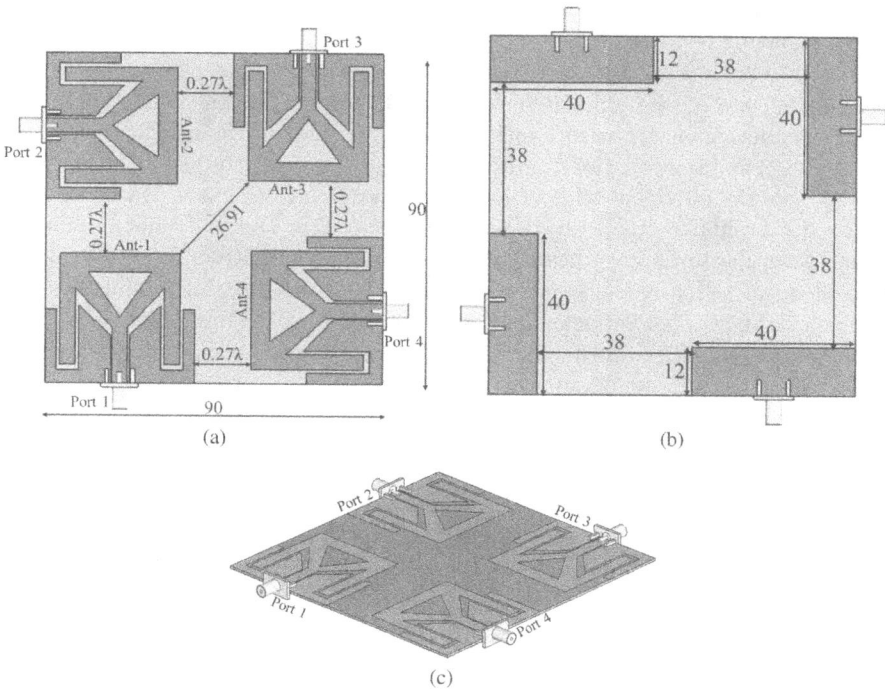

FIGURE 8.10 Structure of the proposed coplanar waveguide ground (CPWG)-fed four-port MIMO antenna without loading the isolating structure: (a) front view, (b) back view, and (c) perspective view.

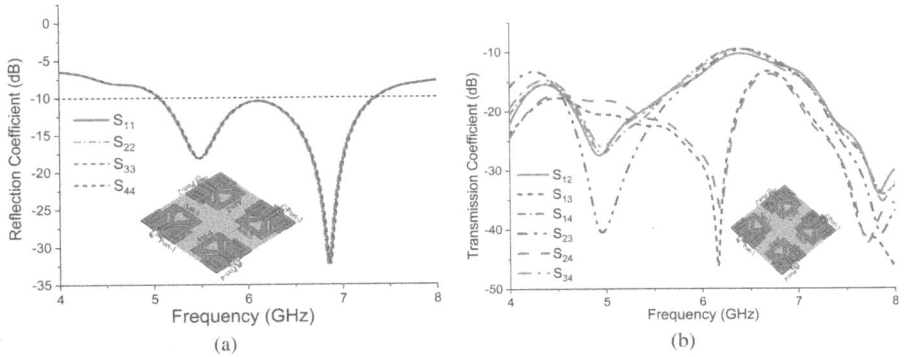

FIGURE 8.11 Scattering parameters of the proposed coplanar waveguide ground (CPWG)-fed four-port MIMO antenna without isolating structure: (a) reflection coefficients and (b) transmission coefficients.

thickness of 0.8 mm (0.014 λ), and it has a total planar size of 90×90 mm^2 (1.65 $\lambda \times$ 1.65 λ), where λ is the free space wavelength measured at a lower resonating frequency (f_L) of 5.5 GHz. Notably, the separation distance between any two antenna elements is 15 mm (0.27 λ). As shown in Figure 8.10b, the back view of the proposed four-port MIMO antenna (without loading the isolating structure) shows four separate partial ground planes, each having a size of 40×12 mm^2. To get a clear view of geometry, a perspective view of the proposed four-port MIMO antenna is shown in Figure 8.10c.

The scattering characteristics including the reflection coefficient curves (S_{11}, S_{22}, S_{33}, and S_{44}) and transmission curves (S_{12}, S_{13}, S_{23}, S_{14}, S_{24}, and S_{34}) of the proposed four-port MIMO antenna without loading the isolating structure, are depicted in Figure 8.11a and b, respectively. In Figure 8.11a, the four identical antenna elements have shown the same impedance bandwidth of 36.14% (5.10–7.35 GHz), which is also identical to the result shown in Figure 8.2 for the designed proposed single antenna element. This validates that the deployment of the four identical antenna elements does not affect the impedance bandwidth of each other. However, as shown in Figure 8.11b, even though the four antenna elements are arranged in a sequential rotational manner in clockwise direction with separated ground, the separation distance between any two adjacent antenna elements of 0.27 λ may have led to poor isolation of approximately 10 dB (especially at 6.5 GHz band), which indicates that without using any isolating structure, all antenna elements are mutually coupled with each other. Notably, this mutual coupling will mitigate the antenna efficiency and deteriorate the far-field patterns of each antenna element. Therefore, one of the crucial challenges of deploying MIMO antennas in wireless devices is to amalgamate maximum antenna elements in an exclusive area with minimum coupling.

8.2.4 DESIGN ANALYSIS OF THE PROPOSED FOUR-PORT MIMO ANTENNA WITH WHIRLIGIG ISOLATING STRUCTURE

Considering the aforementioned analysis, to offer maximum isolation between any two antenna elements, a whirligig isolating structure is meticulously designed and

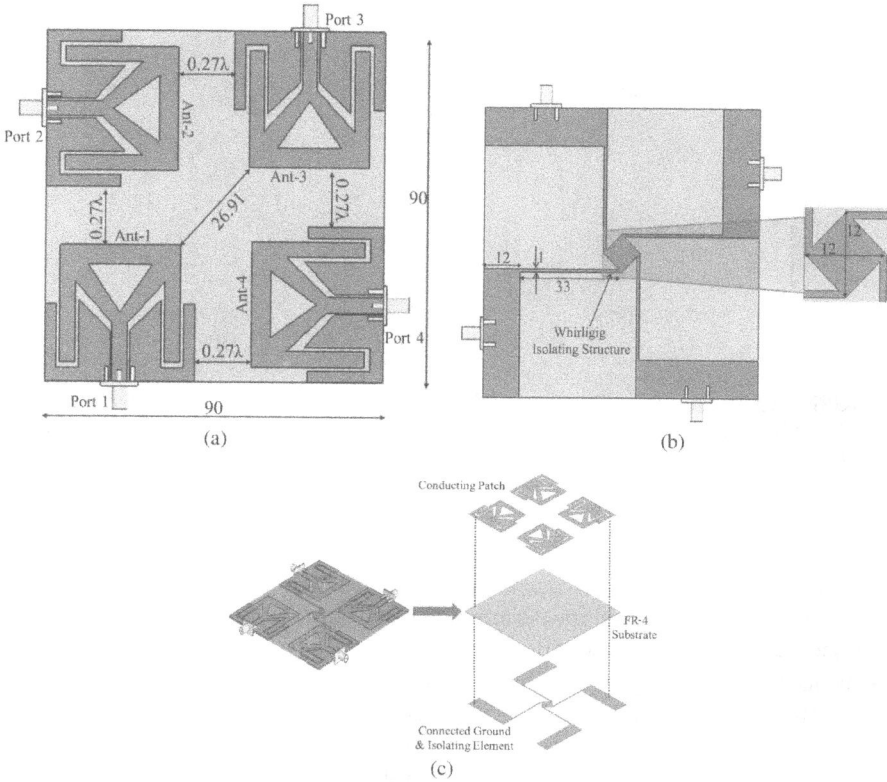

FIGURE 8.12 Structure of the proposed coplanar waveguide ground-fed four-port MIMO antenna with isolating structure: (a) front view, (b) back view, and (c) layered view.

centrally printed at the middle of the bottom layer of the substrate, as shown in Figure 8.12. As illustrated in Figure 8.12, the whirligig structure consists of a centrally located square-shaped structure (of side length 6 mm) with four narrow lines (each 1 mm) that extended (from the four corners) toward the nearest partial ground plane, forming a whirligig shape. Due to the deployment of the whirligig structure, all the partial ground of the antenna elements now share a common ground plane that can nullify/cancel the radiative near-field coupling effects between any two adjacent antenna elements. Notably, the whirligig structure is also optimized to ensure that the operational impedance bandwidths of the four antenna elements remain undisturbed (same as those without the whirligig structure in Figure 8.10a) while contributing toward the enhanced isolation.

Figure 8.13 illustrates the reflection coefficients (S_{11}, S_{22}, S_{33}, and S_{44}) and transmission coefficients (S_{12}, S_{13}, S_{23}, S_{14}, S_{24}, and S_{34}) curves of the proposed CPWG-fed four-port MIMO antenna with the isolating structure. As depicted in Figure 8.13a, even though the isolating structure was loaded into the proposed four-port MIMO antenna, there is no degradation on the impedance bandwidths of all four separate antenna elements, as they have shown identical simulated 10 dB impedance bandwidths of

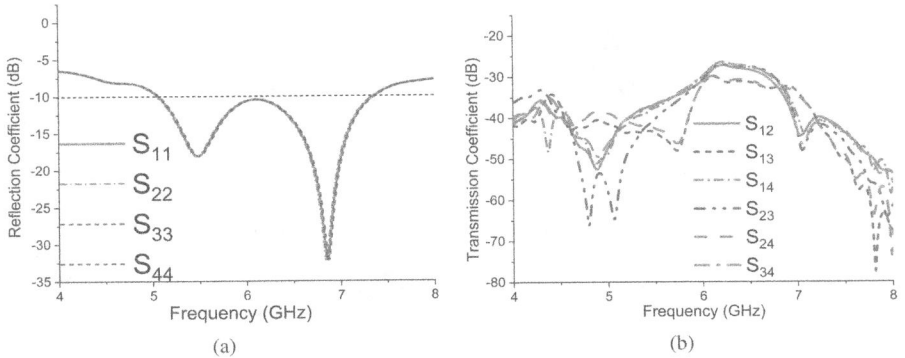

FIGURE 8.13 Scattering parameters of the proposed coplanar waveguide ground-fed four-port MIMO antenna with isolating structure: (A) reflection coefficients and (B) transmission coefficients.

(5.10–7.35 GHz), as compared to those in Figure 8.11a, without loading the isolating structure. Furthermore, as shown in Figure 8.13b, very good isolation levels of greater than 26.5 dB are demonstrated across the operational bandwidths. Thus, it can be validated that the deployment of the isolating structure has improved the isolation level by at least 16.5 dB, as well as not affecting the impedance bandwidth of all four antenna elements, which is one of the main novelties of this proposed four-port MIMO antenna.

To further comprehend how the whirligig isolating structure can enhance the isolation of the proposed CPWG-fed four-port MIMO antenna, its corresponding surface current distribution plots for port 1 (Ant. 1) excitation with and without loading the isolating structure (at 5.5 GHz and 6.8 GHz) are plotted in Figures 8.14 and 8.15, respectively. Notably, this simulation analysis is carried out by exciting one port at a time while the other ports are terminated with 50 Ω load impedance, but for brevity, the two figures only show the excitation of Ant. 1 (with other three ports terminated). As shown in Figure 8.14, during the absence of the isolating structure, Ants. 2–4 are highly coupled by the excited Ant. 1, especially at 5.5 GHz. One can see that the coupling effects with Ant. 4 by the excited Ant. 1 is the largest, because the main radiator of Ant. 4 is pointing at the vertical section of the ICS (with high current density) of the excited Ant. 1. By observing Figure 8.15, once the whirligig isolation structure is loaded, the radiative near-field coupling effects of Ants. 2–4 by the excited Ant. 1 are blocked and counter-balanced (by producing opposite phase current) by the narrow lines connected between the centrally located square-shaped structure and the four partial grounds. Thus, it validates that each antenna element function can yield improved isolation across the bands of interest.

8.2.5 RESULTS AND DISCUSSION OF THE PROPOSED FOUR-PORT MIMO ANTENNA

To validate the above simulation results, the proposed four-port MIMO antenna was manufactured, and its corresponding front view and back view are shown in

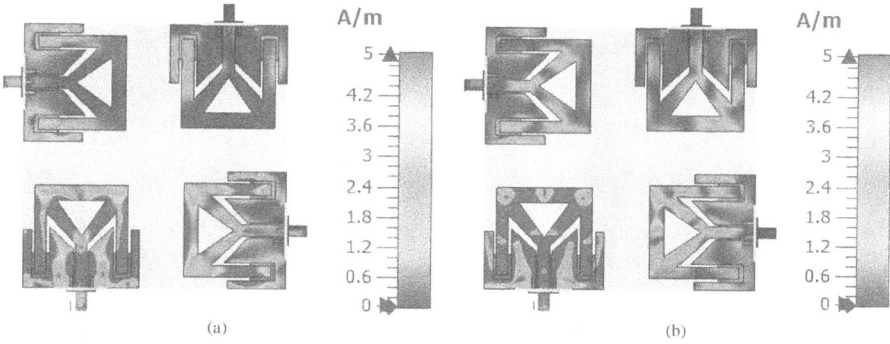

FIGURE 8.14 Surface current distribution of the proposed coplanar waveguide ground-fed four-port MIMO antenna without isolating structure, when Ant. 1 is excited at (a) 5.5 GHz and (b) 6.8 GHz.

Figure 8.16a and b, respectively. Here, the S-parameters are measured using an Agilent N5247A vector network analyzer (VNA), and the measurement of far-field radiation patterns, gain, and efficiency are carried out in an anechoic chamber at a resonating frequency of 5.5 GHz (f_L) and 6.8 GHz (f_H). Due to the symmetry and identical performance of each antenna element, only the scattering and radiation performances of two antenna elements (Ant. 1 and Ant. 2) are considered.

The simulated and measured reflection coefficient (S_{11}) and isolation (S_{12}) are illustrated in Figure 8.17. Here, the simulated and measured f_L are almost the same at 5.5 GHz, except that the measured f_H is slightly shifted to the higher spectrum at 6.8 GHz. Nevertheless, the measured 10 dB impedance bandwidth was 35.46% (5.15–7.37 GHz), which is slightly lower than the simulated one at 36.14% (5.10–7.35 GHz). As for its corresponding isolation level (S_{12}), the measured isolation level was >25 dB, which is lower than the simulated one at >27.5 dB. Notably, the differences between the simulation results and the measured ones could be due to minor

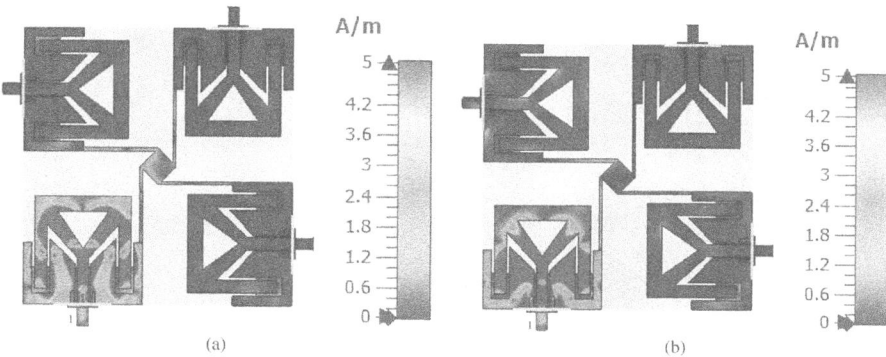

FIGURE 8.15 Surface current distribution of the proposed coplanar waveguide ground-fed four-port MIMO antenna with isolating structure, when Ant. 1 is excited at 5.5 GHz (A) and 6.8 GHz (B).

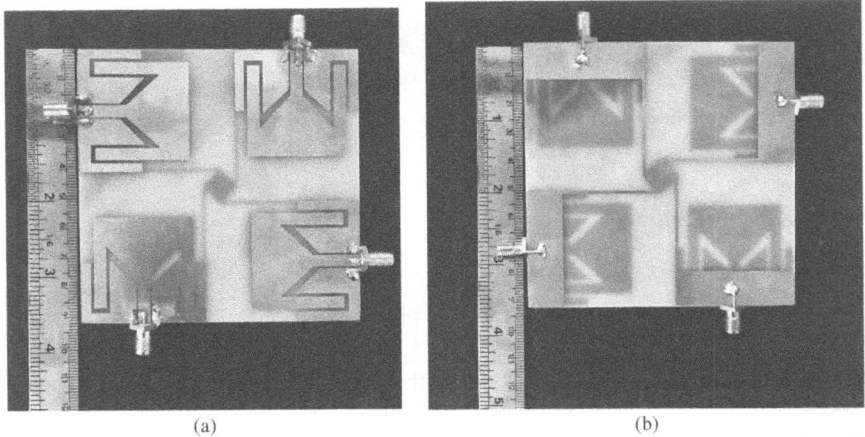

(a)　　　　　　　　　　　　　(b)

FIGURE 8.16 Fabricated prototype of the proposed coplanar waveguide ground-fed four-port MIMO antenna with isolating structure: (a) front view and (b) back view.

fabrication inaccuracies and semiflexible coaxial feeding cables. Nonetheless, the measured impedance bandwidth and isolation level are still very well validated with the simulated ones.

The antenna setup for measuring the gain and radiation pattern in an anechoic chamber is shown in Figure 8.18. The far-field radiation patterns at the two resonant frequencies f_L (5.5 GHz) and f_H (6.8 GHz) across the two principal planes (E-plane and H-plane) are measured under the condition that when one port is excited, other ports are terminated with matched impedance. From Figure 8.19, it is observed that Ants.1–4 applied in the proposed four-port MIMO antenna exhibit nearly omnidirectional radiation at 5.5 GHz and 6.8 GHz while maintaining the difference greater

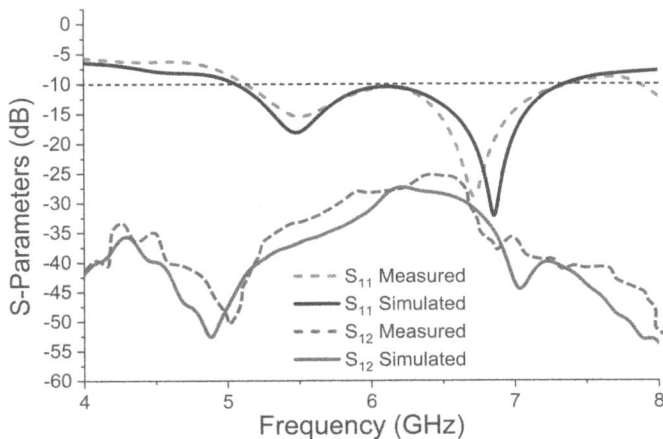

FIGURE 8.17 Simulated and measured scattering parameters of the proposed coplanar waveguide ground-fed four-port MIMO antenna.

FIGURE 8.18 Measurement setup of the proposed coplanar waveguide ground-fed four-port MIMO antenna in an anechoic chamber.

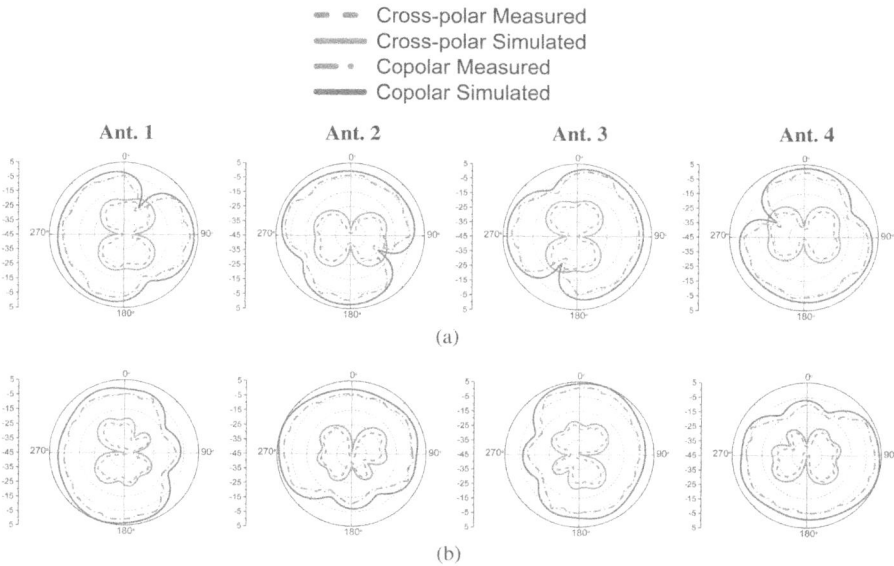

FIGURE 8.19 Measured and simulated 2D radiation patterns of the proposed coplanar waveguide ground-fed four-port MIMO antenna: (a) E-plane at 5.5 GHz, (b) E-plane at 6.8 GHz (c) H-plane at 5.5 GHz, and (d) H-plane at 6.8 GHz. (*Continued*)

(c)

(d)

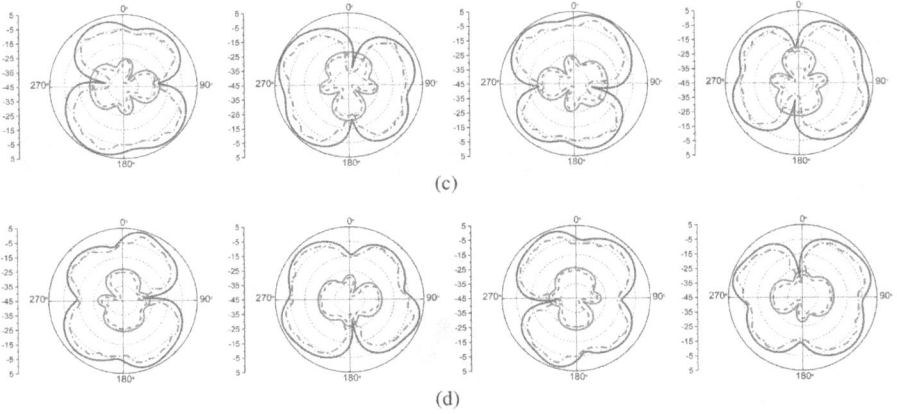

FIGURE 8.19   (*Continued*)

than 10 dB between co/cross-polar patterns. Figure 8.20 shows the simulated 3D radiation patterns of Ants. 1–4 at the two resonant frequencies f_L and f_H. It can be easily visualized that these patterns are sequentially rotating clockwise by 90°, validating the geometry of the proposed MIMO antenna.

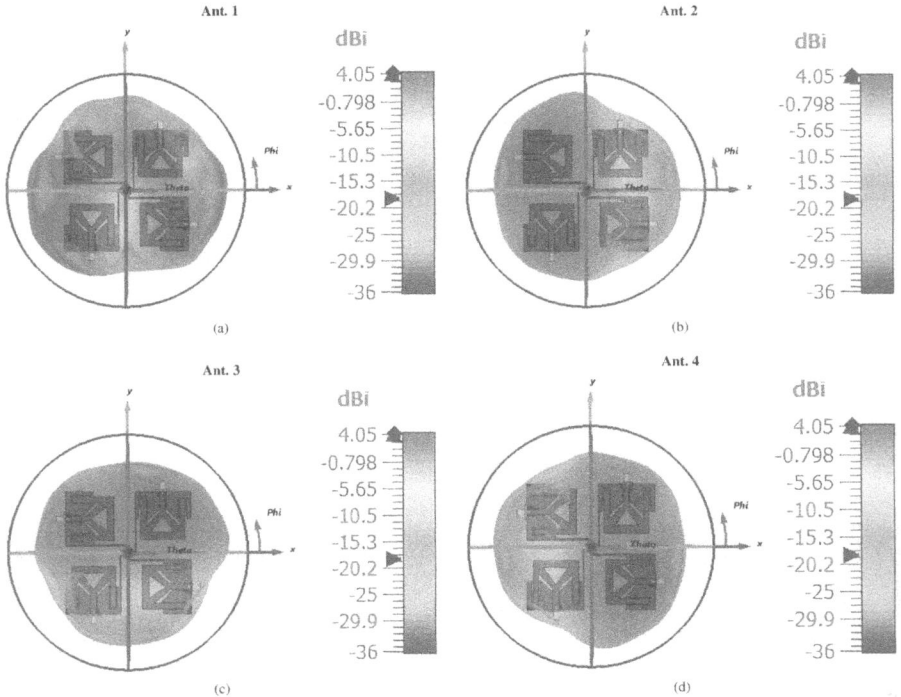

FIGURE 8.20   Simulated 3D radiation patterns of the proposed coplanar waveguide ground-fed four-port MIMO antenna: (a) Ant. 1, (b) Ant. 2 (c) Ant. 3, and (d) Ant. 4.

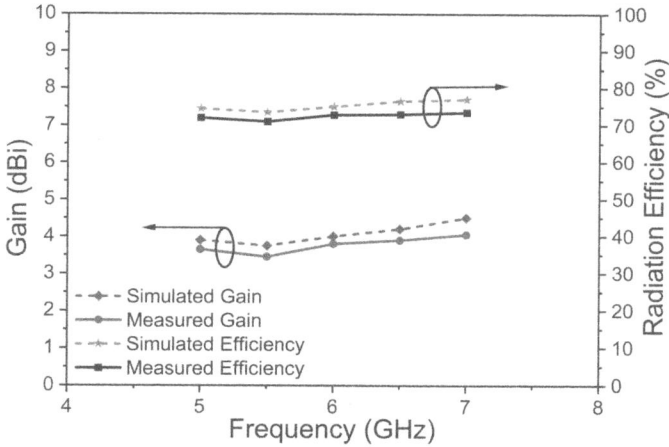

FIGURE 8.21   Simulated and measured gain and radiation efficiency of the proposed coplanar waveguide ground-fed four-port MIMO antenna.

Figure 8.21 illustrated the measured gain and radiation efficiency of the proposed four-port MIMO antenna, and it has shown stable gain and radiation efficiency variation of approximately 3.5–4 dBi and 72–74%, respectively, which are slightly lower than the simulated ones.

Figure 8.22 shows the ECC values of the proposed four-port MIMO antenna that are calculated from far-field radiation patterns using formula mentioned in Ref. [24]. It is noted that the ECC values are less than 0.008 for any combination of antenna elements across the desired bands of interest. Therefore, the proposed four-port MIMO antenna meets the requirement of the standard of ECC <0.5 for the considerable diversity gain (DG) (~10 dB as shown in Figure 8.23) needed by wireless

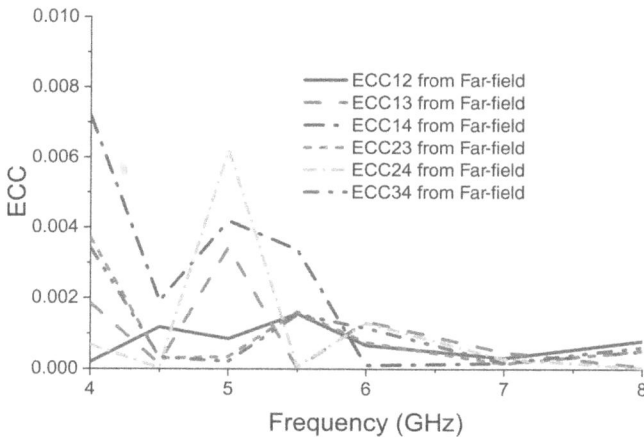

FIGURE 8.22   Envelope correlation coefficient of the proposed CPWG-fed four-port MIMO antenna.

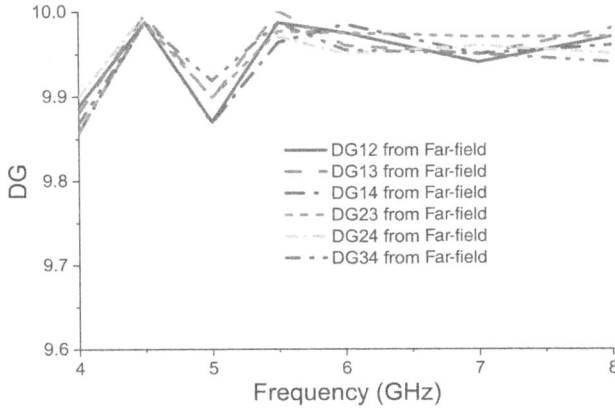

FIGURE 8.23 Diversity gain of the proposed coplanar waveguide ground-fed four-port MIMO antenna.

communication systems [25, 26]. The ergodic channel capacity of the proposed four-port MIMO antenna across the operational bandwidths (5.15–7.37 GHz) is analyzed in Figure 8.24. It is validated that the capacity varies between 20 bps/Hz and 21 bps/Hz, and it is 1.85 times greater than the two-port MIMO antenna type.

8.2.6 Performance Comparision of Proposed Four-port CPWG MIMO Antenna with Existing State-of-the-Art Technology

The performance comparison of the proposed CPWG-fed four-port MIMO antenna with other recent MIMO systems, including antenna elements, size, frequency range, gain, and MIMO diversity parameters, such as isolation and ECC, are compared and illustrated in Table 8.2. In this Table, it can be observed that the proposed CPWG-fed

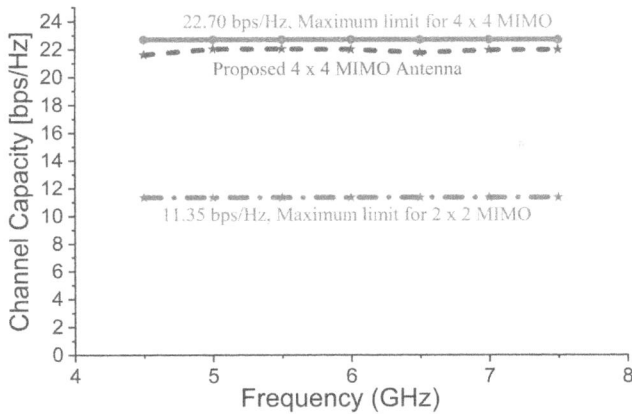

FIGURE 8.24 Channel capacity of the proposed coplanar waveguide ground-fed four-port MIMO antenna.

TABLE 8.2
Performance Comparison of Proposed CPWG-Fed Four-Port MIMO Antenna

Ref	Elements	Feed	Common Ground	IBW (%)/Freq Range (GHz)	Size (mm²)	Gain (dBi)	MIMO Diversity Parameters	
							ECC	Isolation (dB)
[4]	2	Microstrip	Yes	(87.8%)/1.48–3.8	100×65	3.4	0.01	18
[6]	4	Microstrip	No	(12.72%)/5.15–5.85	39×30	2.8	0.02	20
[7]	4	Microstrip	No	(96.22%)/2.2–6.28	45×45	4.0	0.25	14
[8]	4	Microstrip	No	(136.50%)/2–10.6	45×45	4.0	0.003	17
[9]	4	Microstrip	No	(58.56%)/3.20–5.85	40×30	3.45	0.01	17.5
[10]	4	Microstrip	No	(13.06%)/2.36–2.69 (13.71%)/4.89–5.61	70×70	4.0	0.3	21
[11]	4	Probe feed	No	(3.71%)/2.38–2.47	60×60	4.0	–	22
[12]	4	Microstrip	Yes	(4.08%)/2.4–2.5 (38.20%)/4.9–6.5	46×21	3.5	0.3	12
[13]	4	Microstrip	Yes	(67.53%)/2.70–4.94	40×40	4.0	0.1	12
[14]	4	Microstrip	Yes	(136.93%)/2.19–11.7	60×41	6.5	0.25	20
[17]	2	CPW	No	(8.69%)/3.3–3.6 (4.08%)/4.8–5.0	50×50	2.0	0.001	20
[19]	2	CPW	Yes	(122.14%)/2.9–12.0	30×60	6.0	0.003	25
[20]	4	CPW	No	(33.71%)/4.71–6.62	45×45	4.95	0.2	20
[21]	4	CPW	No	(40%)/4.30–6.45	50×50	4.0	0.004	20
Proposed	**4**	**CPW**	**Yes**	**(35.46%)/5.15–7.37**	**90×90**	**4.0**	**0.008**	**25**

four-port MIMO antenna has acceptable dimensions with a wide 10 dB impedance bandwidth covering the Wi-Fi 5/6 bands, high gain, excellent ECC of <0.008, and very high isolation of >25 dB.

8.2.7 CONCLUDING REMARKS

A novel CPWG-fed four-port MIMO antenna has been successfully investigated with designed footprint of 1.65 λ × 1.65 λ × 0.014 λ (at 5.5 GHz). The measured results show that the proposed MIMO antenna can yield a very wide 10 dB impedance bandwidth of 35.46% (5.15–7.37 GHz) that covers both the Wi-Fi 5 (5.15–5.85 GHz) and Wi-Fi 6 (5.925–7.125 GHz) operational bands. Very high isolation of >25 dB was achieved because of the loaded whirligig isolating structure to the four partial grounds. Besides showing very stable gain and radiation efficiencies of approximately 3.5–4 dBi and 72–74%, respectively, very low ECC of <0.008, DG of ~10 dB, and desirable channel capacity of up to 21 bps/Hz are also demonstrated. Therefore, the proposed MIMO antenna can be considered a promising candidate for Wi-Fi 5 and Wi-Fi 6 AP and repeater applications.

8.3 CASE STUDY: WIDEBAND FOUR-PORT MIMO ANTENNA ARRAY WITH HIGH ISOLATION FOR FUTURE WIRELESS SYSTEMS

A four-port MIMO antenna array [27] with wideband and high isolation characteristics for imminent wireless systems functioning in fifth generation (5G) new radio (NR) sub-6 GHz n77/n78/n79 and 5 GHz WLAN bands is proposed. Each array antenna element is a microstrip line-fed monopole type. The novelty of the antenna lies in designing an "EL" slot along with two identical stubs coupled to the partial ground in order to improve the impedance matching and radiation characteristics across the bands of interest. To further attain high port isolation without affecting the compactness and radiation performance of each antenna element, the technique of introducing an innovative and novel unprotruded multislot (UPMS) isolating element (of low profile 2 × 19mm^2) between two closely spaced antenna elements (with edge-to-edge distance of approx. 0.03 λ at 4.6 GHz) is also presented. Besides demonstrating a small footprint of 30 × 40 × 1.6 mm^3, the proposed four-port MIMO antenna array has also shown wide 10 dB impedance bandwidth of 58.56% (3.20–5.85 GHz), high isolation of more than 17.5 dB, and good gain and efficiency of around 3.5 dBi and 85%, respectively, across the bands of interest. Finally, the MIMO performance metrics of the proposed antenna are also analyzed.

8.3.1 OVERVIEW OF EXISTING STATE-OF-THE-ART TECHNOLOGY

Due to the recent development in 5G communication systems functioning at 5G NR sub-6 GHz n77/n78/n79 bands, MIMO technology is hence involved so that larger channel capacity with high data rates (>10 Gbps) can be attained. To meet the aforementioned characteristics, it is required to deploy as many antennas as possible in close proximity. However, enhancing the isolation between closely spaced antennas is quite challenging. Therefore, recently, several types of MIMO antenna arrays with good characteristics including high port isolation, single/dual-band operation, and compact size have been reported in Refs. [28–52] to meet the requirements of WLAN/5G standards. In Ref. [28], a decoupling structure (X-strip) is employed between conventional rectangular patch antennas, which forms an array of five antenna elements with a center-to-center spacing (between two antenna elements) of 0.4 λ. In Ref. [29], a decoupling structure using two thin metallic strips is reported for two planar microstrip-based antennas operating in 2.45 GHz band. With this technique, isolation of better than 20 dB has been achieved. However, the antenna array reported in Refs. [28] and [29] has very large dimensions of 152×60 mm^2 and 70×140 mm^2, respectively. To reduce the overall array size and mutual coupling between the antenna elements, the antennas reported in [30] and [31] have applied parallel-coupled line resonators and ring resonators, respectively, with smaller dimensions of 57×32 mm^2 and 40×40 mm^2. Notably, the antennas reported in Ref. [28–31] uses Roger substrate, which is expensive and, thus, leads to increased manufacturing costs.

To reduce the manufacturing cost, the antennas in Ref. [32–37] use FR-4 substrate, which is less expensive and easily available. In Ref. [32], to reduce the mutual coupling, a PIN diode is placed in the band-stop filter decoupling network and the dimensions of the antenna are only 38×38 mm^2. As for the antenna in Ref. [33], a shorting pin-based interconnected semicircle-enclosed decoupling structure (with a dimension of 200×70 mm^2) is used. Nevertheless, it can only be applied for single-band operation. To accommodate dual/multiband operations, in Ref. [34] reported the technique of applying radiators at different layers to reduce the mutual coupling for 2.4/5 GHz band, whereas in Ref. [35] applied the conventional spatial diversity technique (increasing the distance between the antenna elements) to increase isolation level across the 2.4/5.8 GHz band. Even though the antennas reported in Refs. [34] and [35] have exhibited dual-band operation, they have a large dimension of 110×110 mm^2 and 89×86 mm^2, respectively. To realize reduced antenna array dimensions of 48×48 mm^2 and 30×30 mm^2 while achieving good mutual decoupling between the antenna elements, Ref. [36] reported the use of a thin inductive line and Ref. [37] introduced a decoupling structure composed of a circular patch and four L-shaped branches (placed counterclockwise) printed on the upper surface of the substrate. Nevertheless, Refs. [36] and [37] can only cover a single 5 GHz band operation.

By further observing the open literature, many unconventional techniques have been applied recently to reduce the array antenna size as well as improve the isolation between the antenna elements. In Ref. [38], a transparent antenna using plexiglass as a substrate and operating in a dual-band was reported. As for Ref. [39], the electronic band gap (EBG) structure is applied, and authors in Refs. [40–42] have reported the use of defective ground structure (DGS) that can reduce the antenna array size to 68×38 mm^2. However, these works are not designed for the 5G sub-6 GHz band.

The use of cognitive radio (CR) technology and metamaterials are also expected to improve antenna performance, which plays a vital role in developing 5G systems

[43–45]. The authors in Ref. [43] presented a novel CR technique integrated with the MIMO technology for future 5G communications. In Ref. [44], the authors reported compact 34 × 34 mm^2 MIMO antenna array based on a surface integrated waveguide (SIW) structure with a metallic via hole as an isolating element to operate in 125–300 GHz Terahertz band. Notably, an isolation of 25 dB could be achieved with this technique. In Ref. [45], to reduce the mutual coupling, a novel metasurface structure (with a dimension of 200 × 120 mm^2) is introduced between the antenna elements. However, it supports only a single-band operation.

In Refs. [46–48], the MIMO antenna array with multiple antenna elements (up to 10) for mobile terminal applications was reported for operating in the 5G sub-6 GHz band. Nevertheless, the reported antennas do not cover the complete 5G sub-6 GHz band, and their reflection coefficients are measured at a reference of -6 dB. As for those MIMO arrays designed for laptop/notebook computer applications [49–52], to reduce the mutual coupling, the techniques such as using the system ground as an isolating element [49], loading of the shunt inductor into the antenna [50], and the use of a pie-shaped decoupling structure [51] and a self-decoupled structure [52] have been introduced. Nonetheless, none of the aforementioned antennas has covered the complete 5G NR sub-6 GHz band n77/n78/n79, in which the n77 (3.3–4.2 GHz), n78 (3.3–3.80 GHz), and n79 (4.4–5.0 GHz) require a wideband coverage of at least 3.3 GHz to 5.0 GHz. Therefore, the MIMO systems functioning in 5G NR sub-6 GHz bands are essential for future wireless systems, and its corresponding MIMO array antenna (such as the four-port MIMO array) should possess the characteristics such as being simple in design [53, 54], compact in size, wideband, low profile, lightweight, cost-effective, high port isolation, good impedance matching, and space-efficient capabilities in array functioning.

In this design, a compact-size single monopole antenna with a wideband operation that can well cover the 5G NR sub-6 GHz band and WLAN 5 GHz band is initially introduced. To attain high isolation within a restricted space for a two-port MIMO antenna array, a novel UPMS-isolating element of a low-profile of 2 × 19 mm^2 is deployed between the antenna elements that are closely spaced at an edge-to-edge distance of 0.03 λ, where λ is the free space wavelength at 4.6 GHz. Finally, a four-port MIMO antenna array is also studied. Besides showing the typical results such as bandwidth, gain, and radiation properties, the MIMO and diversity performance parameters of the proposed MIMO array antenna, such as ECC, DG, mean effective gain (MEG), channel capacity loss (CCL), and total active reflection coefficient (TARC), are also analyzed for the validation of MIMO performances.

8.3.2 Controlling Mechanism and Wideband Operation of Single Monopole Antenna

Figure 8.25 illustrates the geometry of the proposed single monopole antenna with optimized parameters. It has a designed footprint of 30 × 25 mm^2 and is printed on a 1.6 mm thick FR-4 substrate (dielectric constant of 4.3 and loss tangent of 0.0025).

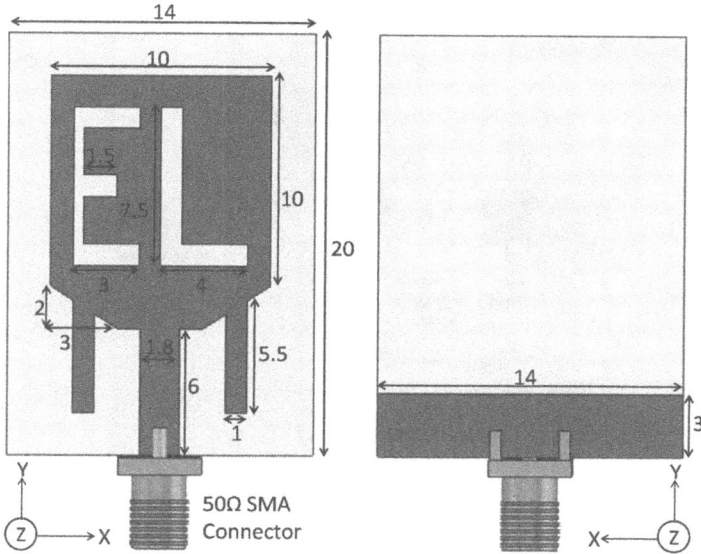

FIGURE 8.25 Geometry of proposed single monopole antenna.

8.3.2.1 Design Steps of Single Monopole Antenna

To begin, a single rectangular patch resonating at 4.6 GHz, whose length and width are calculated from well-known mathematical equations, is optimized using electromagnetic full wave CST Microwave Studio® [55]. As shown in Equations (8.1–8.4), W and L are denoted as the width and length of the patch, h is the height of FR-4 substrate, and ε_{eff}, c, and f_r are the effective permittivity, speed of light (3×10^8 m/s), and resonating frequency, respectively.

$$W = \frac{c}{4f_r\sqrt{\frac{\epsilon_r +1}{2}}} \quad (8.1)$$

$$\epsilon_{eff} = \frac{\epsilon_r +1}{2} + \frac{\epsilon_r -1}{2}\left(\frac{1}{\sqrt{1+12\frac{h}{W}}}\right) \quad (8.2)$$

$$L = \frac{c}{4f_r\sqrt{\epsilon_{eff}}} \quad (8.3)$$

Figure 8.26 shows the simulated design steps/evolution of the proposed single monopole antenna. Here, a single microstrip line-fed rectangular monopole with a partial ground size of 14×3 mm^2 is initially studied, and its corresponding reflection coefficient (S_{11}) shows a simulated 10 dB impedance bandwidth of 4–5.2 GHz.

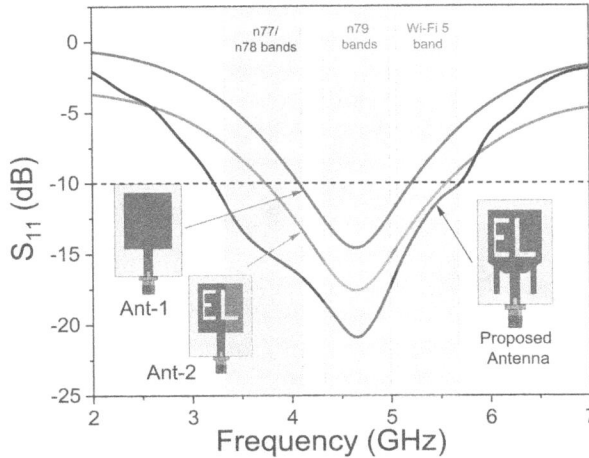

FIGURE 8.26　Simulated design steps of proposed single monopole antenna.

However, this bandwidth is not sufficient to cover the entire 5G sub-6 GHz band. Therefore, to widen the bandwidth, two slots of E and L shapes are etched on the rectangular monopole radiator, which effectively enhances the bandwidth in the range of 3.6–5.5 GHz by modifying the surface current distribution at higher order modes, as depicted in Figure 8.26. Finally, to further enhance the bandwidth so that the proposed single monopole antenna can cover the entire 5G NR sub-6 GHz n77/n78/n79 bands and 5.2 (5.15–5.5) GHz WLAN bands, two high-frequency protruded stubs are introduced at the left and right bottom sections of the monopole. Owing to the use of these two stubs that couple with the partial ground, additional capacitive reactance is mitigated, which improves the bandwidth of the antenna to 3.25–5.6 GHz.

8.3.2.2　Vector Current Distribution of Single Monopole Antenna Element

To further comprehend the wideband mechanism of the proposed single monopole antenna, its corresponding vector current distribution without and with the EL slots are shown in Figure 8.27. The currents flow in only one direction (Figure 8.27a), whereas in Figure 8.2b, the currents are densely populated and flow in various directions. Therefore, merging of these currents will result in reducing the quality factor of the antenna, which in turn widens the bandwidth.

8.3.2.3　Experimental Results

To validate and endorse the proposed single monopole antenna capability for practical utilization, the prototype was fabricated and its radiation performances including radiation patterns, gain, and efficiency were measured in an anechoic chamber.

The S_{11} of the proposed single monopole antenna element prototype was measured using the Rohde & Schwarz ZVH8 VNA. Figure 8.28 illustrates the simulated and measured S_{11} (dB) of the proposed single monopole antenna.

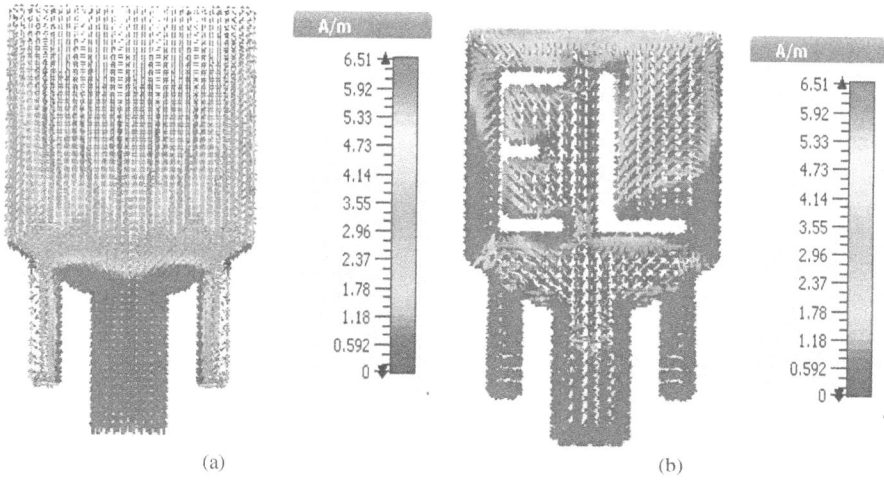

(a)   (b)

FIGURE 8.27   Vector current density of reference antenna (a) without the extra-low (EL) slot and (b) with the EL slot.

Here, a good agreement with a relatively small deviation is observed between the simulated and measured results, which may be due to soldering or fabrication tolerances.

Figure 8.29 depicts the simulated and measured 2D radiation pattern in the E-plane and H-plane at 4.6 GHz with an angle θ varied from -90° to +90°. Here, a near omnidirectional radiation pattern is observed in the E-plane. Moreover, it is notable that the measured and simulated radiation patterns are in good agreement with each other.

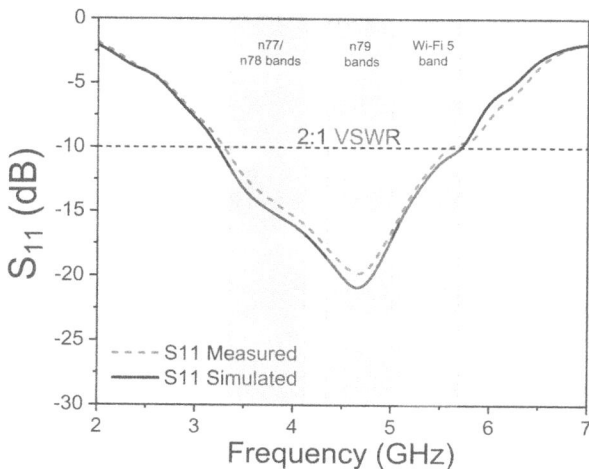

FIGURE 8.28 S_{11} (dB) of single monopole antenna element.

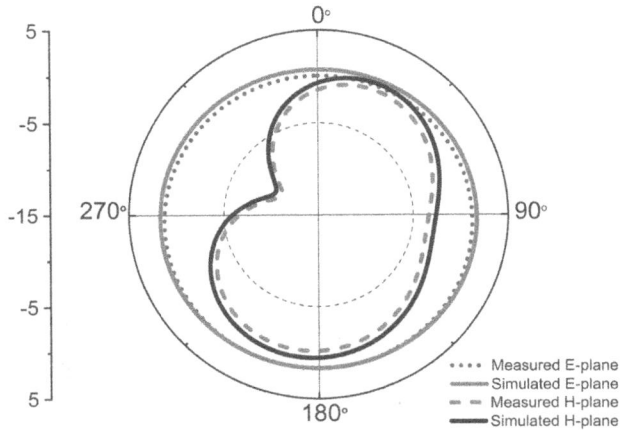

FIGURE 8.29 Radiation pattern of a single monopole antenna in the E-plane and H-plane at 4.6 GHz.

Figure 8.30 demonstrates that the proposed single antenna has a simulated gain of 2.5 dBi and measured gain of 2.45 dBi. Similarly, the maximum simulated efficiency is 82% and maximum measured efficiency is 80%. It is noteworthy that the proposed single monopole antenna has a gain above 2.25 dBi and efficiency above 75% throughout the operating bands.

8.3.3 MIMO ANTENNA ARRAY WITH UPMS ISOLATING ELEMENT

The geometry of a two-port closely placed antenna array, namely, Antenna-1 and Antenna-2, with an edge-to-edge distance of 0.03 λ (at 4.6 GHz) is illustrated in

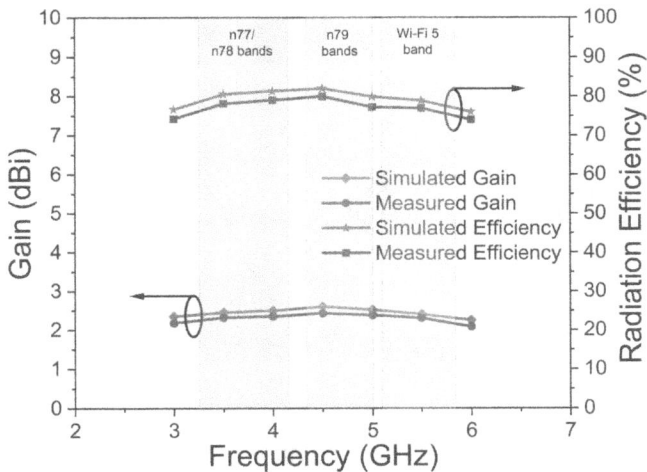

FIGURE 8.30 Gain and efficiency of single antenna element.

Figure 8.31a. The two antennas share a common ground of 30×3 mm^2, as shown in Figure 8.31b. The array is printed on a 1.6 mm thick FR-4 substrate with footprint of 30×25 mm^2. Both Antenna-1 and Antenna-2 are fed by SubMiniature Version A (SMA) connector, individually soldered to a microstrip-fed line 6 mm in length and 1.8 mm in width, which ensures proper 50 Ω impedance matching. For obtaining the corresponding simulated S-parameters, when Antenna-1 is active, Antenna-2 is terminated with 50 Ω impedance load and vice versa. The simulated S-parameters in Figure 8.31c illustrate that there is an apparent increase in bandwidth when the antenna is transformed from a single antenna to a two-antenna array. The bandwidth obtained for the two-port antenna array is 56.69% (3.21–5.75 GHz). However, from the transmission coefficient curve shown in Figure 8.31c, it is noticed that the isolation between Antenna-1 and Antenna-2 is very undesirable.

Therefore, to achieve isolation larger than 15 dB, an UPMS isolating element is deployed between Antenna-1 and Antenna-2 and it is extended from the middle section of the ground plane, as depicted in Figure 8.32b. The UPMS is composed of multiple narrow slots (each of size 0.6×5 mm^2) etched on a 2×19 mm^2 ground-protruded stub.

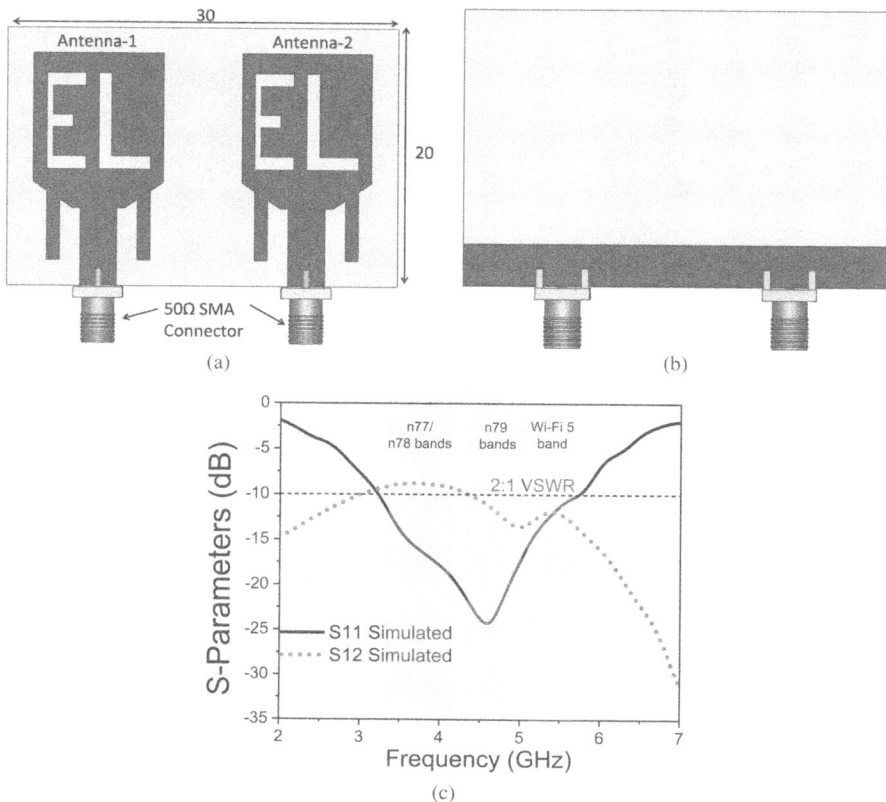

FIGURE 8.31 Geometry of two-port antenna element without an isolating element.

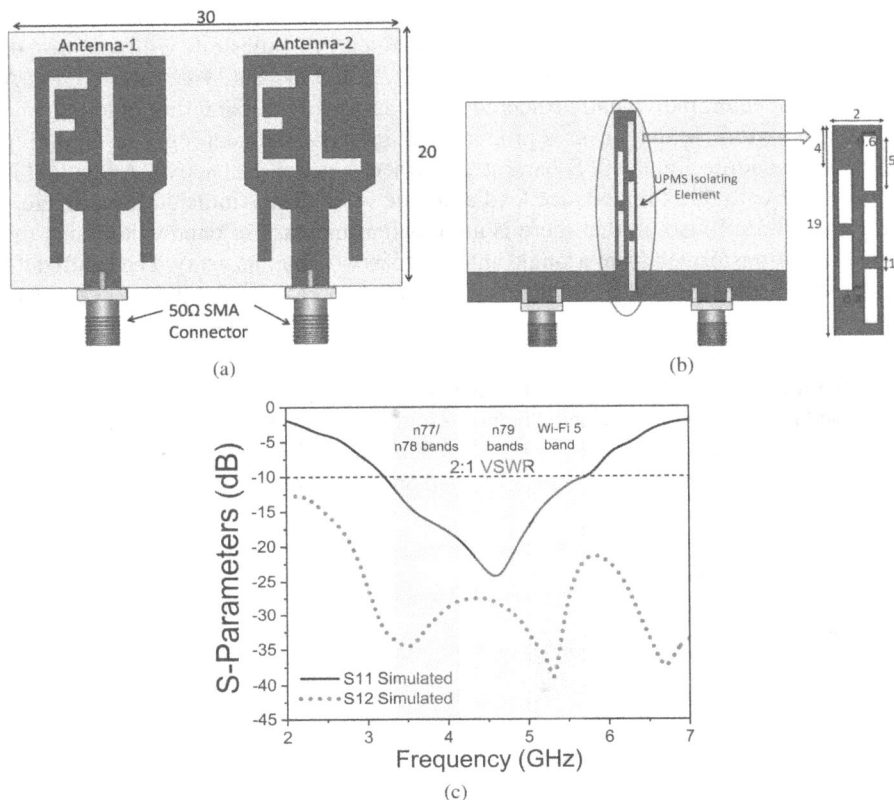

FIGURE 8.32 Geometry of the two-port antenna array element with a unprotruded multislot isolating element.

From Figure 8.32c, the reflection coefficient remains unchanged, whereas the transmission coefficient (S_{12}) is increased from -12 dB to -28 dB at the resonance of 4.6 GHz. The simulated isolation throughout the desired operating bands is better than 20 dB.

8.3.3.1 Reduction Coupling Mechanism of Proposed UPMS Isolating Element

The features of the UPMS isolating element can be further explained by applying two plane waves across the length of isolating element. The transmission coefficients ($S_{12} = S_{21}$) are obtained by simulating the UPMS isolating element under the defined boundary condition in Ref. [56]. Figure 8.33 illustrates the S_{12} (dB) of the proposed UPMS, where it is visualized that the multislot acts as a reflector and supports an S_{12} coefficient below -18 dB throughout the functioning band (3.21–5.75 GHz) by absorbing the surface wave and, therefore, serving as a wide-stop band filter and nullifying the surface wave propagation between the Antenna-1 and Antenna-2. As a result, the electric isolation between Antenna-1 and Antenna-2 has been increased

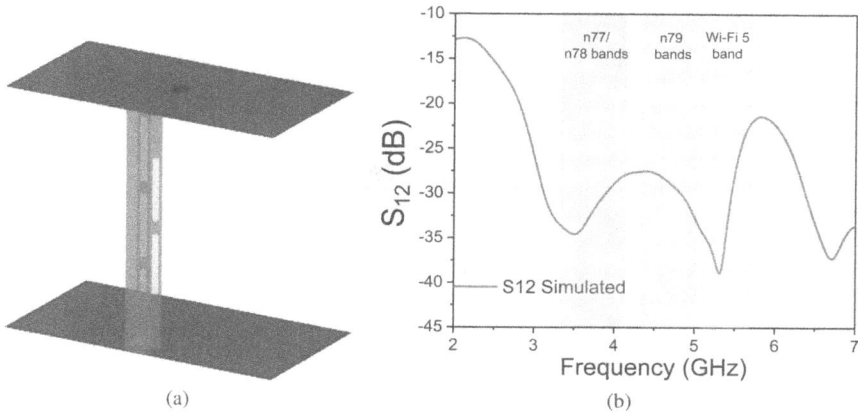

(a)   (b)

FIGURE 8.33 S_{12} analysis of proposed unprotruded multislot isolating element.

without increasing their physical separation. This proves that the proposed UPMS element acts as an isolating element when placed between two antennas.

8.3.3.2 Electric Field Intensity (V/m) of Proposed UPMS Isolating Element

To further validate the reduction coupling mechanism and the claim of increased isolation between Antenna-1 and Antenna-2 due to the proposed UPMS, the electric field intensity (V/m) distribution on the surface of the two-port MIMO antenna array without and with the UPMS isolating element is depicted in Figures 8.34a and b, respectively. For analyzing the effect, Antenna-1 is excited, while Antenna-2 is kept terminated with 50 Ω matched impedance load. Under this condition, in the absence of UPMS, from Figure 8.34a, it is apparent that an electromagnetic energy induced by Antenna-1 is strongly coupled with the radiating edge of Antenna-2 through a gap of 2 mm and the maximum electric field lies on the radiating edges of both antennas. Whereas in the presence of UPMS, as shown in Figure 8.34b, the electromagnetic energy of Antenna-1 is now coupled with the radiating edge of the UPMS, resulting in nullifying the surface wave by alternating the current path, which is now flowing along the UPMS, essentially reducing the mutual coupling between Antenna-1 and Antenna-2.

8.3.4 EXPERIMENTAL RESULTS

The two-port MIMO antenna array was fabricated, and its front and back pictures are depicted in Figure 8.35. Figure 8.36 illustrates the simulated and measured S-parameters of the proposed two-port MIMO antenna array. The result shows that there is an apparent improvement in bandwidth of S_{11} for the two-port MIMO antenna array as compared to the single monopole antenna element. From Figure 8.36, the measured 10 dB impedance bandwidth was 56.69% (3.21–5.75 GHz) and it conforms to the bandwidth requirements of 5G NR sub-6 GHz n77/n78/n79 bands and the unlicensed 5.2 GHz WLAN band. It is also verified that measured

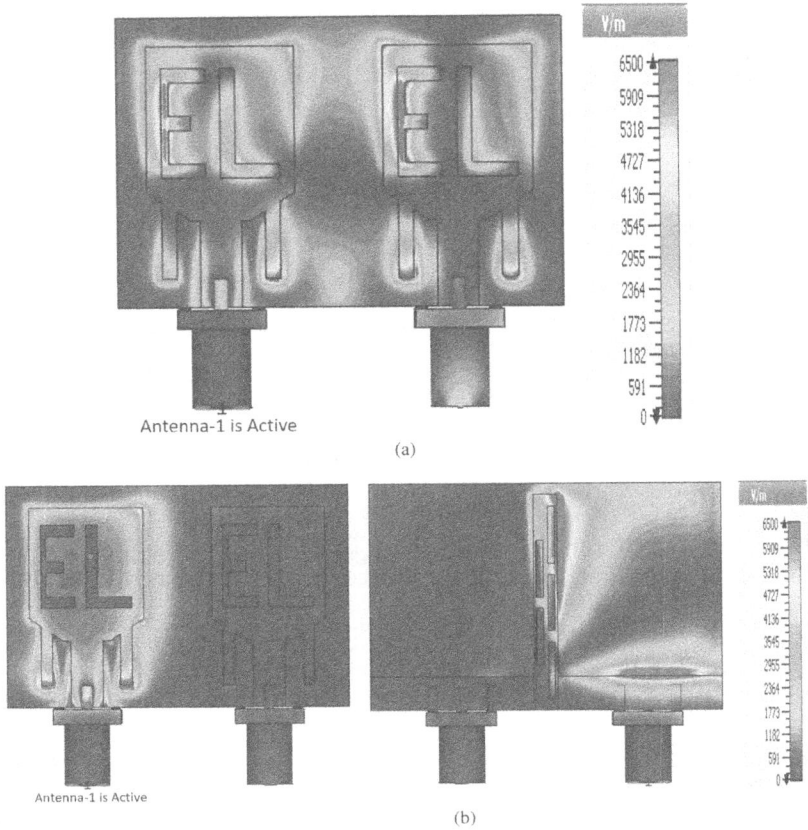

FIGURE 8.34   The electric field (V/m) distribution on the surface of a two-port MIMO antenna array without (a) and with (b) an unprotruded multislot isolating element.

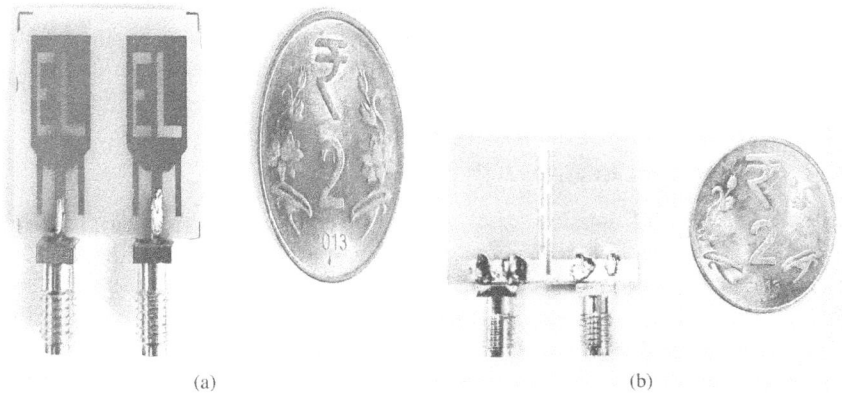

FIGURE 8.35   Fabricated prototype of the proposed two-port MIMO antenna array.

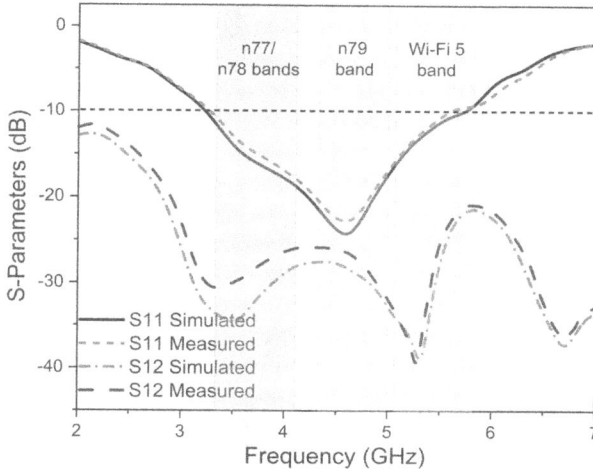

FIGURE 8.36 Simulated and measured scattering parameters of the two-port MIMO antenna array.

isolation was better than 20 dB throughout the bands of interest, which is acceptable as per 5G standards as well as the industry requirement to ignore the detrimental effects of wireless devices operating in the practical environment of a MIMO system. Here, a minimal deviation is observed between the simulated and measured results, which may be due to soldering inaccuracies or fabrication tolerances.

Figure 8.37 shows the 2D radiation pattern of the proposed two-port MIMO array. The patterns in the H-plane are practically the same for both Antenna-1 and Antenna-2, whereas their corresponding E-plane patterns are almost 180° out of phase with each other. This proves that the proposed two-port MIMO antenna array has achieved pattern diversity in the E-plane, which ensures that there is no interference during reception.

Figure 8.38 illustrates the simulated and measured gain and efficiency of the proposed two-port MIMO antenna array, and good agreements are demonstrated

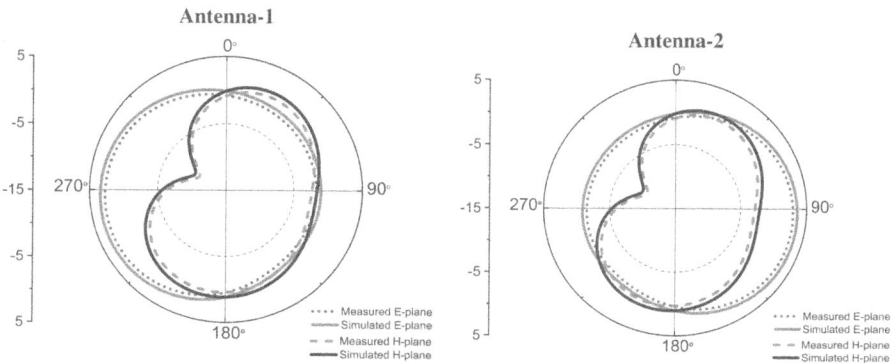

FIGURE 8.37 Radiation patterns of two-port MIMO antenna array in the E-plane and H-plane. (*Continued*)

FIGURE 8.37   *(Continued)*

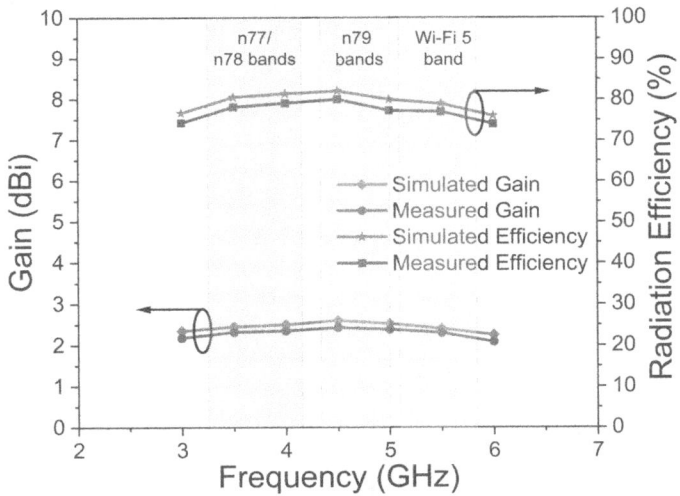

FIGURE 8.38   Gain and efficiency of Antenna-1 in two-port MIMO antenna array.

TABLE 8.3

Comparison of Gain and Efficiency of Antenna-1 in Two-Port MIMO Array with Single Monopole Antenna Element

Operating Band	Gain (dBi) (Single Antenna)	Gain (dBi) (Two-port Antenna)	Efficiency (%) (Single Antenna)	Efficiency (%) (Two-port Antenna)
n77/n78	2.35	2.85	79.05	81.25
n79	2.45	2.90	80	82
Wi-Fi 5	2.40	2.85	77.25	81.00

between them. Here, we have shown only the gain and efficiency results of Antenna-1. There is an increase in the gain and efficiency value as compared with the single monopole antenna element. Furthermore, the gain and efficiency of Antenna-1 in the two-port MIMO antenna is above 2.85 dBi and 81%, respectively, throughout the bands of interest. The performance comparison (measured gain and efficiency) between the single monopole antenna element and Antenna-1 in the proposed two-port MIMO antenna are catalogued in Table 8.3. From the Table 8.3, it is observed that the two-port MIMO antenna has exhibited slightly better antenna performance.

To further verify whether this UPMS isolating element is appropriate for a MIMO array that has more than a two-antenna array element, a four-antenna element MIMO array with UPMS isolating elements was developed and its analysis with complete MIMO performance metrics are discussed in further sections

8.3.5 FOUR-PORT MIMO ANTENNA ARRAY

A four-port MIMO antenna array with a designed footprint of $30 \times 40 \times 1.6 \text{ mm}^3$ is shown in Figure 8.39. Here, Antenna-3 and Antenna-4 are mirror images of Antenna-2 and Antenna-1, respectively, and they have applied the same UPMS isolating element. The performance metrics of the four-port MIMO configuration are extensively discussed later in the chapter.

8.3.5.1 Experimental Results of Proposed Four-Port MIMO Antenna Array

The fabricated prototype of the proposed four-port MIMO antenna array (using FR-4 substrate) is shown in Figure 8.40. Typical antenna and MIMO performance measurements are conducted in an effort to endorse the MIMO capabilities for practical utilization. The detailed examination and analysis of measured results are provided in the following sections.

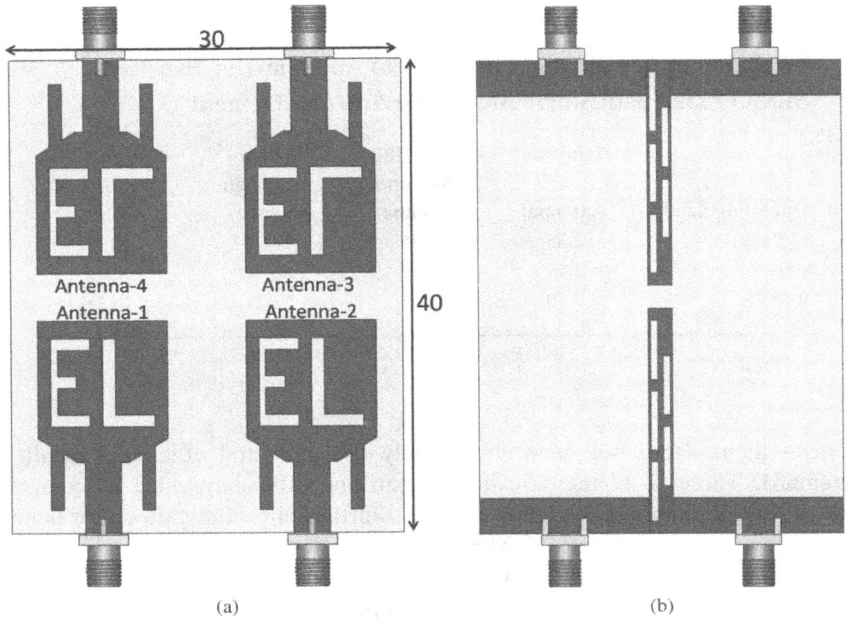

FIGURE 8.39 Proposed four-port MIMO antenna array: (a) top view and (b) bottom view.

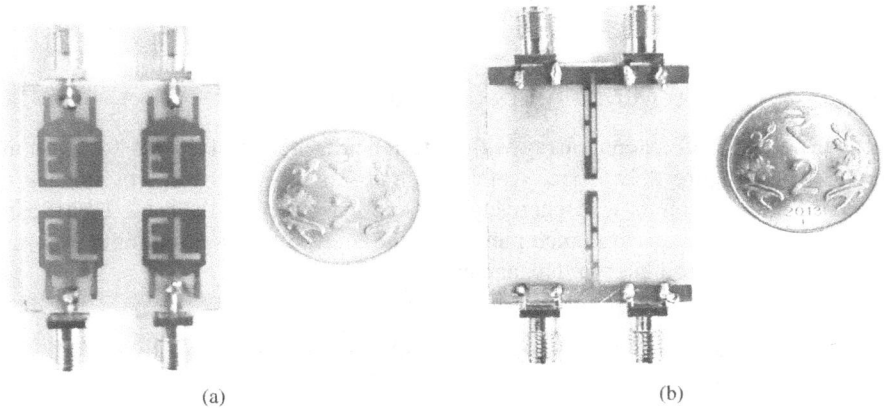

FIGURE 8.40 Fabricated prototype of proposed four-port MIMO antenna array: (a) top layer and (b) bottom layer.

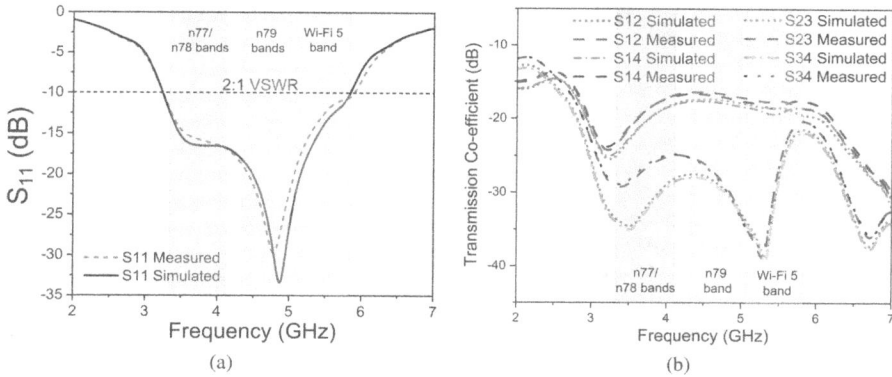

FIGURE 8.41 Simulated and measured scattering parameter curves: (a) reflection coefficient curves and (b) transmission coefficient curves.

8.3.5.1.1 S-parameters

Figure 8.41a demonstrates the simulated and measured S_{11} of the proposed four-port MIMO antenna array. The other S-parameters (S_{22}, S_{33} and S_{44}) are not measured because their results are almost identical to S_{11}.

As compared with the two-port MIMO array (see Figure 8.36), the 4.6 GHz resonance (in two-port MIMO) is now shifted to approximately 4.8 GHz (in four-port MIMO) and the measured S_{11} of the four-port MIMO array has exhibited a slightly wider 10 dB impedance bandwidth of 58.56% (3.20–5.85 GHz). Therefore, the proposed four-port MIMO array can well cover the bandwidth requirements of 5G NR sub-6 GHz n77/78/79 bands and the entire 5 GHz WLAN band (5.15–5.85 GHz). The transmission coefficient investigation is illustrated in Figure 8.41b. It is observed that S_{12} and S_{34} are almost the same and they are well below −20 dB throughout the operating bands of interest, whereas S_{14} and S_{23} are the same and below −17.5 dB. From this investigation, it is realized that the simulated and measured curves have displayed good coherence. Notably, the unexpected variations are due to minor fabrication tolerances or unavoidable conductor loss in the usage of coaxial cables at the time of measurements.

8.3.5.1.2 Radiation Patterns of Proposed Four-Port MIMO Antenna Array

To understand the radiational function of the proposed four-port MIMO antenna array, its corresponding simulated copolar and cross-polar radiation patterns at a resonant frequency of 4.8 GHz are depicted in Figure 8.43. The magnitude of the

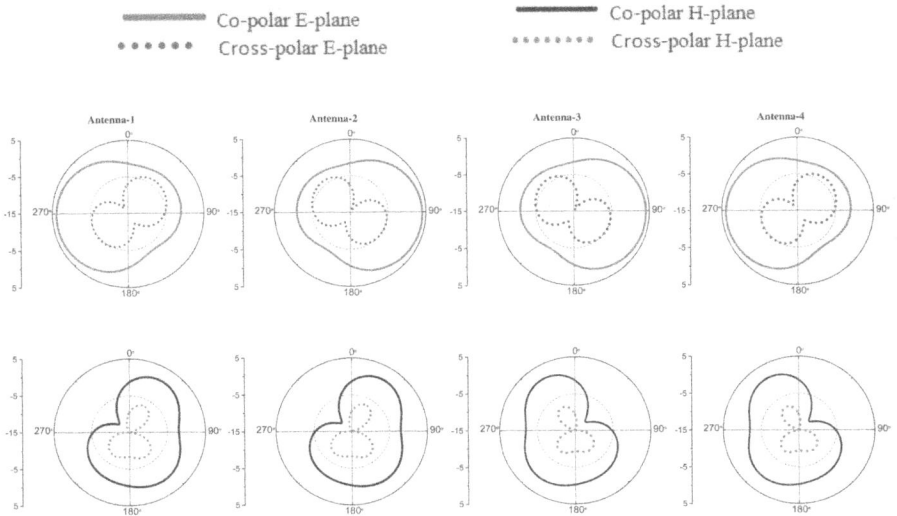

FIGURE 8.42 Simulated copolar and cross-polar radiation pattern at 4.8 GHz.

cross-polar is less than that of the copolar radiation pattern, which indicates that the proposed MIMO antenna array has good isolation (Figure 8.42).

Figure 8.43 illustrates the simulated and measured radiation pattern in the E-plane and H-plane. It is observed that the 2D patterns in the H-plane are similar between Antenna 1 and 2 and between Antenna 3 and 4, whereas the E-plane patterns are 180° out of phase. Similarly, the E-plane patterns are similar between Antenna 1 and 4 and Antenna 2 and 3, whereas the H-plane patterns are 180° out of phase. This proves that the proposed antenna has achieved pattern diversity, which ensures that there is no interference during reception and maintains an omnidirectional radiation pattern in both the E-plane and H-plane.

There is also a very small deviation between the simulated and measured results, and discrepancies, if any, may be due to fabrication inaccuracies or unavoidable conductor losses in the cable used during measurement analysis.

8.3.5.1.3 Gain and Efficiency of Proposed Four-Port MIMO Antenna Array

Figure 8.44 depicts the gain and efficiency of the proposed four-port MIMO antenna array. Here, the gain and efficiency of Antenna-1 are measured only because the other three antenna elements (Antenna-2 to Antenna-4) are of identical structures. In this figure, it is observed that the gain and efficiency results of Antenna-1 measured from the proposed four-port MIMO antenna array are better than the two-port MIMO antenna array. Here, the measured gain and efficiency are well above 3.45 dBi and 84%, respectively, throughout the bands of interest.

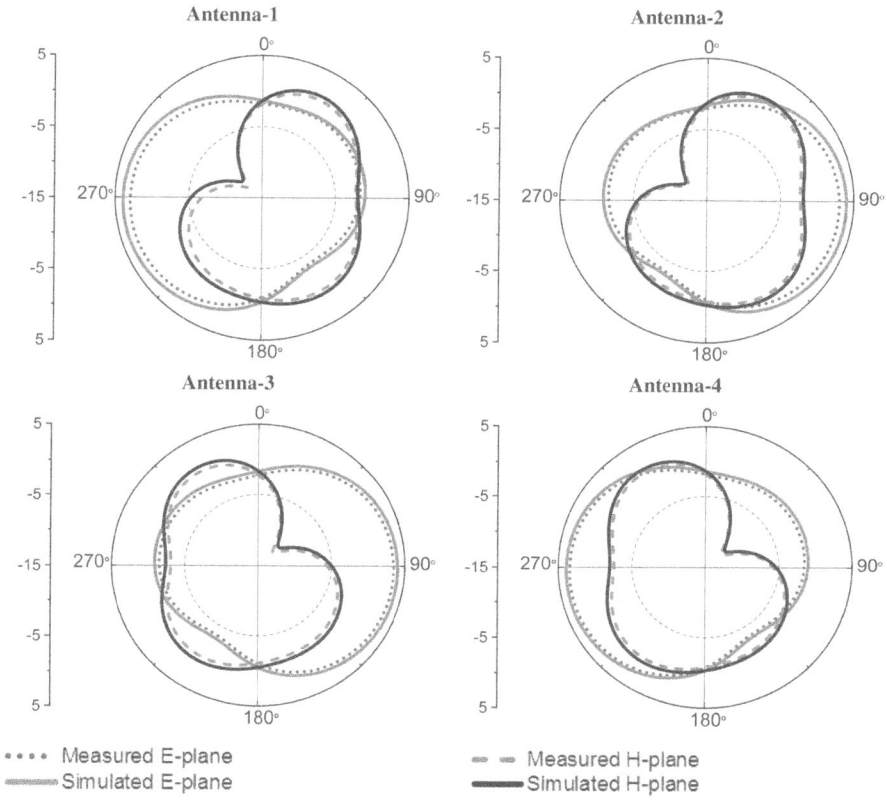

FIGURE 8.43 Simulated and measured radiation patterns in the E-plane and H-plane at 4.8 GHz.

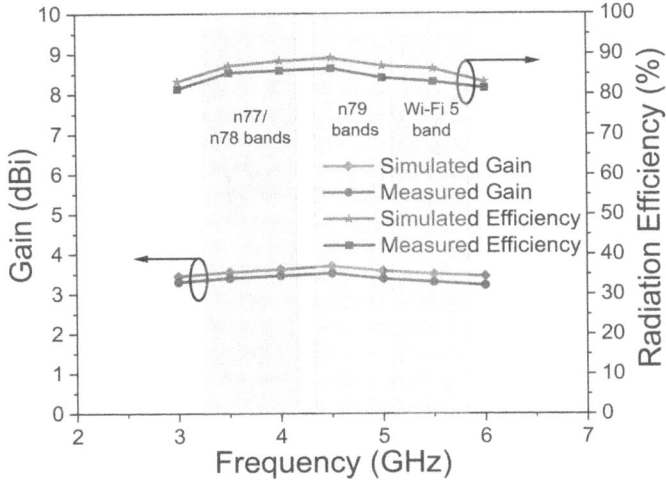

FIGURE 8.44 Simulated and measured gain and efficiency of the proposed four-port MIMO antenna array.

Table 8.4 shows the comparison of measured gain and efficiency of Antenna-1 in the four-port MIMO array and in the two-port MIMO array.

8.3.5.2 Performance Metrics of Proposed Four-Port MIMO Antenna Array

To ensure that the performance of the proposed four-port MIMO antenna array is acceptable for a practical environment, the key performance metrics including ECC, DG, CCL, MEG, and TARC are evaluated. A detailed discussion of the parameters is presented in the subsequent sections.

TABLE 8.4

Comparison of Measured Gain and Efficiency Values of Two-Port Antenna Array with Four-Port Antenna Array

Operating Bands (GHz)	Gain (dBi) (Two-Antenna)	Gain (dBi) (Four-Antenna)	Efficiency (%) (Two-Antenna)	Efficiency (%) (Four-Antenna)
n77/n78	2.85	3.45	81.25	84.55
n79	2.90	3.50	82	85.00
Wi-Fi 5	2.85	3.45	80.50	83.85

8.3.5.2.1   ECC

ECC is one of the key performance metrics of a MIMO antenna system, and it can be calculated from the S-parameters using Equation (8.4).

$$ECC = \frac{\left| S_{ii}^{*} S_{ij} + S_{ji}^{*} S_{jj} \right|^{2}}{\left(1 - \left|S_{ii}\right|^{2} - \left|S_{ji}\right|^{2}\right)\left(1 - \left|S_{jj}\right|^{2} - \left|S_{ij}\right|^{2}\right)} \tag{8.4}$$

However, Equation (8.4) is only accurate when the antenna efficiencies are very high (near 100%). Therefore, another method to determine the ECC value is by applying the far field radiation of the antennas using the Equation (8.5), provided as follows:

$$\rho_{e} = \frac{\left| \iint \left[\overline{F_{1}}(\theta,\varphi) \times \overline{F_{2}}(\theta,\varphi) \right] d\Omega^{2} \right|}{\iint \left| \overline{F_{1}}(\theta,\varphi) \right|^{2} d\Omega \iint \left| \overline{F_{2}}(\theta,\varphi) \right|^{2} d\Omega} \tag{8.5}$$

where "$F_{i}(\theta, \phi)$" is the radiated field of the ith antenna.

Figure 8.45 shows the ECC curves of the proposed four-port MIMO antenna over the bands of interest determined from both the S-parameter equation as

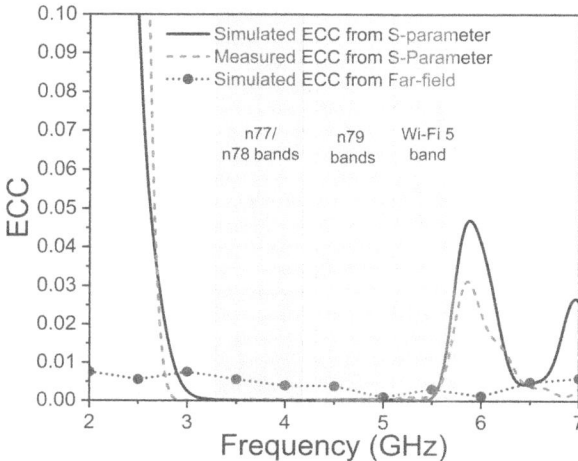

FIGURE 8.45   ECC values from the S-parameter data and from far-field patterns.

well as far-field patterns equation. Here, the plotted ECC curves are between Antenna-1 and Antenna-2 only, and the other curves are not shown due to brevity. In this figure, the ECC values (measured and simulated) determined from the S-parameter equation are well agreed with each other across the bands of interest, showing ECC values of well below 0.05. As for its far-field patterns equation counterpart, it has shown ECC values of well below 0.01 across the bands of interest. Therefore, the investigation of ECC confirms that the proposed MIMO antenna system has achieved high port isolation and excellent diversity performance.

8.3.5.2.2 DG of Proposed Four-Port MIMO Antenna Array

DG exhibits "the loss in transmission power when diversity mechanism is performed on the MIMO system" for the MIMO configuration. The DG is calculated by using Equation (8.6).

$$DG = 10\sqrt{1-\left|\rho_{eij}\right|} \tag{8.6}$$

Figure 8.46 depicts the calculated DG values from both the S-parameters and far-field patterns. It can be observed that the DG values are above 9.8 dB throughout the bands of interest. This ensures a good diversity performance of the proposed four-port MIMO antenna array.

8.3.5.2.3 TARC

The TARC is a parameter used to validate the performance of a MIMO antenna in a single attempt. TARC expresses the total incident power to the radiated power of the MIMO antenna array. TARC is calculated as the square root of the incident power provided by all excitations minus the radiated power and then dividing

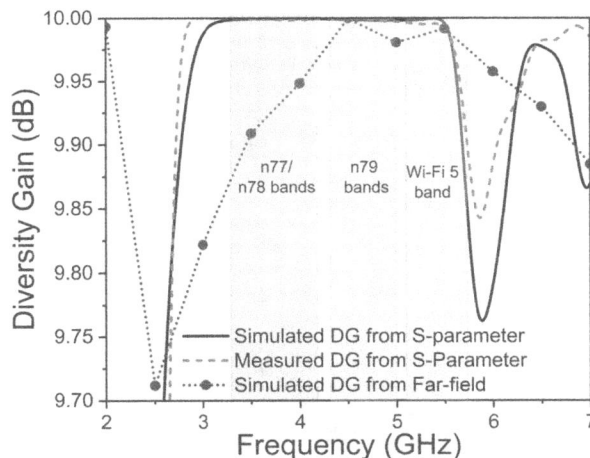

FIGURE 8.46 Diversity gain (DG) of proposed four-port MIMO antenna array.

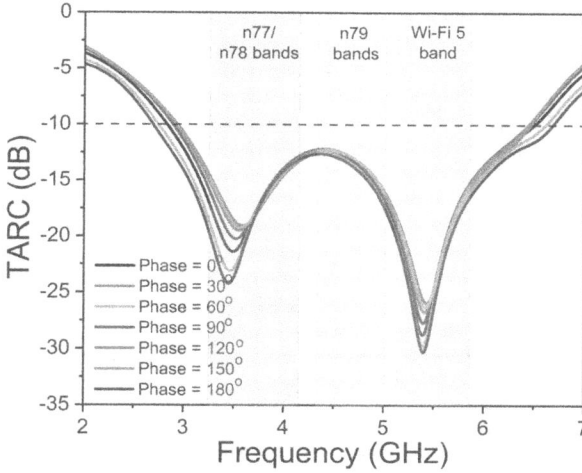

FIGURE 8.47 Total active reflection coefficient curves of the proposed four-port MIMO antenna array.

by the incident power. By assuming lossy antennas, it can be calculated using Equation (8.7):

$$\Gamma = \frac{\sqrt{\left(\left|S_{ii} + S_{ij}e^{j\theta}\right|^2\right) + \left(\left|S_{ji} + S_{jj}e^{j\theta}\right|^2\right)}}{\sqrt{2}} \tag{8.7}$$

where θ is the input phase angle that is varied from 0° to 180° with step increment of 30° and S_{ii} and S_{jj} are the reflection coefficients of the port one and port two, respectively. The TARC values are calculated for Antenna-1 and Antenna-2. Therefore, S_{ii} and S_{ij} represent reflection coefficient S_{11} and S_{22} in (dB) of port 1 and port 2, respectively, whereas S_{ij} and S_{ji} are the port isolation S_{12} and S_{21} in (dB) between the antenna array.

Figure 8.47 depicts the measured TARC values. The TARC values are almost stable throughout the operating bands, which ensures that the proposed MIMO antenna system has achieved high port isolation.

8.3.5.2.4   CCL

The CCL is listed as one of the most important metrics among the MIMO performance parameters. Thus, it is essential to provide the details of the CCL of the proposed four-port MIMO antenna array during the correlation impact. The CCL is calculated by using the following equations:

$$C_{loss} = -log_2 det\left(\Psi^R\right) \tag{8.8}$$

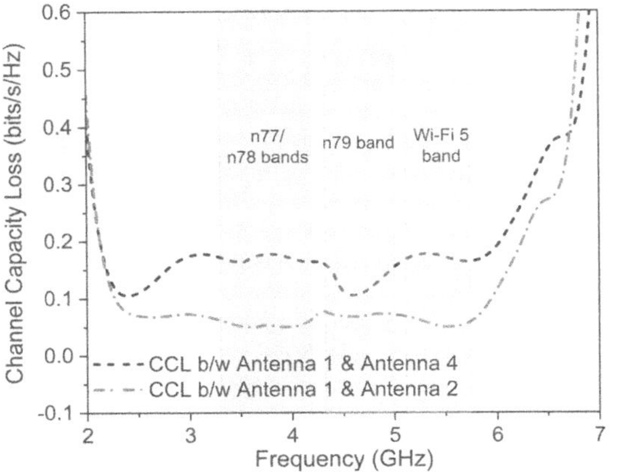

FIGURE 8.48   CCL of the proposed four-port MIMO antenna array.

where, Ψ^R is the correlation matrix of the receiving antenna and is expressed as:

$$\Psi^R = \begin{bmatrix} \Psi_{ii} & \Psi_{ij} \\ \Psi_{ji} & \Psi_{jj} \end{bmatrix} \tag{8.9}$$

where,

$$\Psi_{ii} = 1 - \left(\left| S_{ii} \right|^2 + \left| S_{ij} \right|^2 \right) \tag{8.10}$$

$$\Psi_{ij} = -\left(S_{ii}^{*} \, S_{ij} + S_{ji}^{*} S_{jj} \right) \tag{8.11}$$

$$\Psi_{ji} = -\left(S_{jj}^{*} S_{ji} + S_{ij}^{*} S_{ii} \right) \tag{8.12}$$

$$\Psi_{jj} = 1 - \left(\left| S_{jj} \right|^2 + \left| S_{ji} \right|^2 \right) \tag{8.13}$$

Figure 8.48 demonstrates that for the proposed MIMO antenna system, the achieved CCL value is less than 0.4 bit/s/Hz as per the requirement of the industry for the entire bands of interest that specifies the proposed MIMO antenna system has a high throughput for practical applications.

8.3.5.2.5   MEG

For diversity performance analysis, the MEG is another essential parameter and is expressed as the mean power received by the diversity antenna to the mean power received by an isotropic antenna in affluent multipath fading environment. The MEG is achieved using Equation (8.14):

$$MEG_i = 0.5\mu_{irad} = 0.5\left(1 - \sum_{j=1}^{K} \left| S_{ij} \right|^2 \right) \tag{8.14}$$

TABLE 8.5
MEG Values of Proposed Four-Port MIMO Antenna Array

Frequency (GHz)	MEG-1 (dB)	MEG-2 (dB)	MEG-3 (dB)	MEG-4 (dB)	MEG-1/MEG-2	MEG-3/MEG-4
3.0	-8.53	-8.68	-8.76	-8.60	0.981	1.019
3.5	-6.65	-7.01	-7.08	-6.74	0.947	1.050
4.0	-6.61	-6.69	-6.76	-6.68	0.988	1.011
4.5	-6.69	-7.02	-6.97	-7.03	0.992	0.992
5.0	-7.01	-7.10	-7.02	-7.08	0.987	0.992
5.5	-7.04	-7.08	-7.05	-7.10	0.994	0.991
6.0	-7.35	-7.33	-7.37	-7.42	1.002	0.993

where in Equation (8.14), K is number of antennas, in this case 4, i is the active antenna and η_{irad} is the radiation efficiency of the ith antenna. By expanding Equation (8.14), the MEG values for Antenna-1, Antenna-2, Antenna-3, and Antenna-4 are found using the following equations and the calculated values are given in Table 8.5:

$$MEG_1 = 0.5\left(1-|S_{11}|^2 -|S_{12}|^2 -|S_{13}|^2 -|S_{14}|^2\right) \tag{8.15}$$

$$MEG_2 = 0.5\left(1-|S_{21}|^2 -|S_{22}|^2 -|S_{23}|^2 -|S_{24}|^2\right) \tag{8.16}$$

$$MEG_3 = 0.5\left(1-|S_{31}|^2 -|S_{32}|^2 -|S_{33}|^2 -|S_{34}|^2\right) \tag{8.17}$$

$$MEG_4 = 0.5\left(1-|S_{41}|^2 -|S_{42}|^2 -|S_{43}|^2 -|S_{44}|^2\right) \tag{8.18}$$

To achieve good diversity performance, the MEG values should be in between −3 and −12. From Table 8.5, the MEG values of all four antenna elements are within the well-defined limit. Also visualized from Table 8.5, the ratio of MEG1/MEG2 and MEG3/MEG4 is equal to 1 that confirms that the proposed MIMO antenna system has achieved better diversity performance in a multipath fading environment.

8.3.6 PERFORMANCE COMPARISON OF PROPOSED FOUR-PORT MIMO ANTENNA ARRAY WITH OTHER EXISTING STATE-OF-THE-ART TECHNOLOGY

The performance comparison of the proposed four-port MIMO antenna system with other existing MIMO systems, including operating bands, size, type of isolating element, gain, efficiency, ECC, and DG are displayed in Table 8.6. In Table 8.6, it can be noted that the proposed four-port MIMO antenna possess wide impedance bandwidth with smaller dimensions, followed by excellent ECC and DG values.

TABLE 8.6

Performance Comparison of Proposed Four-Port MIMO Antenna Array against the References

Ref.	Bands (GHz)	Dimensions (mm³)	Isolating Element	Gain (dBi)	Efficiency (η)	ECC	DG
[28]	5	2.53 λ × 1.0 λ × 0.05 λ	X-shaped strip	13.8	NG	NG	NG
[29]	2.45	0.56 λ × 1.12 λ × 0.012 λ	Metallic strip	2.86	NG	< 0.3	NG
[30]	5.7	1.09 λ × 0.61 λ × 0.03 λ	Parallel coupled line resonator	6.43	85	<0.05	NG
[31]	2.4	0.32 λ × 0.32 λ × 0.01 λ	Ring resonator	1.69	NG	<0.05	NG
[33]	5	3.33 λ × 1.16 λ × 0.026 λ	Shorting pin-based decoupling structure	5.4	72.8	0.09	9.5
[34]	2.4/5	0.87 λ × 0.87 λ × 0.01 λ	Radiators at different layers	50	90	0.02	NG
[35]	2.4/5.8	0.71 λ × 0.67 λ × 0.01 λ	Spatial diversity	1.62	NG	<0.05	9.95
[36]	3.6/5	0.57 λ × 0.57 λ × 0.014 λ	Thin inductive lines	4.9	NG	<0.16	NG
[37]	5	0.5 λ × 0.5 λ × 0.013 λ	Circular patch and L-shaped strip	4.35	75	<0.15	9.90
[38]	2.4/3.6	0.83 λ × 0.83 λ × 0.01 λ	Not used	4.3	88	0.002	9.99
[39]	2.4/5.2	0. 408 λ × 0.23 λ × 0.01 λ	EBG structure	3	NG	>0.07	9.8
[40]	2.4	0.54 λ × 0.30 λ × 0.012 λ	H-Shaped DGS	2.2	84	0.0002	NG
[41]	2.95	0.48 λ × 0.48 λ × 0.015 λ	DGS structure and slotted microstrip resonator	5.63	NG	NG	NG
[14]	2.5/4.5	0.59 λ × 0.27 λ × 0.012 λ	DGS and inverted U-shaped strip	NG	NG	NG	NG
[42]	125-300	-	SIW and metasurface	40	90	NG	NG
[45]	2.5	1.6 λ × 0.96 λ × 0.175 λ	Metasurface structure	9	NG	NG	NG
This Work	**n77/n78/ n79 and WLAN**	**0.64 λ × 0.48 λ × 0.025 λ**	**Un-protruded multislot**	**3.5**	**85**	**<0.05**	**9.98**

Note: Electrical lengths are calculated at a lower frequency.

Abbreviations: DGS = defective ground structure, EBG = electronic band gap, NG = not given, SIW = surface integrated waveguide.

8.3.7 CONCLUDING REMARKS

A four-port MIMO antenna array with an innovative and novel UMPS isolating element was studied and analyzed successfully. The antenna is very compact in size, has a dimension of $40 \times 30 \times 1.6$ mm³, and operates in 5G NR sub-6 GHz n77/n78/ n79 and 5 GHz WLAN band. Each MIMO antenna of the proposed design exhibited gain and efficiency of 3.5 dBi and 85%, respectively. Good radiation patterns were observed at the center frequency in both the E-plane and H-plane. The MIMO performance of the proposed antenna was evaluated by calculating the metrics such as ECC, DG, CCL, TARC, and MEG over the operating band. It was found that the ECC values were below 0.05 and DG values were above 9.8 dB throughout the operating bands, which are well within the acceptable limits. The other MIMO parameters like CCL, TARC, and MEG were also well within the acceptable limits, which indicates good MIMO performance. Therefore, because of its good radiation characteristics, better diversity performance, and compactness, the proposed four-port MIMO antenna array is a good candidate for assimilation into the future wireless devices operating in the 5G NR sub-6 GHz and 5 GHz WLAN bands, as well as suitable for large antenna packing density for better transmission and reception quality of signals.

REFERENCES

1. Kulkarni, J., Sim, C. Y. D. & Desai, A. *et al.* "A compact four Port ground-coupled CPWG-fed MIMO antenna for wireless applications." *Arabian Journal for Science and Engineering*, vol. 47, pp. 14087–14103, 2022. https://doi.org/10.1007/s13369-022-06620-z.
2. Toni, A., Carrascosa, M., Bellalta, B., Pretel, I. & Etxebarria, I. "Channel load aware AP/Extender selection in home Wi-Fi networks using IEEE 802.11 k/v." *IEEE Access*, vol. 9, pp. 30095–30112, 2021.
3. Bellalta, B. "IEEE 802.11ax: High-efficiency WLANS." *IEEE Wireless Communications*, vol. 23, no. 1, pp. 38–46, 2016.
4. Zixian, Y., Yang, H. & Cui, H. "A compact MIMO antenna with inverted C-shaped ground branches for mobile terminals." *International Journal of Antennas and Propagation*, vol. 2016, pp. 1–6, 2016.
5. Hadri, D. E., Zakriti, A., Zugari, A., Ouahabi, M. E. & Aoufi, J. E. "High isolation and ideal correlation using spatial diversity in a compact MIMO antenna for fifth-generation applications." *International Journal of Antennas and Propagation*, vol. 2020, pp. 1–10, 2020.
6. Cheng, Y., Liu, H., Sheng, B. Q. & Zhu, L. "A compact 4-element MIMO antenna for terminal devices." *Microwave and Optical Technology Letters*, vol. 62, pp. 2930–2937, 2020.
7. Anitha, R., Vinesh, P. V., Prakash, K. C., Mohanan, P. & Vasudevan, K. "A compact quad element slotted ground wideband antenna for MIMO applications." *IEEE Transactions on Antennas and Propagation*, vol. 64, no. 10, pp. 4550–4553, 2016.
8. Tripathi, S., Mohan, A. & Yadav, S. "A compact Koch fractal UWB MIMO antenna with WLAN band-rejection." *IEEE Antennas and Wireless Propagation Letters*, vol. 14, pp. 1565–1568, 2015.
9. Kulkarni, J., Desai, A. & Sim, C. Y. D. "Wideband four-port MIMO antenna array with high isolation for future wireless systems." *AEU – International Journal of Electronics and Communications*, vol. 128, pp. 1–14, 2021.
10. Malviya, L., Rajib, K. P. & Machavaram, V. K. "A 2 × 2 dual-band MIMO antenna with polarization diversity for wireless applications." *Progress In Electromagnetics Research C*, vol. 61, pp. 91–103, 2016.
11. Ramachandran, A., Pushpakaran, S. V., Pezholil, M. & Kesavath, V. "A four-port MIMO antenna using concentric square-ring patches loaded with CSRR for high isolation." *IEEE Antennas and Wireless Propagation Letters*, vol. 15, pp. 1196–1199, 2015.

12. Soltani, S., Lotfi, P. & Murch, R. D. "A dual-band multiport MIMO slot antenna for WLAN applications." *IEEE Antennas and Wireless Propagation Letters*, vol. 16, pp. 529–532, 2016.

13. Sarkar, D. & Srivastava, K. V. "A compact four-element MIMO/diversity antenna with enhanced bandwidth." *IEEE Antennas and Wireless Propagation Letters*, vol. 16, pp. 2469–2472, 2017.

14. Liu, X. L., Wang, Z. D., Yin, Y. Z., Ren, J. & Wu, J. J. "A compact ultra-wideband MIMO antenna using QSCA for high isolation." *IEEE Antennas and Wireless Propagation Letters*, vol. 13, pp. 1497–1500, 2014.

15. Desai, A., Bui, C. D., Patel, J., Upadhyaya, T., Byun, G. & Nguyen, T. K. "Compact wideband four element optically transparent MIMO antenna for mm-wave 5G applications." *IEEE Access*, vol. 8, pp. 194206–194217, 2020.

16. Simons, R. N. *Coplanar Waveguide Circuits, Components, and Systems*, John Wiley & Sons, New York, 2001.

17. Du, C. & Zhao, Z. "A CPW-fed dual-band MIMO antenna with enhanced isolation for 5G application." *Progress in Electromagnetics Research*, vol. 98, pp. 11–20, 2020.

18. Yang, Y., Chu, Q. & Mao, C. "Multiband MIMO antenna for GSM, DCS, and LTE indoor applications." *IEEE Antennas and Wireless Propagation Letters*, vol. 15, pp. 1573–1576, 2016.

19. Kumar, R. & Pazare, N. "A CPW-fed stepped slot UWB antenna for MIMO/diversity applications." *International Journal of Microwave and Wireless Technologies*, vol. 9, no. 1, pp. 151–162, 2017.

20. Saadh, A. W., Ramaswamy, P. & Ali, T. "A CPW fed two and four element antenna with reduced mutual coupling between the antenna elements for wireless applications." *Applied Physics A*, vol. 127, no. 2, pp. 1–18, 2021.

21. Saadh, A. W., Khangarot, S., Sravan, B. V., Aluru, N., Ramaswamy, P., Ali, T. & Pai, M. M. "A compact four-element MIMO antenna for WLAN/WiMAX/satellite applications." *International Journal of Communication Systems*, vol. 33, no. 14, e4506, 2020.

22. Sharawi, M. S. "Current misuses and future prospects for printed multiple-input, multiple-output antenna systems." *IEEE Antennas and Propagation Magazine*, vol. 59, no. 2, pp. 162–170, 2017.

23. Yin, W., Chen, S., Chang, J., Li, C. & Khamas, S. K. "CPW fed compact UWB 4-element MIMO antenna with high isolation." *Sensors*, vol. 21, no. 8, p. 2688, 2021.

24. Kulkarni, J., Desai, A. & Sim, C.-Y. "Two port CPW-fed MIMO antenna with wide bandwidth and high isolation for future wireless applications." *International Journal of RF and Microwave Computer-Aided Engineering*, vol. 31, e22700, 2021. https://doi.org/10.1002/mmce.22700

25. Sharawi, M. S. "Printed multi-band MIMO antenna systems and their performance metrics." *IEEE Antennas and Propagation Magazine*, vol. 55, no. 5, pp. 218–232, 2013.

26. Najam, A. I., Duroc, Y. & Tedjini, S. "Multiple-input multiple-output antennas for ultra-wideband communications." *Intech Open*, vol. 10, pp. 209–236, 2012.

27. Kulkarni, J., Desai, A. & Sim, C.-Y. D. "Wideband four-Port MIMO antenna array with high isolation for future wireless systems." *AEU – International Journal of Electronics and Communications*, vol. 128, 153507, 2021, https://doi.org/10.1016/j.aeue.2020.153507.

28. Zou, X.-J., Wang, G.-M., Wang, Y.-W. & Li, H.-P. "Decoupling antenna array with X-shaped strip." *International Journal of RF and Microwave Computer-Aided Engineering*, vol. 29, pp. 1–7, 2018.

29. Abbosh, A., Al-Rizzo, H., Yahya, S. & Al-Wahhamy, A. "Decoupling and MIMO performance of two planar monopole antennas with protruded strips." *Microwave and Optical Technology Letters*, vol. 60, pp. 2712–2718, 2018.

30. Mark, R., Singh, H. V., Mandal, K. & Das, S. "Reduced edge-to-edge spaced MIMO antenna using parallel coupled line resonator for WLAN applications." *Microwave and Optical Technology Letters*, vol. 61, pp. 2374–2380, 2019.
31. Isaac, A. A., Al-Rizzo, H., Yahya, S., Al-Wahhamy, A. & Tariq, S. Z. "Miniaturised MIMO antenna array of two vertical monopoles embedded inside a planar decoupling network for the 2.4 GHz ISM band." *IET Microwaves, Antennas & Propagation*, vol. 14, no. 1, pp. 132–140, 2020.
32. Thummaluru, S. R., Kumar, R. & Chaudhary, R. K. "Isolation and frequency reconfigurable compact MIMO antenna for wireless local area network applications." *IET Microwaves, Antennas & Propagation*, vol. 13, no. 4, pp. 519–525, 2019.
33. Henridass, A. & Gulam, N. A. M. "CPW fed circularly polarized wideband pie-shaped monopole antenna for multi-antenna techniques." *COMPEL - The International Journal for Computation and Mathematics in Electrical and Electronic Engineering*, vol. 37, no. 6, pp. 2109–2121, 2018.
34. José, A., Tirado, M., Hildeberto, J.-A., Arturo, R.-M., Luis, A. V.-T. & Ricardo, G.-V. "Four ports wideband drop-shaped slot antenna for MIMO applications." *Journal of Electromagnetic Waves and Applications*, vol. 34, no. 9, pp. 1159–1179, 2020.
35. Birwal, A., Singh, S., Kanaujia, B. K. & Kumar, S. "Low-profile 2.4/5.8GHz MIMO/ diversity antenna for WLAN applications." *Journal of Electromagnetic Waves and Applications*, vol. 34, no. 9, pp. 1283–1299, 2020.
36. Xiaohua, T., Weimin, W., Yongle, W., Yuanan, L., Ahmed, A., Kishk, A. & Heng, W. "Enhancing isolation and bandwidth in planar monopole multiple antennas using thin inductive line resonator." *AEU – International Journal of Electronics and Communications*, vol. 117, p. 153094, 2020.
37. Yang, M. & Zhou, J. "A compact pattern diversity MIMO antenna with enhanced bandwidth and high-isolation characteristics for WLAN/5G/Wi-Fi applications." *Microwave and Optical Technology Letters*, vol. 62, no. 6, pp. 2353–2364, 2020.
38. Desai, A., Upadhyaya, T., Palandoken, M. & Gocen, C. "Dual band transparent antenna for wireless MIMO system applications." *Microwave and Optical Technology Letters*, vol. 61, no. 7, pp. 1845–1856, 2019.
39. Maturi, T. & Harikrishna, B. "Electronic band-gap integrated low mutual coupling dual-band MIMO antenna." *International Journal of Electronics*, vol. 107, no. 7, pp. 1166–1176, 2020.
40. Acharjee, J., Mandal, K. & Mandal, S. K. "Reduction of mutual coupling and cross polarization of a MIMO/diversity antenna using a string of H-shaped DGS." *AEU – International Journal of Electronics and Communications*, vol. 97, pp. 110–119, 2018.
41. Biswas, S., Ghosh, C. K., Banerjee, S., Mandal, S. & Mandal, D. "High port isolation of a dual polarized microstrip antenna array using DGS." *Journal of Electromagnetic Waves and Applications*, vol. 34, no. 6, pp. 683–696, 2020.
42. Ghosh, C., Mandal, B. & Parui, S. K. "Mutual coupling reduction of a dual-frequency microstrip antenna array by using U-shaped DGS and inverted U-shaped microstrip resonator." *Progress in Electromagnetics Research C*, vol. 48, pp. 61–68, 2014.
43. Thummaluru, S. R., Ameen, M. & Chaudhary, R. K. "Four-Port MIMO cognitive radio system for mid-band 5G applications." *IEEE Transactions on Antennas and Propagation*, vol. 67, no. 8, pp. 5634–5645, 2019.
44. Alibakhshikenari, A. & Virdee, B. S. "Study on isolation and radiation behaviours of a 34 × 34 array-antennas based on SIW and metasurface properties for applications in terahertz band over 125–300 GHz." *Optik*, vol. 206, pp. 1–11, 2019.
45. Yin, B., Feng, X. & Gu, J. "A metasurface wall for isolation enhancement: Minimizing mutual coupling between MIMO antenna elements." *IEEE Antennas and Propagation Magazine*, vol. 62, no. 1, pp. 14–22, 2020.

46. Shi, H., Zhang, X., Li, J., Jia, P., Chen, J. & Zhang, A. "3.6-GHz eight-antenna MIMO array for mobile terminal applications." *AEU – International Journal of Electronics and Communications*, vol. 95, pp. 342–348, 2018.

47. Hu, W. *et al.* "Dual-band eight-element MIMO array using multi-slot decoupling technique for 5G terminals." *IEEE Access*, vol. 7, pp. 153910–153920, 2019

48. Li, Y., Sim, C., Luo, Y. & Yang, G. "Multiband 10-antenna array for sub-6 GHz MIMO applications in 5-G smartphones." *IEEE Access*, vol. 6, pp. 28041–28053, 2018.

49. Kulkarni, J. & Seenivasan, R. "Design and development of novel monopole antenna for WLAN/WiMAX/MIMO applications in the laptop computers." *Circuit World*, vol. 45, no. 4, pp. 257–267, 2019.

50. Lee, C. & Su, S. "Small-sized, tri-band, two-antenna system aimed for 4 × 4 Gbps notebook applications." *2018 International Symposium on Antennas and Propagation (ISAP)*, Busan, South Korea, pp. 1–2, 2018.

51. Su, S., Lee, C. & Hsiao, Y. "Compact two-inverted-F-antenna system with highly integrated π-shaped decoupling structure." *IEEE Transactions on Antennas and Propagation*, vol. 67, no. 9, pp. 6182–6186, 2019.

52. Chang, W., Su, S. & Tseng, B. "Self-decoupled, 5G NR77/78/79 two-antenna system for notebook computers." *2019 Electrical Design of Advanced Packaging and Systems (EDAPS)*, Kaohsiung, Taiwan, pp. 1–3, 2019.

53. Kulkarni, J. "Multi-band printed monopole antenna conforming bandwidth requirement of GSM/WLAN/WiMAX standards." *Progress in Electromagnetics Research Letters*, vol. 91, pp. 59–66, 2020.

54. Kulkarni, J., Kulkarni, N. & Desai, A. "Development of "H-shaped" monopole antenna for IEEE 802.11a and HIPERLAN 2 applications in the laptop computer." *International Journal of RF and Microwave Computer-Aided Engineering*, vol. 30, no. 7, pp. 1–14, 2020.

55. Computer Simulation Technology (CST) Studio Suit 2017. https://www.cst.com/2017.

56. Kumar, A., Ansari, A. Q., Kanaujia, B. K. & Kishor, J. "A novel ITI-shaped isolation structure placed between two-port CPW-fed dual-band MIMO antenna for high isolation." *AEU – International Journal of Electronics and Communications*, vol. 104, pp. 35–43, 2019.

9 Software-Defined Radio, Receiver, and Transmitter Analysis

9.1 INTRODUCTION

The rapidly advancing development of components for digital signal processing and the rapid rise of processing power enables more and more new concepts in the implementation of so-called software-defined radios (SDR). The SDR approach includes transmitter and receiver concepts, whereas the signal processing is largely done in programmable devices, such as field programmable gate array (FPGA) and digital signal processors (DSP). Thereby, components that have been typically implemented in hardware (e.g. amplifiers, mixers, filters, modulators, etc.) are instead implemented by means of software on a digital signal-processing platform. Hardware becomes replaced by software. SDR can be defined as a communication system, where the major signal processing components, typically realized in hardware, are instead replaced by digital algorithms written in software (FPGA). An SDR can facilitate radio interoperability and also enable spectrum reuse. Figure 9.1 shows a typical SDR architecture. The SDR platform has the advantage of being able to quickly change a receiver or transmitter to meet new requirements. Thus, it is possible to keep these products up-to-date for a long period. A typical example is the mobile telephone according to the various GSM and LTE standards. Even inexpensive amateur radios use more and more of this modern technology because of the increasing integration of available DSP and FPGA modules; the performance enhancement of microprocessors accelerates the application in the price segment also. The SDR approach makes use of time-discrete signal processing by sampled signals. Such sampled systems were first published in 1985 [1, 2].

The typical receiving frequency range is 9 KHz to 50 MHz, given a typical sampling frequency of 120MHz, with a resolution equal to 16 bit. Figure 9.2 shows a typical block diagram of SDR.

9.1.1 GENERAL ADVANTAGES AND BENEFITS OF SDR

- **Ease of design**: Reduces design-cycle time, quicker iterations.
- **Ease of manufacture**: Digital hardware reduces costs associated with manufacturing and testing radios.
- **Multimode operation**: SDR can change modes by loading appropriate software into memory.
- Use of advanced signal processing techniques allows for implementation of new receiver structures and signal processing techniques.

DOI: 10.1201/9781003331018-9

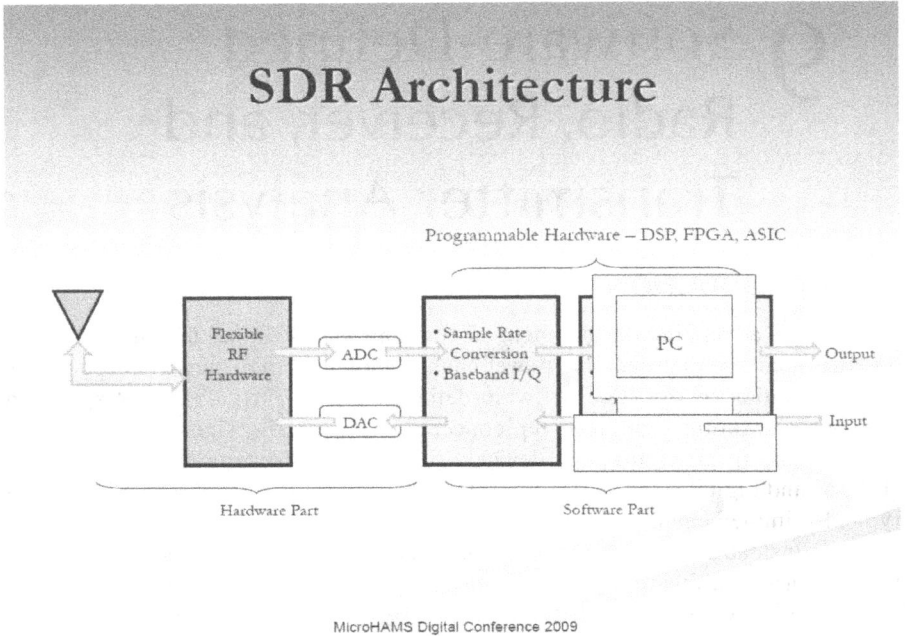

FIGURE 9.1   Typical SDR architecture.

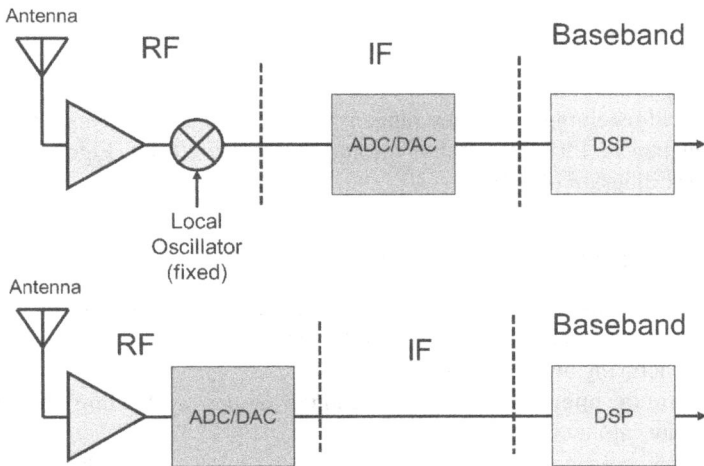

FIGURE 9.2   Typical block diagram of SDR.

- Fewer discrete components, such as the digital processors, can implement functions including synchronization, demodulation, error correction, decryption, etc.
- Flexibility to incorporate additional functionality, as modifications can be done in the field to correct problems and to upgrade.
- Flexible/reconfigurable due to its reprogrammable units and infrastructure.
- Reduced obsolescence as it is multiband/multimode offering ubiquitous connectivity, where different standards can co-exist.
- Enhances/facilitates experimentation.
- Brings analog and digital worlds together due to the full convergence of digital networks and radio science, networkable with simultaneous voice, data, and video.

The following technologies facilitate SDR systems:

- Antennas
  - Receiver antennas better achieve wideband performance than transmitter antennas
  - New fractal and plasma antennas are expected in smaller sizes and wideband capabilities.
- Waveforms
  - Management and selection of multiple waveforms.
  - Cancellation carriers and pulse-shaping techniques.
- Analog-to-digital converters (ADCs)
  - High ADC sampling speed.
  - ADC bandwidth could be digitized instantaneously.
- Digital signal processing
  - Number of transistors doubles every 18 months.
  - More specific purpose DSPs and FPGAs.
- Amplifiers
- Batteries
  - More and more power needed (need to focus on more efficient use of power).
  - Fuel-cell development for handhelds.
- Cognition
  - A key aspect will be to understand how multiple cognitive radios (CRs) work with each other.
- Design tools
- Terrain databases
  - Interference prediction, environment awareness.

SDR can be a promising solution to many present-day problems related to communication standards. Myriad standards exist for terrestrial communications: cell phone standards change every year, satellite ground stations listen to multiple spacecrafts, frequency domain is owned, and uncertainty about how to accommodate ultra-wide band (UWB) issues, etc. SDR can facilitate a flexible radio system that

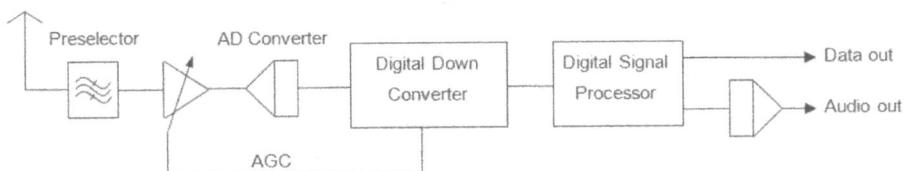

FIGURE 9.3 Direct sampling software-defined radio receiver [4].

allows communication standards to migrate and flexible methods for reconfiguring the radio using software that allows ability to communicate via different protocols at different times.

9.1.2 DIRECT SAMPLING RECEIVER

Figure 9.3 shows a form of a direct sampling receiver. Here, the input signal is immediately converted after the preselection without any mixing. The frequency range of that receiver is practically in the baseband, suitable for a shortwave receiver. This front-end concept is an almost completely digital realization and has no need of analog synthesizers and mixers. When the digital part is driven by a low-noise crystal oscillator, the phase noise performance of the whole receiver is excellent. The frequency drift of the crystal oscillator can be monitored by an external reference and numerically corrected in the DSP. For very high frequency (VHF) to super high frequency (SHF) frequency ranges, a simple heterodyne down-converter can be used [4]. When the suppression of the image frequency band is an issue, the following arrangement shown in Figure 9.3 is proposed:

9.1.3 DIRECT-CONVERSION RECEIVER

The arrangement in Figure 9.4 is applicable for frequencies much higher than the ADC sampling rate. It uses a down-conversion stage with two mixers, driven by a local oscillator with two outputs in quadrature. This allows an image frequency

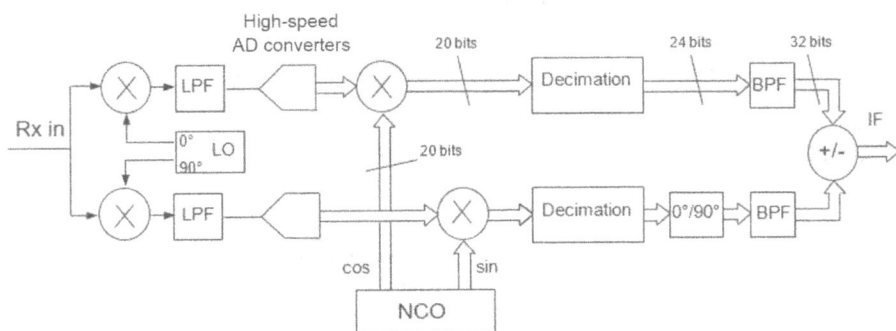

FIGURE 9.4 Direct conversion receiver with an image rejection front end [4].

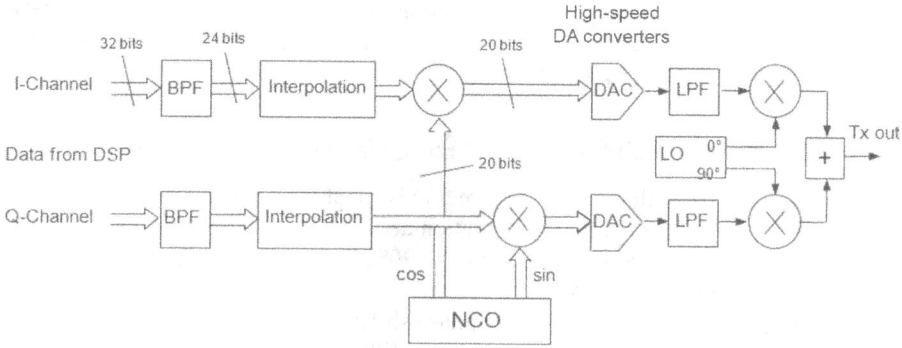

FIGURE 9.5 Transmitter with additional analog image rejection up-converter [4].

suppression of 20 to 40 dB, which can be further improved by a calibration algorithm in the digital part. An advantage of this arrangement is that the local oscillator can consist of a simple low-noise fixed-frequency oscillator, whereas the frequency tuning is done by the digital numerical controlled oscillator (NCO). The tuning range is then, of course, limited to < $f_s/2$ (f_s: sampling frequency). Intermediate frequencies in the range of 20–140 MHz are usual, depending on the ADC. The low-pass filters (LPFs) in the front end are used to limit the frequency spectrum to prevent aliasing in the ADC and to reject the image frequency [4].

For the transmit path, the diagram in Figure 9.4 must be reversed. The low-rate data from DSP is bandpass-filtered and up-sampled in the interpolation block (Figure 9.5). The output of the interpolation blocks contains the complex baseband signal with the high sample rate f_s. It is then numerically mixed up in a desired range < $f_s/2$ by an NCO and digital mixers, and is then directed into a pair of digital-to-analog converters (DACs) to get an analog I/Q signal, which then can be further up-converted. The sum of the I- and Q-signals represent the transmit signal with a substantial suppression of the image frequency.

For a transmit frequency range below 60 MHz, the simplified arrangement of Figure 9.6 is recommended. The I- and Q-channels are added within the digital part, resulting in an almost ideal sideband rejection of >100 dB. Modern 16-bit DACs offer

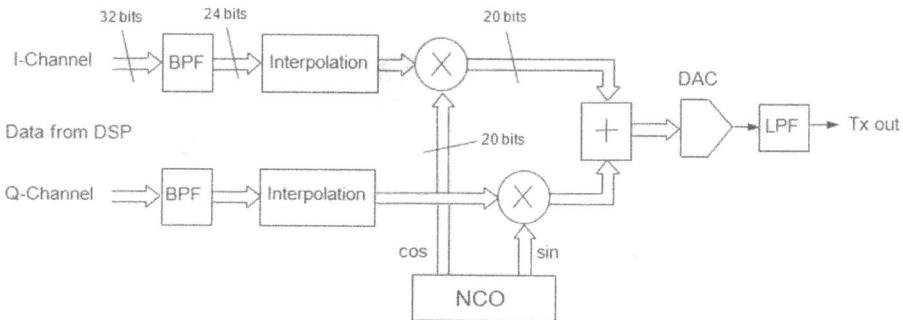

FIGURE 9.6 Simplified arrangement for a frequency range up to 60 MHz [4].

a spurious free dynamic range (SFDR) from up to 80 dB for frequencies <30 MHz. The LPF following the DAC is needed to reject the alias frequencies and, therefore, should have a cutoff frequency of fc < $f_s/2$.

9.2 THE IMAGE REJECTION MIXER/QUADRATURE MIXER

An analog image rejection mixer (Figure 9.7) is capable of attenuating the image by 30–40 dB. Criteria for the image attenuation are amplitude and phase errors in both branches, the most critical element being the 90° phase shifter, mainly for wideband intermediate frequency (IF) [5].

SDR technology allows moving the phase shifter from analog into digital, where it can be realized to near ideal by means of a Hilbert transformer.

To get a maximum reduction of the sample rate, the spectrum of interest must be down-converted to a frequency range near 0 Hz. The usage of a simple mixer yields unwanted frequency band (image) overlapping the desired band (Figure 9.8).

The problem stated in Figure 9.8 exists in a continuous time analog system, and also in a discrete time digital system, and can be solved when the signals are represented in the complex form and the Euler formula as:

$$\bar{S}(t) = A(t) \cdot e^{j\omega t} = A(t) \cdot \left[\cos(\omega t) + j \cdot \sin(\omega t) \right]$$

In 1928, Ralph Hartley, an American engineer, and inventor, known for his contribution to information theory, presented a mixer architecture (Figure 9.9) that allows suppression of the image, the so-called direct conversion receiver, also called the quadrature mixer [4].

The mathematical analysis of this circuit starts from the following conditions:

$$\text{Desired signal: } S_{des} = S(t) \cdot \cos(\omega_2 t)$$

$$\text{Image signal : } S_{img} = R(t) \cdot \cos(\omega_1 t)$$

$$\omega_2 - \omega_0 = \omega_{IF;} \, \omega_1 - \omega_0 = -\omega_{IF}$$

$$S(t) = R(t) = 1$$

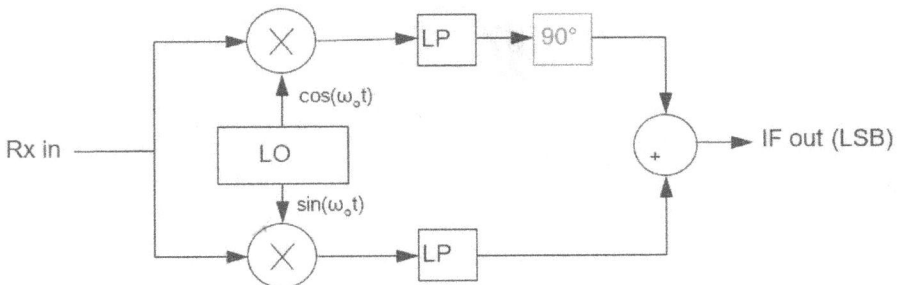

FIGURE 9.7   Image rejection mixer [5].

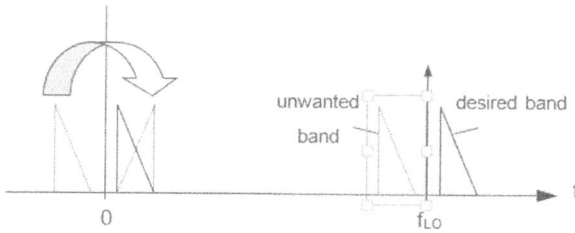

FIGURE 9.8   Problem of overlapping at down-conversion to 0 Hz [5].

I channel: with these assumptions, the signal on node A can be written as:

$$S(t) \cdot \cos(\omega_2 t) \cdot \cos(\omega_o t) + R(t) \cdot \cos(\omega_1 t) \cdot \cos(\omega_o t)$$

$$= \frac{1}{2} \cdot S(t) \cdot \cos(\omega_2 t - \omega_o t) + \frac{1}{2} \cdot R(t) \cdot \cos(\omega_1 t - \omega_o t) \tag{9.1a}$$

$$= \frac{1}{2} \cdot S(t) \cdot \cos(\omega_{IF} t) + \frac{1}{2} \cdot R(t) \cdot \cos(\omega_{IF} t) \text{ (Low pass filtered)}$$

Q channel: on node B, when respecting sin(-x) = -sin(x):

$$S(t) \cdot \cos(\omega_2 t) \cdot \sin(\omega_o t) + R(t) \cdot \cos(\omega_1 t) \cdot \sin(\omega_o t)$$

$$= \frac{1}{2} \cdot S(t) \cdot \sin(\omega_2 t - \omega_o t) - \frac{1}{2} \cdot R(t) \cdot \sin(\omega_1 t - \omega_o t) \tag{9.1b}$$

$$= \frac{1}{2} \cdot S(t) \cdot \sin(\omega_{IF} t) - \frac{1}{2} \cdot R(t) \cdot \sin(\omega_{IF} t) \text{ (Low-pass filtered)}$$

After the -90° phase shift, node C:

$$\frac{1}{2} \cdot S(t) \cdot \cos(\omega_{IF} t) - \frac{1}{2} \cdot R(t) \cdot \cos(\omega_{IF} t) \tag{9.1c}$$

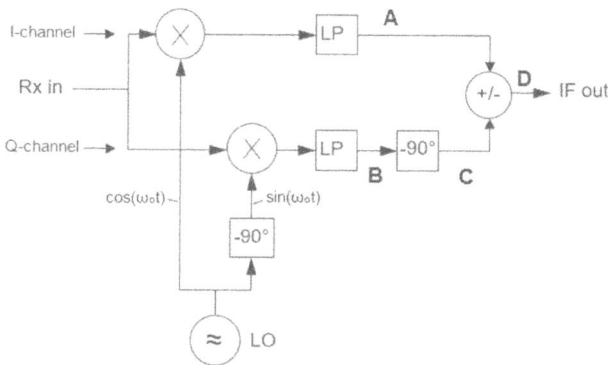

FIGURE 9.9   Hartley architecture [4].

After the addition of node A minus node C follows:

$$IF_{USB}(t) = S(t) \cdot \cos(\omega_{IF}t) \qquad (9.1d)$$

The outcome of subtraction in place of addition is:

$$IF_{LSB}(t) = R(t) \cdot \cos(\omega_{IF}t) \qquad (9.1e)$$

The selection between the upper sideband (USB) and lower sideband (LSB) can be accomplished by simply changing the sign of the Q channel output.

It is obvious that unless the gain and phase of the I- and Q-channels are ideally matched, the image will not be fully cancelled. The ratio between the image and the desired signal is designated as the image rejection ratio (IRR). This important quantity can be derived from the previously mentioned formulas by introducing the gain error ε and the phase error θ, which is assumed to be concentrated in the 90° phase shifter:

Output on node A:

$$\frac{1}{2} \cdot (1+\varepsilon) \cdot A(t) \cdot \cos(\omega_{IF}t) + \frac{1}{2} \cdot (1+\varepsilon) \cdot B(t) \cdot \cos(\omega_{IF}t) \qquad (9.2a)$$

Output on node C:

$$-\frac{1}{2} \cdot (1+\varepsilon) \cdot A(t) \cdot \cos(\omega_{IF}t + \theta) + \frac{1}{2} \cdot (1+\varepsilon) \cdot B(t) \cdot \cos(\omega_{IF}t + \theta)$$

$$= -\frac{1}{2} \cdot (1+\varepsilon) \cdot A(t) \cdot \left[\cos(\omega_{IF}t) \cdot \cos(\theta) - \sin(\omega_{IF}t) \cdot \sin(\theta)\right] \qquad (9.2b)$$

$$+\frac{1}{2} \cdot (1+\varepsilon) \cdot B(t) \cdot \left[\cos(\omega_{IF}t) \cdot \cos(\theta) - \sin(\omega_{IF}t) \cdot \sin(\theta)\right]$$

From this base, the IRR can be derived according to Ref. [6] and results in the following formula:

$$IRR = \frac{1 - 2(1+\varepsilon) \cdot \cos\theta + (1+\varepsilon)^2}{1 - 2(1+\varepsilon) \cdot \cos\theta + (1+\varepsilon)^2} \qquad (9.3)$$

$$\varepsilon = 10^{p[dB]/20} \quad \text{p:amplitude error in dB}$$

whereas ε is the amplitude error, and θ is the phase error. Figure 9.10 shows an IRR calculation as a function of amplitude (p) and phase errors (θ).

To achieve an IRR of 80 dB requires an amplitude balance of p = 0.001 dB and a phase matching of θ = 0.01°, both are almost impossible to achieve using analog technology, in particular for volume production [5].

IRRs of more than 100 dB can easily be realized by using digital processing algorithms.

IRR [dB]

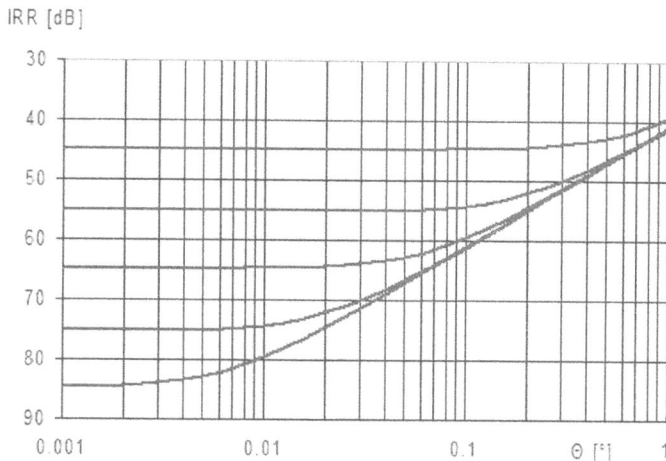

FIGURE 9.10 Image rejection ratio calculation as a function of amplitude (p) and phase errors (Θ) [5].

9.3 THE SAMPLING THEOREM

Harry Nyquist and Claude Shannon, two prominent figures in the field of informa-tion theory and the field of communication engineering defined an important rela-tionship, the sampling theorem:

If a function x(t) contains frequencies less than B [Hz], then it is completely deter-mined by giving its ordinates at a series of points spaced 1/(2B) seconds apart. **Ref. [1]**

In other words, a sample with a sampling rate f_s is capable of generating a digi-tal replica of a band-limited analog spectrum with $f_s/2$ as its highest frequency. A violation of this theorem will lead to aliasing, the folding back of all components $> f_s/2$, about the half sample rate $f_s/2$. Two different operation modes are known for digitalization: the baseband and the band-filter mode. The input spectrum baseband mode fulfills the Nyquist condition and is, therefore, limited to $f_s/2$, the so-called first Nyquist window.

The following frequency slices with a width of $f_s/2$ are denoted as the second, third, etc., windows. Band-limited signals, within higher windows, are mapped down into the first window. This process is referred to as under-sampling. It is important that the band-limited signal B lies totally inside the range:

$$B = n \cdot \frac{fs}{2} .... (n-1) \cdot \frac{fs}{2} \text{ whereas n is the window } (1, 2, ...) \qquad (9.4)$$

Figure 9.11a shows a band-limited spectrum between $f = 0$ and $f = f_s/2$. Frequencies within the even windows are represented with its reverse frequency position, whereas those from the odd windows are not reversed [5].

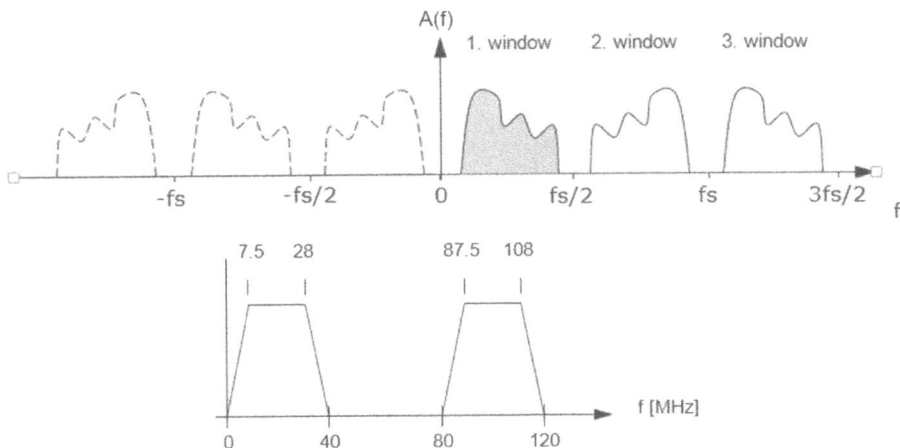

FIGURE 9.11 (a) Aliasing on sampled signals [5] and (b) the frequency allocation of the under-sampled FM band [5].

The oversampling technique can be used by an ADC to digitize signals much higher than its sampling frequency f_s. Modern devices are capable to process signal up to $n = 4$–8.

This is illustrated in an example for an frequency modulation (FM) receiver for $f_L = 87.5$ and $f_H = 108$ MHz (Figure 9.11b).

First, we must determine the possible range of n:

$$2 \leq n \leq \left\lfloor \frac{f_H}{f_H - f_L} \right\rfloor, \lfloor \ \rfloor \text{denotes the floor function} \tag{9.5}$$

The results are: n = 2, 3, 4, 5.

Next, we calculate f_s for an optimum shape of the bandpass filter using the formula:

$$fs(n) = 2 \cdot \sqrt{\frac{f_H \cdot f_L}{n \cdot (n-1)}} \tag{9.6}$$

For n = 3, f_s = 79.3725 MHz.

9.4   ANALOG-TO-DIGITAL CONVERSION

The key element in an SDR receiver is the ADC. It defines the most critical parameters of the receiver, as dynamic range, intermodulation, and sensitivity.

9.4.1   THE ADC

This device is placed ideally as close as possible to the antenna. For VHF (>50 MHz) to SHF frequencies, however, there is a need to place an analog down-converter in front of the ADC. All signal-processing components following the ADC are widely

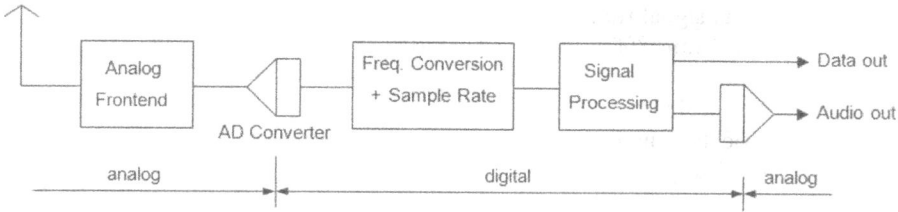

FIGURE 9.12 General block diagram of a software-defined radio receiver analog front end [4].

free from tolerances, noise, unwanted couplings, and profit from high reproducibility and zero drift. The software-configurable hardware components and DSP algorithms allow maximum product flexibility [4].

The analog front-end in Figure 9.12 may consist of preselector filters, a preamplifier and, if needed, a frequency converter. The preconditioned signal then passes the ADC, which produces a sampled discrete time representation of the analog, the time invariant input signal. For receiving frequencies in the VHF range and higher, an analog frequency converter stage is needed.

The evolution toward higher resolutions and sampling rates in the past decade was impressive. In 1998, the first 12-bit, 65 Msps ADC was available, and since 2010, a broad range of 16-bit ADCs are available with sampling rates up to 250 Msps.

The calculation of the SNR for an ideal ADC (N >5-bit) is equal to

$$SNR = \frac{V_{eff}}{V_n} \text{ whereas } V_{eff} \text{ denotes the full scale rms voltage } \frac{V_{FS}}{2 \cdot \sqrt{2}} \quad (9.7)$$

V_n represents the root mean square (rms) error between the sampled and the analog input signal, which is, according to Ref. [7]:

$$V_n = \frac{V_{LSB}}{\sqrt{12}} \text{ and } V_{LSB} = \frac{V_{FS}}{2^N - 1} \quad (9.8)$$

Therefore, the SNR is given by:

$$SNR = \frac{V_{eff}}{V_n} = \frac{V_{LSB} \cdot 2^N}{2 \cdot \sqrt{2}} \cdot \frac{\sqrt{12}}{V_{LSB}} = 2^N \cdot \sqrt{\frac{3}{2}} \text{ or}$$

$$SNR_{dB} = 20 \log_{10}\left(\sqrt{\frac{3}{2}}\right) + N \cdot 20 \log_{10}(2) = 1.76 dB + N \cdot 6.02 dB \quad (9.9)$$

An approximation of the SNR from a real ADC is given in Ref. [8]:

$$SNR_{real} = -10 \cdot \log_{10}\left[\overbrace{\left(2\pi \cdot f_a \cdot t_{jrms}\right)^2}^{\text{Clock Jitter}} + \overbrace{\frac{2}{3}\left(\frac{1+e}{2^N}\right)^2}^{\text{Quantization Noise}} + \overbrace{\left(\frac{2 \cdot \sqrt{2} \cdot V_{NOISE}}{2^N}\right)^2}^{\text{Input Noise}}\right] \quad (9.10)$$

f_a: Input signal frequency (Hz).

t_{jms}:   RMS Jitter (60 f_s for the AD9446).

e: Average differential non-linearity (approximately 0.4).

N: Physical number of bits.

V_{NOISE}:   Effective input noise, generated by the converter.
         (typically 1.9 [LSB] for a 16-bit converter)

For f = 30 MHz and N = 16 bits, this formula delivers a result of SNR_{real} = 81.4 dB.

By setting t_{jrms} = 0, e = 0, and V_{NOISE} = 0 in this expression, the ideal SNR_i is achieved.

$$SNR_{ideal} = 20 \cdot \log_{10}\left(\sqrt{\frac{3}{2}} \cdot 2^N \right) = 20 \cdot \log_{10}\left(\sqrt{1.5}\right) + N \cdot 20 \cdot \log_{10}(2) \qquad (9.11)$$

this gives the same result as above:

$$SNR_{ideal} = 1.76 \text{ dB} + N \cdot 6.02 \text{ dB}$$

It is important to note, that the above value for the SNR is valuable only for a bandwidth of $f_s/2$, the so-called Nyquist bandwidth.

Practical values for real SNR are 75 dB for 14 bits and 80 dB for 16 bits. The effective number of bits (ENOB) is then:

$$ENOB = \frac{SNR_{real}}{SNR_{ideal}} \cdot N \qquad (9.12)$$

Figure 9.13 shows the comparison between the real and ideal ENOB. The next step to 18bits of resolution would gain only a small fraction from one bit when applying the existing technology, as noise is the limiting factor.

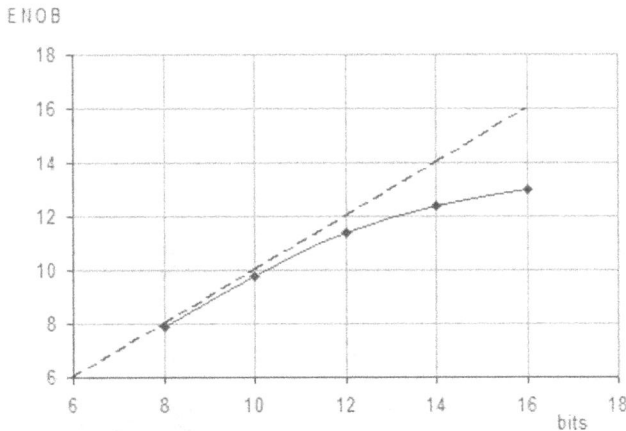

FIGURE 9.13 Comparison of effective number of bits (ENOB) ideal (dashed line) with ENOB real (solid line).

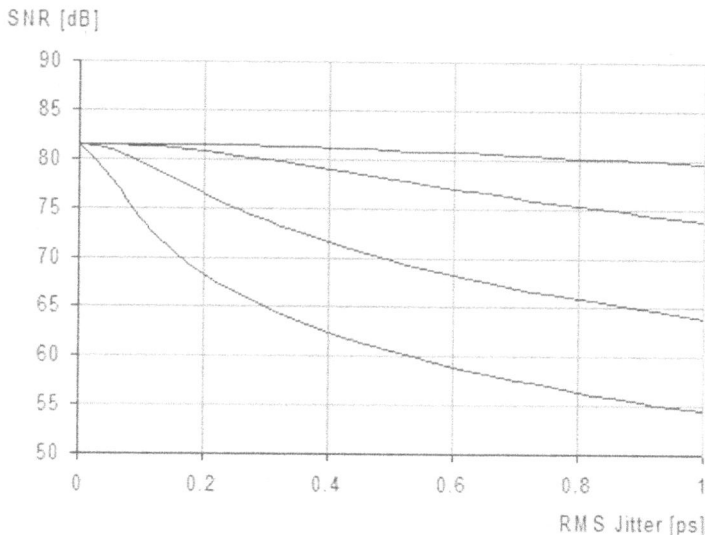

FIGURE 9.14   SNR degradation by jitter of the clock source (AD9446).

High resolution ADCs offer very low intrinsic jitter, as low as 60 f_s. This demands for a clock source with very low phase jitter. The degradation in SNR is dependent on the frequency of input signal and the clock source jitter (Figure 9.14). Therefore, the realization of receiver concepts with signal frequencies higher than the clock frequency by under-sampling suffers a loss in SNR and dynamic range. Also, it is not recommended to use buffers between the clock oscillator and the ADC, as these devices can add a large amount of jitter.

Even special clock distribution circuits add output jitter in the order of 100–400 f_s. Synergy Microwave Corp., offers crystal oscillators with an extremely low broadband jitter of <10 fs_{rms} (100 Hz–10 MHz).

The noise figure of an ADC is dependent on V_{FS}, f_s, R_{in}, and SNR_{eff}.

$$F_{dB} = 174\,dB + 10 \cdot \log_{10}\left(\left(\frac{V_{FS}}{2 \cdot \sqrt{2}}\right)^2 \cdot \frac{1}{R_{in}}\right) + 30\,dB - 10 \cdot \log_{10}\left(\frac{f_s}{2}\right) - SNR_{eff} \quad (9.13)$$

The noise figure of an AD9446 with V_{FS} = 3.2 Vpp, R_{in} = 800 Ω, f_s = 80 MHz, and SNR_{eff} = 81 dB results in F = 19 dB. In the above formula, R_{in} is the only variable that can be altered by more than one decade and, therefore, has the biggest influence on the noise figure. Analyzing the influence of R_{in} gives the following results:

The properties of the AD9446 for f_s = 80 MHz, B = 2,500 Hz, and SNR = 81 dB are:

To achieve an overall noise figure of 10 dB for example, the preamplifier gain is much higher at R_{in} = 50 Ω than at 800 Ω, and, also, the drive level of the amplifier is at 800 Ω easier to achieve:

TABLE 9.1
Properties of the AD9446 with the values given in the text

R_{in}	P_{in_max}	MDS	F
50 Ω	14.1 dBm	-109 dBm	31 dB
200 Ω	8.1 dBm	-115 dBm	25 dB
800 Ω	2.0 dBm	-121 dBm	19 dB

TABLE 9.2
Properties of the AD9446 with the values given in the text

R_{in}	G_{Preamp}	IP3 AD-conv	IP3 $_{total}$
50 Ω	21.9 dB	55 dBm	18.1 dBm
200 Ω	15.9 dB	49 dBm	24.1 dBm
800 Ω	9.9 dB	43 dBm	30.0 dBm

The IP3 (third-order intercept point) of the preamplifier is assumed to be 43 dBm.

The use of the popular operational amplifiers as preamplifier is not recommended, as these devices use resistive feedback, resulting in noise figures >15 dB. The best solutions are passive broadband transformers with a winding ratio of 1:4. Most of the high speed ADCs are specified with an input impedance of ≈ 1 kΩ and a shunt capacitance of a few pF, but only a few devices can effectively be driven with that high impedance because the majority of the ADCs are operating without an input buffer. To drive an unbuffered sample and hold circuit demands, a source impedance as low as 50 or 100 Ω is required.

The specification of the IP3 needs to be done with caution, as the IP3 is not a physical quantity. It is calculated under the quasi-assumption that the third-order distortion product is cubical and, therefore, increases by 3 dB per dB increase of signal power (Figure 9.15a and b). This is not applicable for ADC where the inter modulation distortion (IMD) products at a lower level are larger.

The specification of the IP3 is only meaningful when the power level of the used carriers is also specified, normally corresponding to -7 dB full scale.

The non-linearities of ADCs are specified by the differential non-linearities (DNL) and integral non-linearities (INL) (Figure 9.16). Differential non-linearity is given by errors in the individual amplitude steps. These errors are spread over the whole transfer characteristic. This type of non-linearity mainly produces noise in addition to the given quantization noise.

Integral non-linearities are distortions in the transfer function. They are responsible for intermodulation products. Due to the internal segmentation in most modern high-speed converters, maximum distortion occurs when the input signal periodically passes the transition between two converter sections. The peak in inter modulation in Figure 9.15a at −40 dBm is caused by such a transition error. These intermodulation effects can be reduced by adding an out-of-band noise spectrum,

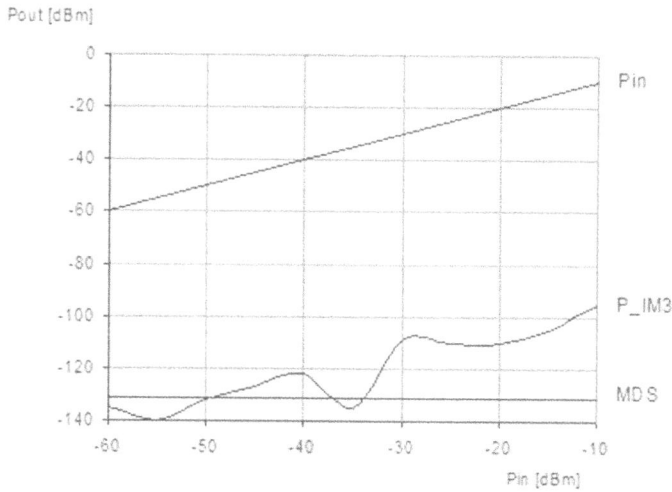

(a)

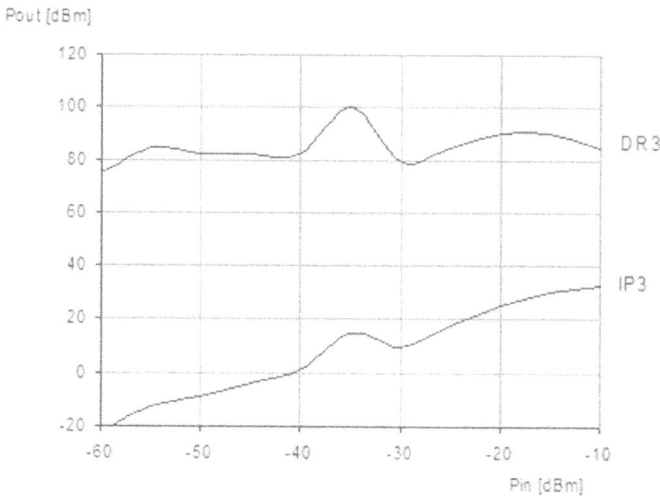

(b)

FIGURE 9.15 (a) The typical third-order intermodulation and (b) the IM3-free dynamic range and associated IP3. (Ref. [5].)

called dithering (Figure 9.17). Applying noise affects the discontinuities, making them no longer exactly periodic and, therefore, reducing intermodulation.

In communications receivers, there are normally enough stochastic signals on the ADC input. In particular, strong broadcast stations with signal strengths of -40 dBFS to -30 dBFS can reduce higher order intermodulation.

For system design, the final receiver bandwidth must be known. All SNR discussions previously discussed are based on a bandwidth of $f_s/2$. In most cases, the final

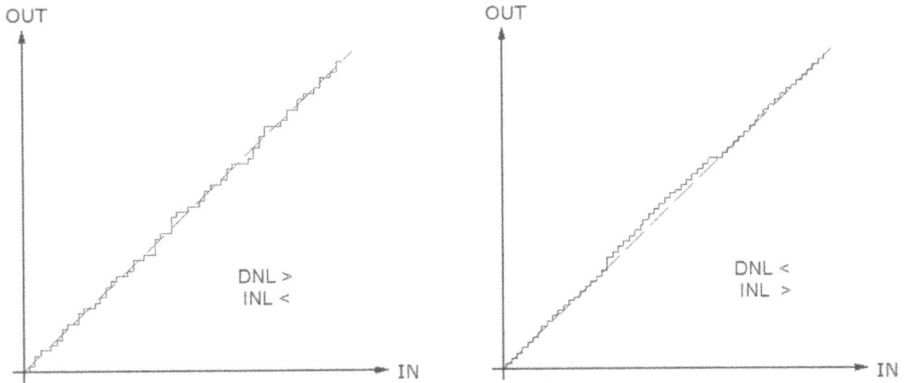

FIGURE 9.16   Differential (left) and integral non-linearities (right) [5].

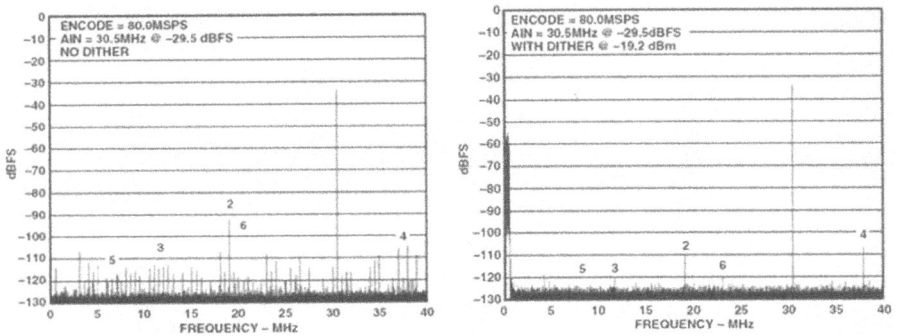

FIGURE 9.17   Influence of dithering on intermodulation. (From Analog Devices Inc. [8].)

bandwidth is much smaller and, thus, the SNR is increased. This effect is also called as processing gain (G_p):

$$G_p = 10 \cdot \log_{10}\left(\frac{f_s}{2 \cdot B}\right) \qquad (9.14)$$

When we assume a narrow band telephony application with a bandwidth of 2,700 Hz and a sample rate of 80 Msps, the resulting SNR is:

$$G_p = 10 \cdot \log_{10}(8 \cdot 10^7 / 5400) = 41.7 \ dB \qquad (9.15)$$

$$SNR = SNR_{AD} + G_p = 81 \ dB + 41.7 \ dB \rightleftarrows\rightleftarrows\rightleftarrows = 122.7 \ dB \qquad (9.16)$$

This SNR value is reached at full scale and, therefore, also represents the dynamic range of the ADC.

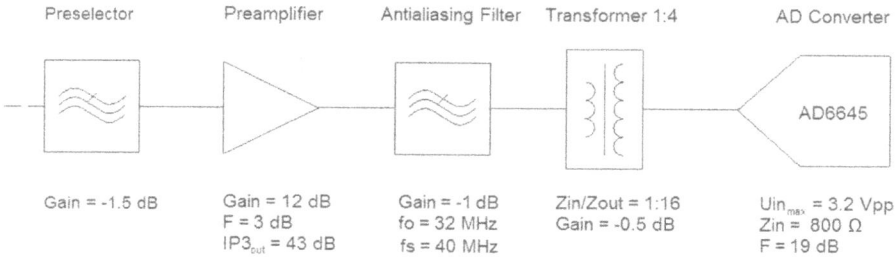

Preselector	Preamplifier	Antialiasing Filter	Transformer 1:4	AD Converter
Gain = -1.5 dB	Gain = 12 dB F = 3 dB $IP3_{out}$ = 43 dB	Gain = -1 dB fo = 32 MHz fs = 40 MHz	Zin/Zout = 1:16 Gain = -0.5 dB	Uin_{max} = 3.2 Vpp Zin = 800 Ω F = 19 dB

FIGURE 9.18 Front-end system design.

The resulting specifications from the front-end system design shown in Figure 9.18 are:

- Gain (to ADC): 9 dB
- Noise figure: 11 dB
- IP3: 30 dBm
- DR3 (dynamic range of 3decibels): 105 dB
- MDS (minimum detectable signal): -128.6 dB
- Uin_{max}: -7 dBm
- DR (dynamic range): 121.6 dB

It is important to place the antialiasing filter after the amplifier to prevent broadband noise from degrading the noise figure of the ADC via aliasing. The transformer must show an IP3 >50 dBm at a signal level of 2 dBm over the whole operating range. The DR3 may be lower in reality due to the effects discussed previously and shown in Figure 9.18.

9.4.2 THE DIGITAL DOWN-CONVERTER

The digital down-converter (DDC) includes (Figure 9.19):

- An NCO.
- A complex IQ-mixer to convert the IF down to approximately 0 Hz (zero IF).
- Several decimation filter stages for reducing the sampling rate.
- Final low-pass finite impulse response (FIR) filters.

The output data rate of the ADC (i.e. the product of sample rate times the resolution) is in the range of 1–2 Gbit/s. This data stream consists of the whole spectrum from DC up to the half sample rate. For most communications purposes, only a small fraction of this bandwidth is of interest, unless the implementation of several receivers in the system is wanted in special cases. Therefore, it needs one or a number of frequency conversions with a subsequent reduction of the sample rate. This function block is designated to act as a DDC. The preconditioned digital signals are then processed at a lower data rate in the signal processing unit.

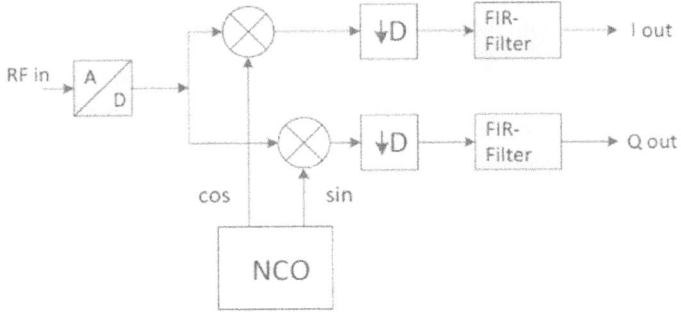

FIGURE 9.19   Block diagram of a digital down-converter.

The task of the DDC is to perform the digital down-conversion, decimation of the channel rate, baseband IQ generation, channel filtering, and offset cancellation, using commercially available application-specific integrated circuits (ASICs), or a programmable hardware in the form of field programmable gate arrays (FPGAs). Subsequently, further processing as demodulation, clock and carrier synchronization, decryption, audio processing, spectrum analysis, etc., can be performed by the DSP [5].

The main task of the DDC is to reduce the sample rate by decimation and to increase the SNR by integration. The most effective way is to down-convert the bandwidth of interest to an intermediate frequency near 0 Hz by using the Hartley architecture described in Figure 9.9.

DDCs are realized either in FPGAs or as a customer-specific integrated circuit (CSIC). FPGAs can be developed by using a high-level language, such as Hardware Description Language (VHDL), and can, thus, be tailored exactly to the needs for a dedicated application. Figure 9.20 shows the typical architecture of a DDC. Its disadvantage is the high power consumption up to 10 W and the large housing dimensions. On the other hand, CISCs use much more development resources and are, therefore, only suitable for mass products.

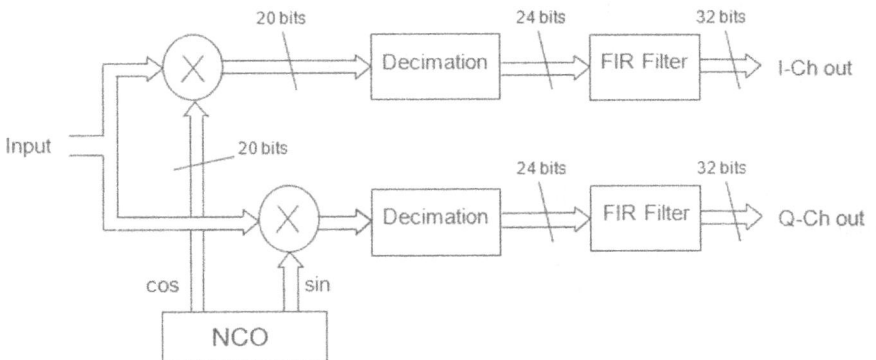

FIGURE 9.20   Architecture of a digital down-converter [4].

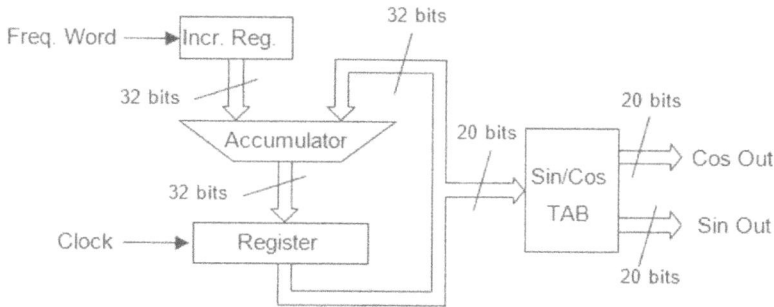

FIGURE 9.21   Principle of the numerical controlled oscillator [4].

The major components of a DDC are the mixer, the NCO, the decimation filters, and the final FIR filters. Other parts, such as automatic gain control (AGC) or demodulators may also be integrated.

The mixer is realized by a signed integer multiplier from the 16-bit input and the 20-bit NCO signal. The product of 36 bits may be truncated to 20–24 bits for the decimation stage.

The FIR filter shown in Figure 9.20 is used for optimal SNR and impulse response.

The NCO shown in Fig. 9.21 accumulates the register content and the increment register and writes it back into the register, generating a ramp function until the register overflows. The 32-bit wide register value is then converted to a Cos- and Sin-function by a read only memory (ROM)-table. The frequency generated by this NCO can be calculated as

$$f_{NCO} = \frac{K \cdot f_{clock}}{2^{32}}$$ where K is the frequency word in the increment register.

At a clock frequency of 80 MHz, the frequency resolution per step is:

$$f_{clock} / 2^{32} = 0.0186 \; Hz$$

The 20-bit wide-output signals allow an SFDR of >120 dB. The ROM-table should have a length of 2^{20} values. This may be too high for realization. For that reason, several methods are known to reduce the ROM-table:

- To use only the values from $0–\pi/2$, all others are deviated from this first quadrant.
- To use only a restricted quantity of most significant bits (MSBs) from the accumulator output and generate the intermediate values by either linear or polynomial interpolation.

Inaccurate values generate phase noise and, therefore, must be avoided. Similar to the ADCs, a small amount of phase or amplitude jitter injected between the phase accumulator and the ROM-table will reduce the SFDR by about 10–15 dB.

The principle of the decimation stage is shown in Figure 9.22.

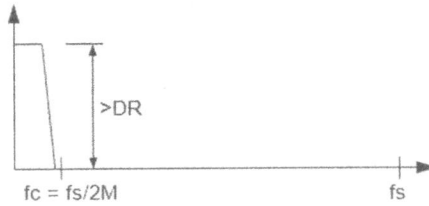

FIGURE 9.22   Conditions for a decimation filter (M = decimation factor) [5].

If we implement a decimation factor of M = 128, then we need a filter with a stop-band frequency of $f_s/256$ and a stop-band attenuation with at least the dynamic range of the whole receiver (i.e. the SNR of ADC plus the process gain). Such a filter must be split into several sections, as the FIR filter's lack on performance when $f_s/fc > 20$ [5].

The realization of nth order FIR filters needs n+1 MAC operation (MAC = multiply and cumulate). The multipliers are especially unpractical, as they need a lot of processing cells in the FPGA. The solution is to replace the high order FIR filters with cascaded integrator comb (CIC) filters. This class of filters uses a structure consisting of a number of integrators and combs and is in-between a rate changer (Figure 9.23). It was documented by Eugene Hogenauer [9].

The integrator sections are similar to first-order infinite impulse response (IIR) filters with unity gain coefficients, therefore, they are not stable by themselves. A single integrator can be described by the simple relation:

$$y(t) = x(t) + y(t-1) \text{ or its transfer function.}$$

$$H_1(z) = \frac{1}{1 - z^{-1}} \tag{9.17}$$

The comb section runs at a reduced sample rate of f_s/R, where R is the decimation factor. A simple comb section can be described by:

$$y(t) = x(t) - x(t-M)$$

$$H_c(z) = 1 - z^M \tag{9.18}$$

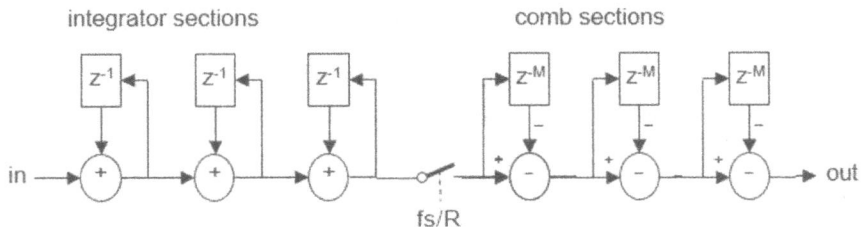

FIGURE 9.23   Structure of a Hogenauer CIC decimator filter [5].

The transfer function of an N-stage CIC filter shown in Figure 9.23 is:

$$H(z) = H_I^N(z) \cdot H_C^N(z) = \left(\frac{1-z^{-M}}{1-z^{-1}} \right)^N \tag{9.19}$$

The magnitude response is then:

$$|H(f)| = \left| \frac{\sin(\pi \cdot M \cdot f)}{\sin(\pi \cdot f / R)} \right|^N \quad N = \text{order}, M = 1, R = \text{decimation factor} \tag{9.20}$$

The factor M can be used to place the zeros of the transfer function. Usually, the values for M are 1 or 2. An M = 2 produces the first zero at $f_s/2$.

The dashed lines in Figure 9.24a show the passband from 0... f_s/R and the first aliasing bands:

$$B_{alias} = \frac{k}{M} \cdot fs \text{ for k} = 1, 2 \ldots R/2 \tag{9.21}$$

It must be noted that aliasing frequencies cannot be removed by subsequent filters. Therefore, the decimation filters must be carefully designed, particularly the first filter in the chain. Whereas the filter in Figure 9.23 has an aliasing-free dynamic range (AFDR) = 50 db, this characteristic value rises up to >100 dB by using N = 5 and R = 16 (see Figure 9.24).

Several CIC filters can cascade to reach the desired decimation and frequency response. Special attention must be given to the gain factor of CIC filters.

The gain factor of an unscaled CIC filter is: $G = (M \cdot R)^N$

The filter shown in Figure 9.24a has an associated gain factor of $16^5 = 1.048 \cdot 10^6$. Therefore, the result must be scaled down after each section by a factor of 16 bits or

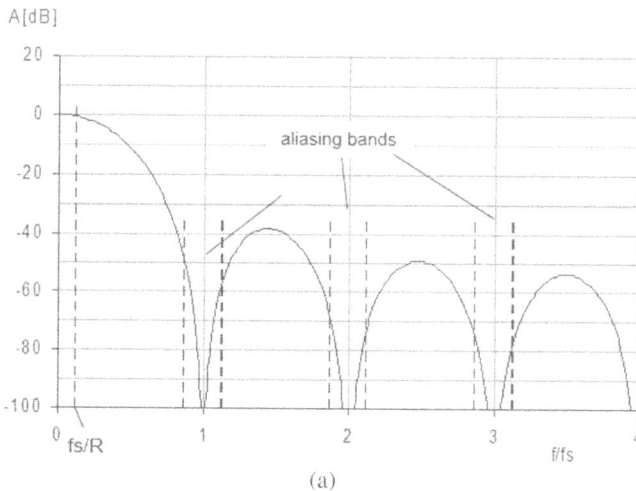

(a)

FIGURE 9.24 (a) Characteristics of CIC filter, M = 1, R = 8, N = 3 [5]. (b) Characteristics of CIC filter, M = 1, R = 16, N = 5 [5]. (*Continued*)

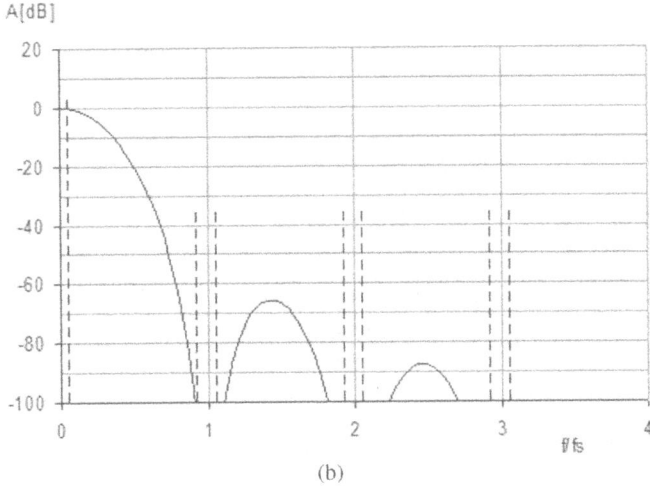

(b)

FIGURE 9.24   (*Continued*)

4 bits. To avoid this unpleasant effect caused by the recursive form of the integrator section, a non-recursive approach is given in Ref. [10].

By exchanging the integrator and comb sections, an interpolation takes place instead of the decimation. This function is used to generate transmission signals where the sample rate must be increased ("up-sampled"). At the interpolation rate, R, the rate changer is filling R-1 zero values into the integrator section. The cascading of several interpolators allows interpolation factors as high as 4,096 (Figure 9.25).

9.5   DEMODULATION ALGORITHMS

For amplitude modulation (AM) and FM demodulation, only an LPF is required, the phase shifter is not needed. For the AM demodulation, the envelope E(t) can be restored as to:

$$E(t) = \sqrt{I(t)^2 + Q(t)^2} = \sqrt{\left[A(t)\cdot\cos(\omega_o t)\right]^2 + \left[A(t)\cdot\sin(\omega_o t)\right]^2} = A(t) \quad (9.22)$$

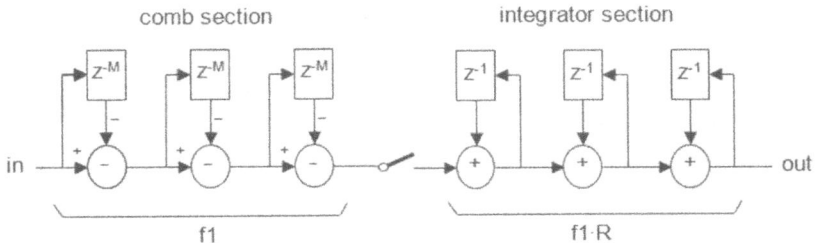

FIGURE 9.25   Interpolator for the transmit path (N = 3) [5].

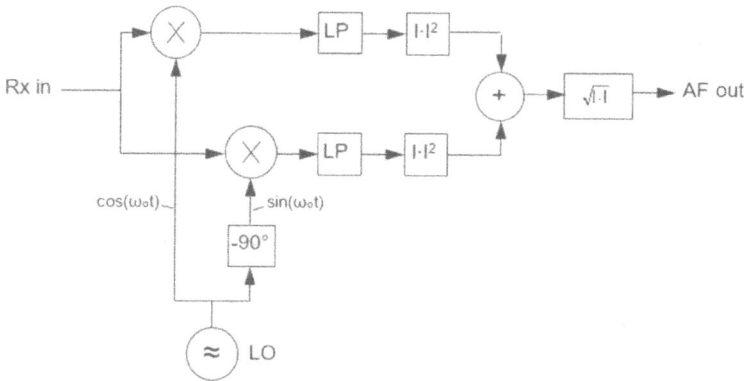

FIGURE 9.26 Direct conversion AM receiver.

whereas A(t) is the amplitude and ω_0 the carrier frequency. Due to the fact, that

$$\cos^2(\omega_o t) + \sin^2(\omega_o t) = 1 \tag{9.23}$$

E(t) is independent from ω_0. Thus, there is no need to synchronize the receiver to the AM carrier frequency, as it is required for the analog synchrodyne receiver. When using a proper algorithm, there is no interference tone noticeable, even when the receiver is detuned up to ±1 kHz. A typical Direct conversion AM receiver is shown in figure 9.26.

For FM demodulation, the arrangement shown in Figure 9.27 is appropriate. To obtain the frequency deviation from a frequency modulated signal, two operations are needed:

$$\phi(t) = arctan\left(\frac{I(t)}{Q(t)}\right) \text{with } \varphi \text{ as the instantaneous phase} \tag{9.24}$$

$$F(t) = \frac{d\phi(t)}{dt} \cdot arctan\left(\frac{I(t)}{Q(t)}\right) \text{with F as the demodulated FM signal} \tag{9.25}$$

The limiter in Figure 9.27 is needed to suppress any variations in amplitude. For this purpose, the signal value is multiplied by the reciprocal value of its instantaneous envelope value. The low-pass filtered DC component of the output signal can be used for an automatic frequency control (AFC).

9.6 SDR REALIZATION EXAMPLE

An SDR shortwave transceiver was developed. This project was started in 2005 with the aim to develop a digital transceiver by using the newest generation of DSP technology.

In another design [11], an analog sampler front end is used and the digital signal processing is performed by a personal computer.

Direct Conversion Receiver   Limiter   FM Demodulator

Rx in

LP

E   E(t)

Z⁻¹

X

cos(ωₒt)   sin(ωₒt)

-90°

$$E(t) = \sqrt{I(t)^2 - Q(t)^2}$$

≈ LO

AF out

arctan

+

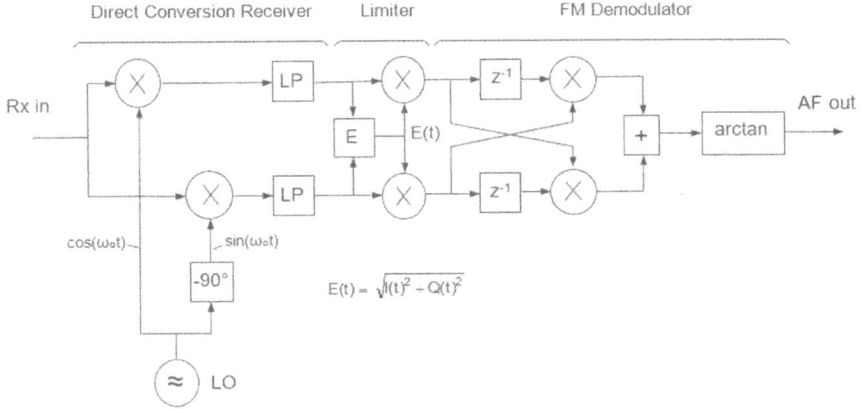

FIGURE 9.27 Direct conversion FM receiver.

The RX- and TX chains are controlled by a clock frequency of 73.728 MHz and the internal DSP clock is 249.912 MHz, resulting in 3.39 ns per instruction or 1.77 GFLOP (floating point operations) due to the parallel operation capability of the SHARC (The Super Harvard Architecture Single-Chip Computer) DSPs. The exchange rate between the DDC, DSP, and the digital up-converter (DUC) is 32 ksps. With this rate, the DSP can perform more than 9,600 instructions in between two sample interrupts (31.25 μs).

The transceiver uses chips from Analog Devices Inc., as shown in the block diagram in fig. 9.28 [5].

The main tasks of the DSP at the full rate of 32 ksps are:

• RX filtering for all modes of operation and for up to four parallel receiving channels.
• Conditioning the TX-signal, including the adaptive predistortion and ALC.

Tasks at 16 ksps are:

• Audio signal processing.
• Pre-emptive AGC.
• Notch filters and audio equalizing.
• Pass-band tuning.

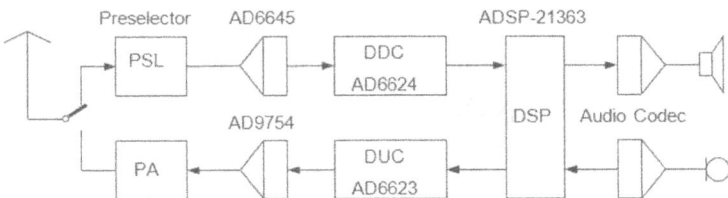

Preselector   AD6645   ADSP-21363

PSL   DDC   DSP   Audio Codec
          AD6624

AD9754

PA   DUC
          AD6623

FIGURE 9.28 Chipset used in the ADT-200A [5].

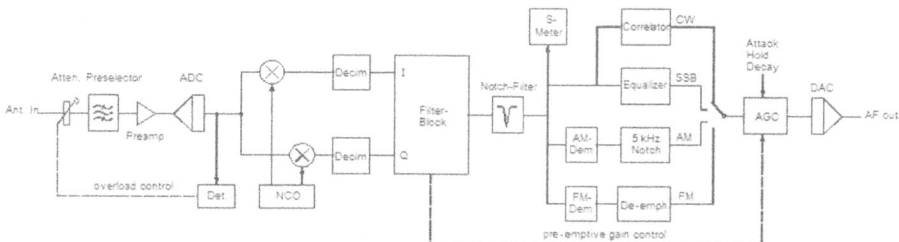

FIGURE 9.29 Block diagram of the receiver part of the ADT-200A [5].

Tasks at 8 ksps are:

- FFT for spectrum monitoring.
- Continuous wave (CW)-shaper and -decoder.

A more detailed block diagram is given in Figure 9.29. An attenuator 0–35 dB in 5 dB steps is placed on the receiver input. It is controlled either manually or by a detector on the wideband input of the DDC, which increases one step when the signal passes the -1 dB full scale limit for >1 second. The attenuator is decreased when the signal is below -8 dBFS for 5 seconds. This measure protects the ADC from being overloaded and automatically adapts the receiver dynamic range on the actual signal levels.

A pre-selector is used especially to improve the second-order intermodulation and to eliminate strong signals from broadcast stations. It is equipped with tunable second-order bandpass filters with a bandwidth of 7–9% of the center frequency.

The possible dynamic ranges with B = 500 Hz depending on the attenuator setting are:

The numerical representation of the signals on the ADC is:

$$U_{in} = 0V \rightarrow 0 \times 0000 = 0$$

$$U_{in} = 2.5V \rightarrow 0 \times 3FFF = 16,383 \qquad (9.26)$$

TABLE 9.3
Possible Dynamic Range Values [5]

Attenuator [dB]	MDS [dBm]	Max. Input [dBm]	Dyn. Range [dB]	F [dB]
0	-137	-18	119	10
5	-132	-13	119	15
10	-127	-8	119	20
15	-122	-3	119	25
20	-117	2	119	30
25	-112	7	119	35
30	-107	12	119	40
35	-102	17	119	45

After the DDC, having a process gain of 35.7 dB at B = 10 kHz and an output resolution of 24 bits:

$$\text{Min.Value} \rightarrow 0 \times 000003 = 3 \,(\text{due to quantization noise})$$

$$\text{Max.Value} \rightarrow 0 \times 1\text{FFFFF} = 2{,}097{,}151 \,(\text{dynamic } 117 \text{ dB}) \qquad (9.27)$$

The integer values are converted in the DSP into a 32-bit floating-point format and scaled down by 12 bits:

$$\text{Min.Value} \rightarrow 0.000721$$

$$\text{Max.Value} \rightarrow 504.0 \qquad (9.28)$$

The 32-bit floating-point single precision format according to IEEE 754 consists of a sign bit (s), 8 bits of exponent (e), and 23 bits of mantissa (m). The numerical value x is given by:

$$x = s \cdot m \cdot b^e \,(b = 2) \qquad (9.29)$$

The possible range that can be covered is:

$$\text{Smallest value} = \pm 1.4 \cdot 10^{-45}$$

$$\text{Biggest value} = \pm\, 3.4 \cdot 10^{38} \qquad (9.30)$$

The resolution is only dependent on the mantissa and, therefore, independent of sign or exponent. The resolution can be assumed as:

$$r = 20 \cdot \log\left(2^{23}\right) = 138\,\text{dB} \qquad (9.31)$$

This value will be even worse when a lot of rounding errors are accumulated. For that reason, the SHARC-signal processors from Analog Devices Inc., are switchable to the 40-bit extended single precision format by setting a flag. This gives critical operations, such as filter- or FFT-processing, a significant improvement in accuracy and, therefore, has only a minor effect on the dynamic range.

The phase noise properties of SDR transceivers using the radio frequency (RF)-sampling method are, in general, superior to the analog systems, in particular when close to the carrier. The phase noise of a receiver (Figure 9.30) is determined by the phase noise of the oscillator, the aperture jitter, quantization noise, and input noise of the ADC.

9.7 FILTERS

The filters can be realized either in the time domain or frequency domain. The frequency domain implementation requires a fast Fourier transformation (FFT), a

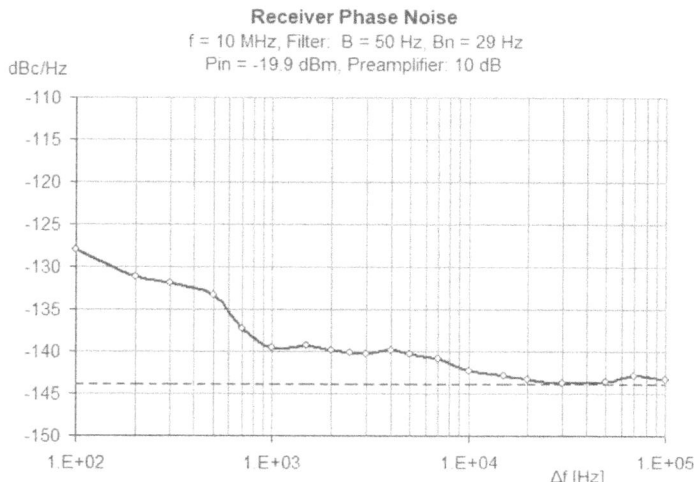

FIGURE 9.30   Receiver phase noise [5].

truncation of unwanted bins, and then an inverse FFT (IFFT). This process allows an easy variation of bandwidth and steep filters with a shape factor near 1.0. On the other hand, low shape factors cause an unfavorable step response, caused by the Gibbs' phenomenon [12], which leads to an unnatural sound and intersymbol interferences at data transmission.

For filters in the time domain, the FIR filter is well suited. It is worthwhile to invest time for a careful design. The following criteria must be considered:

- An optimal trade-off between the filter shape and associated impulse response. For a bandwidth of 2.5 kHz, a shape factor (-6 dB/-60 dB) from 1:1.25 is a good choice for speech transmission.
- A linear phase in the passband, i.e. a constant group delay.
- An optimal relationship between the lower and upper cutoff frequency. As a rule, the product from the lower and upper cutoff frequency (-6 dB) should be near to 5×10^5 Hz² for a natural sound at speech transmission.
- A low time delay. This is essential, especially for data transmission, such as PACTOR (Pacific Area Communication Teletype Over Radio, a digital communication protocol designed for reliable data transmission over radio channels).
- A high stop-band attenuation (>100 dB).

Bandpass filters are required for single side band (SSB) and continuous wave (CW) reception, when the suppression of the unwanted side band is needed. For AM and FM however, LPFs can be applied, whereas the resulting receiver bandwidth is twice the filter bandwidth.

Another desirable characteristic of the filters is to take over the ±90° phase shift. This spares an extra Hilbert transformation filter. All these listed properties can be fulfilled by an Nth order FIR filter. We will start with $N = 255$. The following formulas deliver the $N+1$ coefficients ($h_I[n]$, $h_Q[n]$) for the different I- and Q-channel filters:

$$h_I(n) = \frac{1}{\pi \cdot n}\left[\sin\left(\frac{2\pi \cdot n \cdot f_H}{f_s} \right) - \sin\left(\frac{2\pi \cdot n \cdot f_L}{f_s} \right) \right] n = -128 \ldots +127$$

$$h_o(n) = \frac{1}{\pi \cdot n}\left[\cos\left(\frac{2\pi \cdot n \cdot f_H}{f_s} \right) - \cos\left(\frac{2\pi \cdot n \cdot f_L}{f_s} \right) \right] \tag{9.32}$$

where f_L is the lower cutoff frequency, f_H the higher cutoff frequency, and f_s is the sampling rate. The magnitude frequency response of this filter is given by:

$$|H(f)| = \sum_{n=1}^{\frac{N}{2}} 2 \cdot h(n) \cdot \sin\left(\frac{2\pi \cdot n \cdot f}{f_s} \right) \tag{9.33}$$

The abrupt transitions from the coefficients on both ends yield to big sidelobes as shown in Figure 9.31a. The poor out-of-band selectivity can be improved by windowing. This means, that the coefficients are multiplied by a window function with the aim to reduce the step transition on both ends of the coefficient chain without degrading the filter shape. The simplest window function is a triangle with the shape $w(0) = 0$, $w(N/2) = 1$ and $w(N) = 0$. There are many more known sophisticated window functions. One example is the Hamming window with the function:

$$w(n) = 0.54 - 0.46 \cdot \cos\left(\frac{2\pi \cdot n}{N} \right) \quad \text{for n = 0, 1, \ldots N}-1 \tag{9.34}$$

The filters shown in Figure 9.31a–c are simulated by Matlab. The two sidebands are clearly visible, the frequency = 0 is equal to the LO frequency. It also demonstrates the influence of different window functions. By optimizing the sidelobes, the shape factor is adversely affected. Whereas the shape factor with the Hamming window is 1:1.22, it rises to 1:1.28 for the Blackman-Harris window. At the same time, the side-lobe suppression increases from 60 dB up to 115 dB.

The filters shown above are designed for a passband of 300–2,700 Hz. If a design demands the lower cutoff frequencies, then aliasing on the frequency of 0 Hz will occur. To avoid this, the usage of a low IF in the range of 1–1.5 kHz is recommended. This method will also allow implementation of passband tuning (PBT) to move the whole filter up and down for several hundred hertz.

The introduction of a low intermediate frequency also offers the advantage of the phase transition between the two sidebands being clearly separated. Figure 9.32 shows the amplitude and phase response from the simulated filter, but with an IF = 1 kHz.

The lower diagram in Figure 9.32 shows the phase response of the Q-channel filter. The phase is shifted by $\pi/2$ in the lower sideband and by $-\pi/2$ in the upper sideband, whereas the I-channel has a zero phase offset in both sidebands. Thus, the function of a Hilbert transformer is emulated by this type of filter.

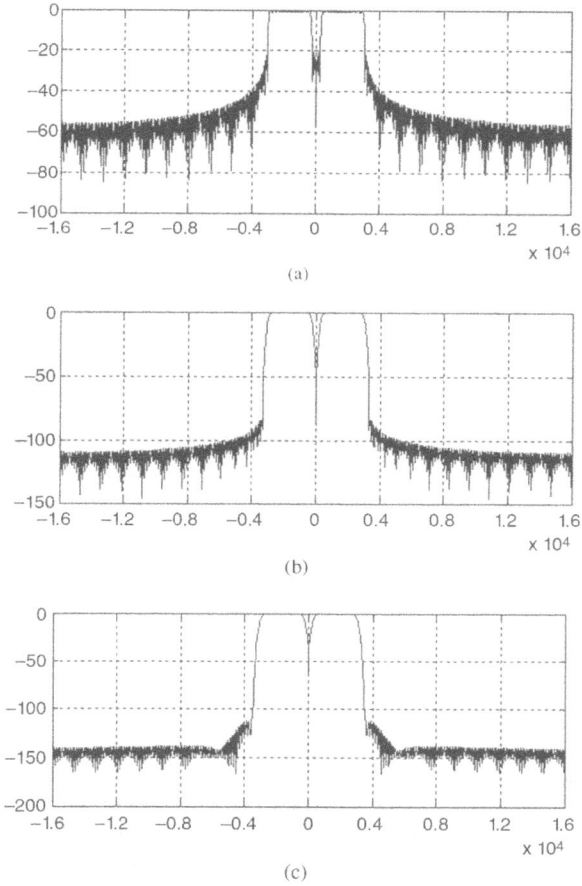

(a)

(b)

(c)

FIGURE 9.31   The magnitude of frequency response without windowing [5] (A), with the Hamming window [5] (B), and with the Blackman-Harris window [5] (C).

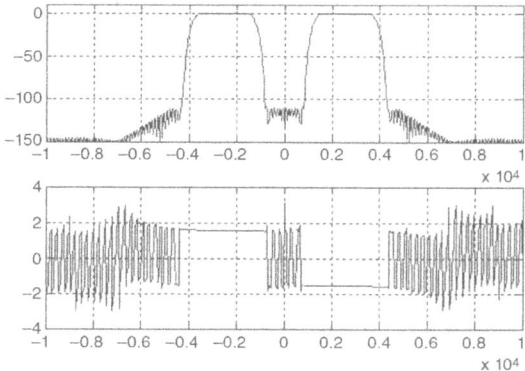

FIGURE 9.32   Amplitude and phase of the Q-channel filter with an IF = 1 kHz [5].

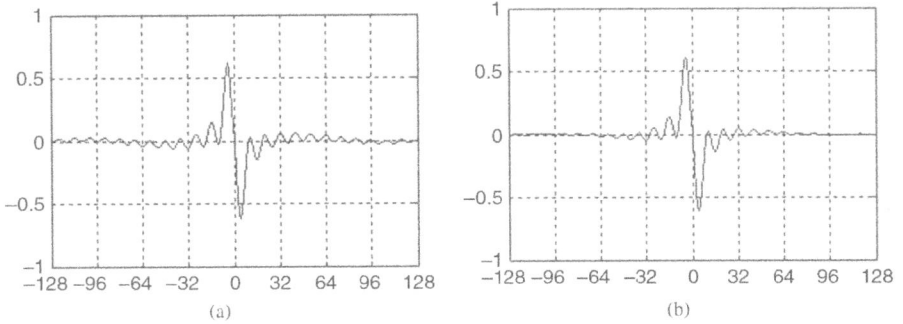

FIGURE 9.33 Impulse response without (a) and with (b) a window [5].

The realization is that such a set of I- and Q-filters with a 40-bit floating point SHARC DSP require a processing time of 1 μs, resulting in a reduction of the unwanted sideband of >100 dB, enough for the needs in most communications receivers.

The impulse response, showed in Figure 9.33 is equal to $h_I(n)$ and $h_Q(n)$, which represent the filter coefficients. The signal time delay of this filter is, therefore, 128 samples, or:

$$\text{Td} = \frac{n}{2 \cdot f_s}, \text{ whereas n is the filter order.} \tag{9.35}$$

The block diagram of the CW/SSB filter section is given in Figure 9.34. The complex input signal is heterodyned by an IF of 1 kHz. In front of the main filters is a first noise blanker followed by the rate decimator n1. The decimator is used to adapt the sampling rate in relation to the filter bandwidth to fulfill the condition f_s/B <20. The main filters are followed by the second noise blanker and the sideband selector. The result is a single sideband signal, shifted by a 1 kHz intermediate frequency. This is split again into I- and Q-channels and passes the post filters. Due to the phase shift capability of these filters, a complex signal results again. The decimator n2 serves to further decrease the sampling rate, especially for the narrow CW filters. For that

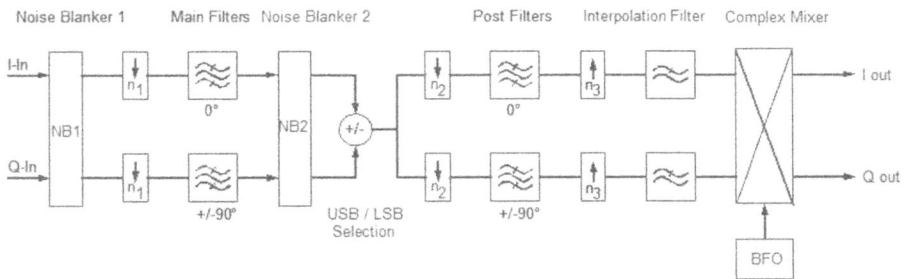

FIGURE 9.34 Block diagram of the filter section [5].

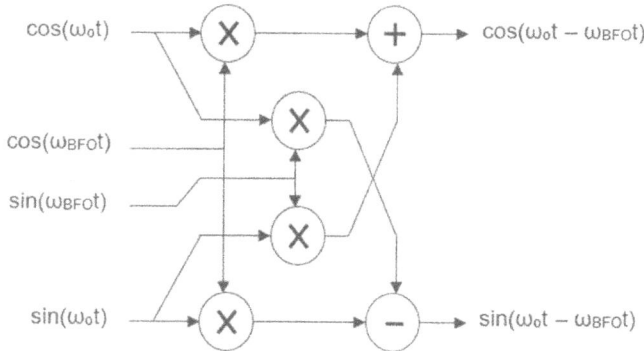

FIGURE 9.35 Complex mixer maintaining quadrature.

reason, the sampling rate, as low as 4 ksps, must be interpolated up to the final sampling rate of 16 ksps by n3. An interpolation filter avoids aliasing. The final complex mixer maintains the quadrature, but suppresses the sum of the inputs and the beat frequency oscillator (BFO) frequencies.

The complex mixer can be accomplished by four multipliers and two adders. The principle is presented in Figure 9.35:

All SSB filters are showed in Figure 9.36 in a single plot for comparison. The filters with the smallest bandwidth deviate from the rectangular form because the f_s/B ratio <20 is violated. On the other hand, the shape factor of the four largest filters is decreased, as they are processed with 32 ksps in contrast to all others with 16 ksps [5].

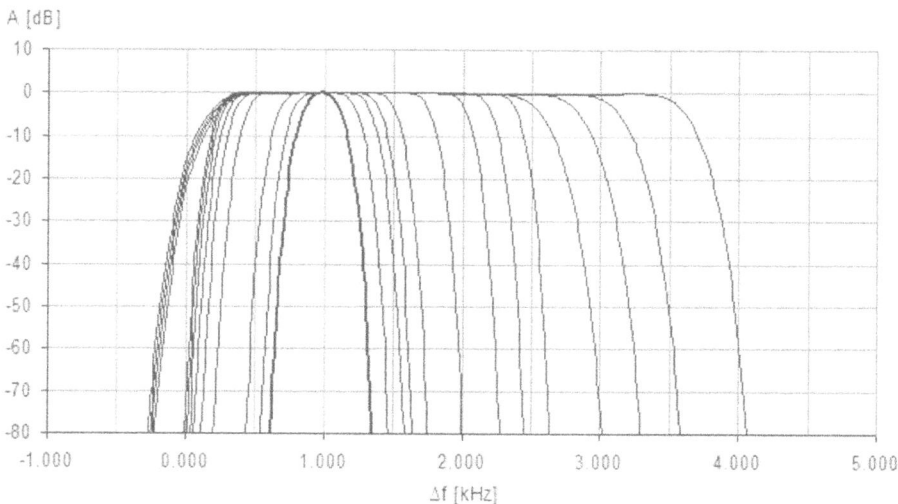

FIGURE 9.36 Measured frequency response of single sideband filters [5].

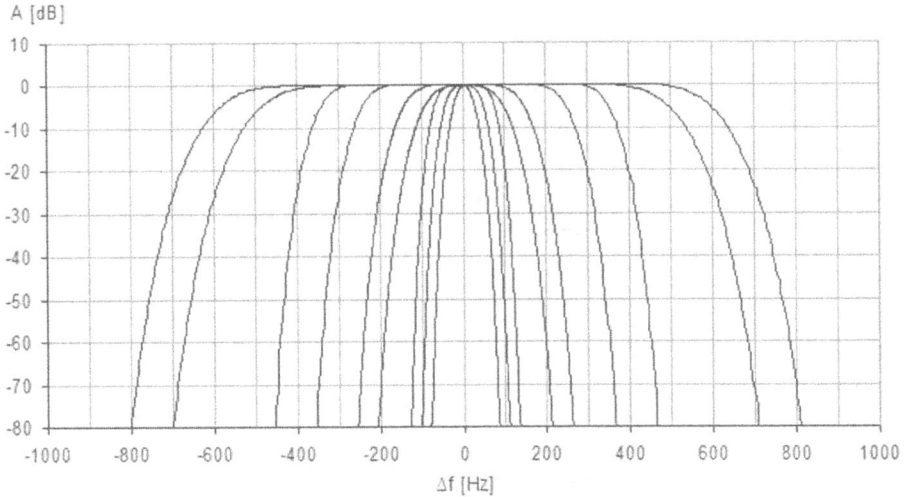

FIGURE 9.37 Measured frequency response of continuous wave filters [5].

The following table shows characteristic data. The group delay is denoted as the whole delay between the antenna input and receiver output.

The CW filters are a little more complex. To realize bandwidths as small as 50 Hz, a drastic reduction of the sample rate is required. The narrowest bandwidth of the main filter is 300 Hz. This filter acts also as antialiasing for the subsequent sample rate of 4 ksps. All CW filters are operating on an IF of 1 kHz, normalized to 0 Hz in Figure 9.37. The pitch frequency can be varied by the BFO.

TABLE 9.4
SSB Filter Characteristics [5]

B [Hz]	fu [Hz]	fo [Hz]	Shape Factor	Group Delay [ms]
300	850	1,150	2.29	20
500	750	1,250	1.78	20
700	650	1,350	1.55	20
1,000	420	1,420	1.38	20
1,200	330	1,530	1.32	20
1,500	280	1,780	1.26	20
1,800	240	2,040	1.21	20
2,000	220	2,220	1.19	20
2,200	210	2,410	1.18	20
2,400	190	2,590	1.32	12
2,700	170	2,870	1.29	12
3,000	160	3,160	1.26	12
3,500	140	3,640	1.22	12

TABLE 9.5
CW Filter Characteristics [5]

B [Hz]	Shape Factor	Sample Rate [ksps]	Group Delay [ms]	Noise Bandwidth [Hz]
50	2.96	4	44	29.1
100	2.02	4	44	73.9
150	1.64	4	44	128.6
200	1.95	8	28	158.9
300	1.63	8	28	262.0
500	1.39	8	28	461.5
750	1.19	8	28	667.0
1,000	1.35	16	20	913.6
1,200	1.30	16	20	1,118.1

The shape factor SF can be estimated using the empirical formula:

$$SF = 1 + (0.0249 \cdot fs/B) \tag{9.36}$$

This relation is true for these presented FIR filters with an order of 255 and the Blackman-Harris window.

The group delay is the whole delay between the antenna input and receiver output. The noise bandwidth was numerically integrated from the amplitude response.

9.8 NOISE BLANKER

The noise blanker used in this design (Figure 9.38) was specially developed to suppress impulse noise and rampant amplitude changes.

This algorithm is effective in cancelling sharp noise impulses. The factor k sets the headroom about the filtered mean input value until the shaping is effective. When a>b, the input signal is scaled down to the level of b and is subsequently increased by the factor p for each sample (Figure 9.39a and b). Therefore, p increases from 1 with a delta of 0.5 dB per sample. Thus, a steep amplitude rising will be shaped with a ramp of 0.5 dB per sample, or 16 dB per millisecond. We must consider that this process is simply a gain variation and, therefore, linear and free from harmonic distortion.

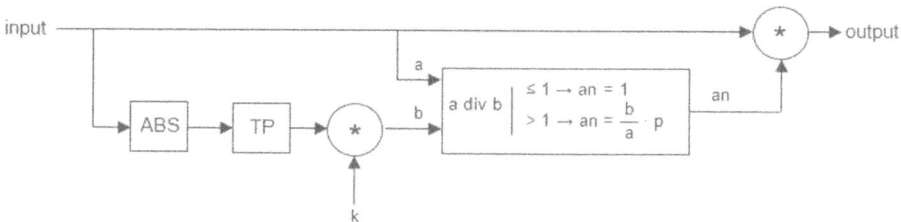

FIGURE 9.38 Block diagram of a noise blanker [5].

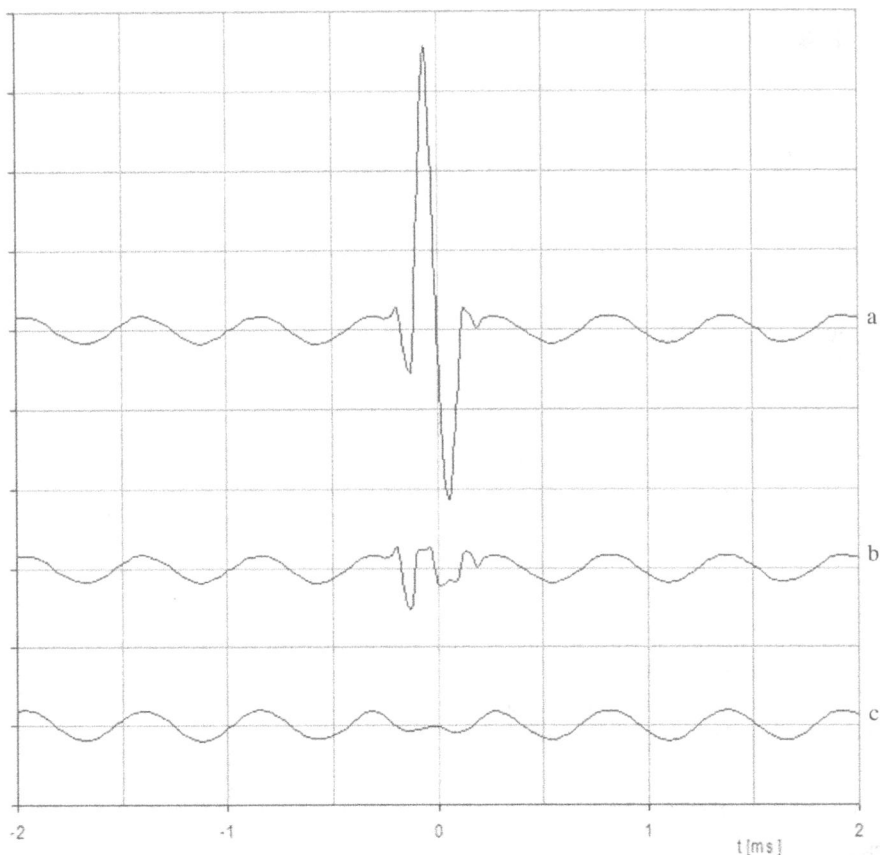

FIGURE 9.39 Capability of the noise blanker (NB): at the input of NB (a), at the output (b), and after the main filter (time shifted) (c) [5].

9.9 AGC

AGC (Figure 9.40) is often untended in receiver designs. The result is an unnatural or popping audio [13]. In software-defined receivers, most parts are free of distortion, so the AGC is the only part that has a big influence on speech quality.

AGC design in analog receivers is difficult, as the control chain is distributed over different stages with filters in between, causing a time delay. In SDR designs, it is much easier to build an AGC algorithm, as the dynamic range of the digital part is much greater than that of the front end and, therefore, the control can be concentrated on the end of a receiver chain.

In the literature, most AGC algorithms are based on feedback systems, which are potentially unstable and tend to overshoot. Thus, the AGC in the ADT-200A is built as a feed-forward system. Moreover, to prevent pops on fast rising signal amplitudes, a pre-emptive part is added to the normal control signal. This part is picked up

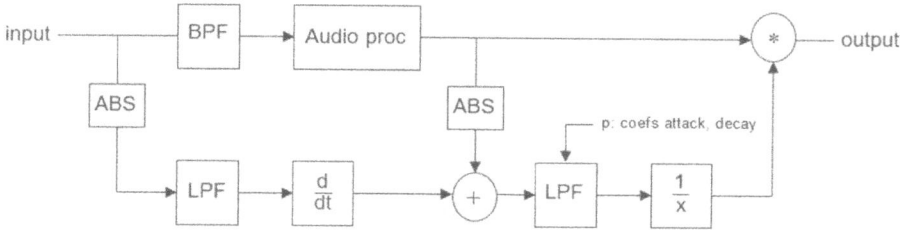

FIGURE 9.40 Simplified block diagram of the pre-emptive AGC [5].

in front of the last filter and, therefore, it receives amplitude steps a few milliseconds (ms) before they reach the AGC. To ensure that the pre-emptive part is only active at strong amplitude steps, its envelope is differentiated before it is added to the normal AGC.

Figure 9.41 shows the reaction when a -10 dBm signal is switched on, while the AGC is set to the noise level of -120 dBm. Graph A denotes the pre-emptive part and Graph B shows the resulting AGC control signal. Note that due to the filter delay, the signal reaches the AGC after 8 ms, when the AGC control signal is already set. This method prevents overshooting, even when the attack time is set to a value >5 ms.

The combination from the noise blanker and the pre-emptive AGC results in an effective audio levelling system, maintaining a natural sound, even under severe conditions. As the AGC control function is placed on the end of the processing chain, the volume is held constant, even when the notch filter is on, or the audio equalizer is set to the minimum or maximum (Figure 9.42).

An unpleasant characteristic of most AGC algorithms is the dependence of the recovery time from the level change. For example, when a step of -20 dB has a recovery time to -6 dB of 30 ms, a step of -100 dB needs 160 ms to reach the same -6 dB audio output (see Figure 9.43).

To solve this problem, we must control the time constant in reference to the step size. This needs knowledge of the high-level A at the end of the attack period and the

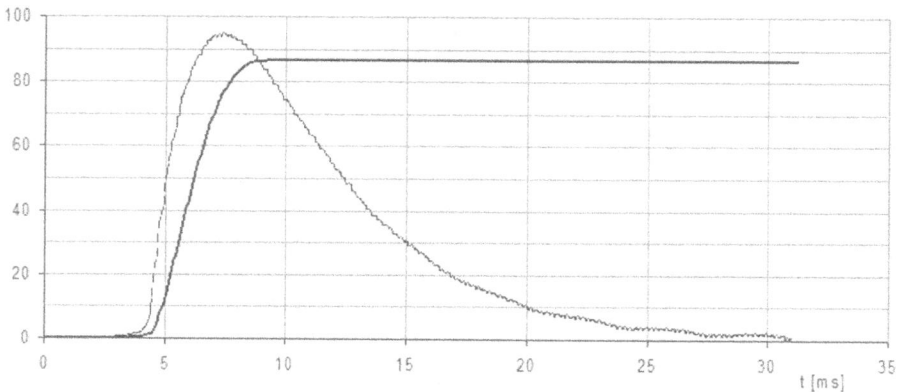

FIGURE 9.41 Step response of the AGC on a level change of 110 dB [5].

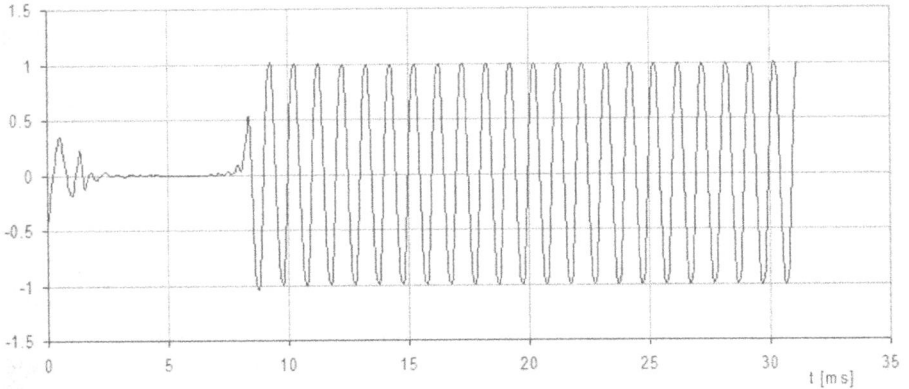

FIGURE 9.42   Audio signal, measured at the AGC output at a setup of 110 dB [5].

low-level B at the beginning of the decay period, following the hold time. Then the coefficient p of the arrangement in Figure 9.33 will become:

$$p = k \cdot \log_2 \left(\frac{A}{B} - 1 \right) \tag{9.37a}$$

$$k = m \cdot \left(1 - e^{-\frac{1}{f_s \cdot tau}} \right) \tag{9.37b}$$

where "f_s" is the sample rate, "tau" the nominal time constant, and "m" a factor. This simple relationship can be implemented with a few lines of code. The result is presented in Figure 9.44. It is clearly visible that the bigger the change in signal, the faster is the AGC reaction, with the goal to reach the -6 dB point always at the same time.

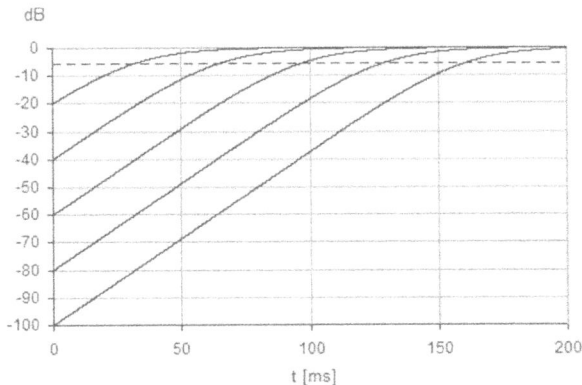

FIGURE 9.43   AGC recovery time for 20–100 dB amplitude steps [5].

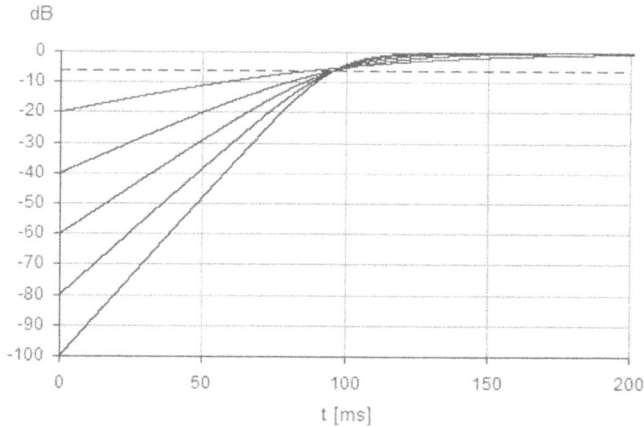

FIGURE 9.44 Compensated automatic gain control recovery time [5].

9.10 THE S-Meter

Due to the fact, that there is no gain control between the input attenuator and the filtered receiving signal, the S-Meter (Figure 9.45) can profit from the full dynamic range of the receiver. It is realized by a digital peak envelope detector with a fast ramp-up and slow decay, making it capable of showing the real peak value of received voice signals. In addition, it has a numerical display for dBm and dBµV, which are calibrated as root mean square indications.

The S-Meter has a measuring range from -148 dBm up to +17 dBm, including the full range of the input attenuator, whose setting is compensated so that the meter always displays the signal level at the receiver input. The insertion loss of the pre-selector filter is digitally compensated by a calibration table, thus, the meter accuracy is ±1 dB over the full range.

9.11 ADAPTIVE TRANSMITTER PREDISTORTION

This section demonstrates how the receiving part in a transceiver can be used to linearize the transmit signal. The non-linearities in a power amplifier (PA) are caused by different categories of distortion products:

- **Harmonic distortions**: Those can be cancelled by an LPF.
- **AM-to-AM distortions**: A non-linear envelope distortion by saturation, producing intermodulation.

FIGURE 9.45 S-Meter, indicating S6.

- **AM-to-PM distortions**: Producing a phase modulation, depending on the envelope.
- **Memory effect**: The sum of non-linearities is also dependent from the past, i.e. distortions on an increasing envelope are different from those of a decreasing envelope.

There are different predistortion procedures known [14] and they can be roughly categorized into feed-forward and feedback systems. Only the latter one shall be discussed here. For SDR transceivers, the baseband feedback method is favored, as the transmit signal is already available in a complex I/Q form. For the feedback path, the receiver can be used, synchronized to the transmit frequency. The comparison of the undistorted and the distorted transmit signal is then carried out in the baseband at the low system sample rate.

The error signal results by comparing the two complex signals after scaling and delaying for correct time correlation. This signal contains both the AM/AM and the AM/PM errors. As they are influencing each other, a long convergence time will result. Therefore, it is better to separate the AM and PM components by converting the Cartesian into a polar representation, which allows the two components to separate.

When the undistorted signal is $\overline{X}$ and r(t) is the instantaneous amplitude, then we can write:

$$\overline{X} = \mathrm{r}(t) \cdot e^{j\omega t} = r(t) \cdot \left(\cos(\omega t) + j \cdot \sin(\omega t) \right) \qquad (9.38)$$

and in the polar representation:

$$\left| \overline{X} \right| = r(t)\theta = arctan\left(\frac{\cos(\omega t)}{\sin(\omega t)} \right) \qquad (9.39)$$

The principle of an adaptive AM/AM predistorter, based on a look-up table (LUT) is presented in Figure 9.46. The function "mag" delivers the scalar envelope from the undistorted baseband transmit signal. The output of the non-linear power amplifier (NLPA) is fed back via the receiver and delivers the distorted baseband envelope (b), after passing the second function "mag" and a multiplier by the factor c. As the propagation time through the PA and receiver is substantial, the undistorted signal must be delayed because signal (a) is time-equivalent to signal (b). Then the envelope-dependent instantaneous error $p(r_w)$ is defined as:

$$\mathrm{p}(r_w) = \frac{a}{b} \qquad (9.40)$$

where r_w represents the physical write address of the LUT. The LPF preceding the DUC serves to limit the noise, introduced by the discontinuities of the LUT output. The selection of the filter bandwidth is not easy. When it is too small, the effectiveness of the predistortion moves toward higher modulation frequencies, when it is too broad, the adjacent channel power ratio (ACPR) will be reduced. The bandwidth must be determined as a trade-off between these two effects.

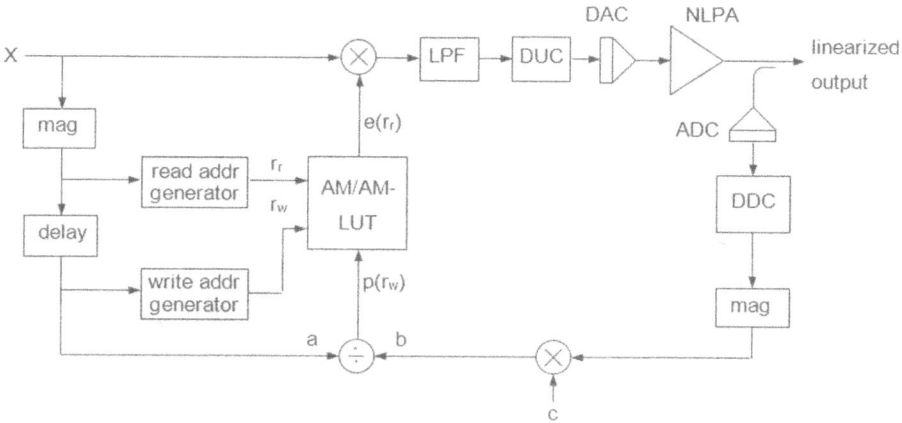

FIGURE 9.46   Baseband representation of a feedback adaptive AM/AM predistortion.

By updating the AM/AM-LUT, one must consider the following points:

a. The error value $p(r_w)$ must be filtered by an exponential average before it is written to the table. An applicable arrangement is shown in Figure 9.47.
b. Some provision must be taken when only a few addresses are updated in case the envelope is occasionally synchronous to the sample rate. To remedy this situation, values between two newly updated points must be periodically interpolated.
c. When the LUT is trained by a low-level modulation signal, the untrained higher values must be extrapolated to avoid a discontinuity in the correction.
d. When the transmitter is modulated by a single carrier or FM, frequency shift keying (FSK), PM or phase shift keying (PSK), then the predistortion must be switched off, as the carrier remains constant.
e. The step size in the LUT must be small so that the non-linearities remain nearly constant, a number of ≥ 128 steps is sufficient.

The predistortion algorithm must be able to react to any load variations on the PA that affect the loop gain.

Figure 9.48a shows the typical voltage gain (G_v real) of a MOS-FET PA, measured with 50 Ω terminations, whereas Figure 9.48b represents the voltage gain correction factor p to achieve a linear gain (G_v linear).

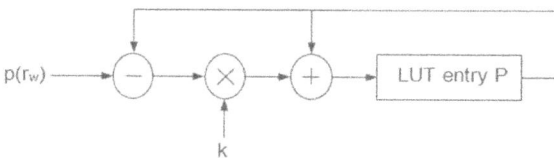

FIGURE 9.47   Exponential averager for LUT update values (k = filter coefficient).

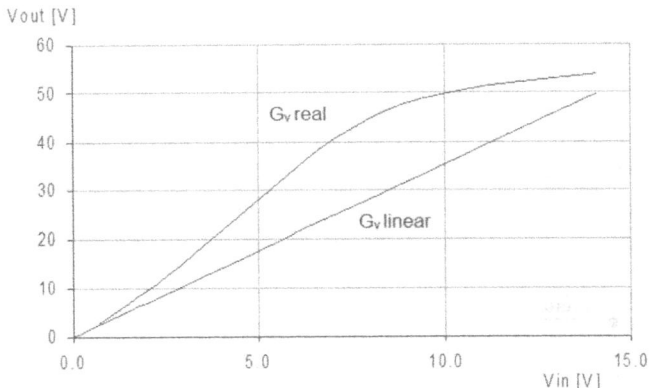

(a)

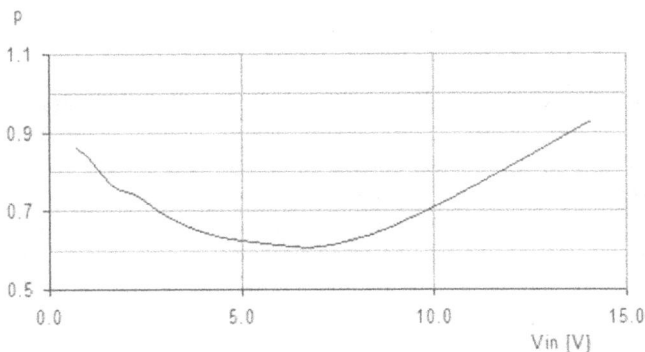

(b)

FIGURE 9.48   (a) Typical power amplifier voltage gain and (b) gain correction factor (p).

Generally, the same procedure as for AM/AM distortions can also be applied to correct AM/PM distortions. However, it is somewhat more delicate, as the phase of the RF signal is now involved. The phase may be affected also by the harmonic filter or a complex load impedance (Figure 9.49).

To get the phase values, an arc tangent function is needed. This algorithm can be implemented by using power series approximations:

$$\arctan(x) = \frac{\pi}{2} - \frac{1}{x} + \frac{1}{3x^3} - \frac{1}{5x^5} + \frac{1}{7x^7} - \frac{1}{9x^9} + \frac{1}{11x^{11}} - \dots \qquad (9.41)$$

This series converges rapidly for x ≥1.4. For lower values of x, the substitution arcsin(a) is needed:

$$\arcsin(a) = a + \frac{a^3}{6} + \frac{3a^5}{40} + \frac{15a^7}{336} + \frac{105a^9}{3,456} + \frac{945a^{11}}{42,240} + \dots \text{ whereas, } a = \frac{x}{\sqrt{1+x^2}} \qquad (9.42)$$

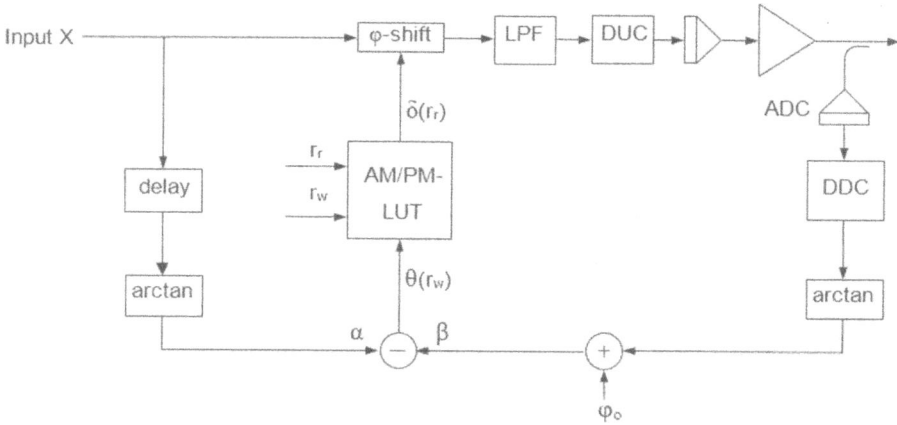

FIGURE 9.49 Adaptive AM/PM predistortion.

By using series with seven terms, a maximal error of ±0.03° can arise. The optimized DSP code needs 40 instructions, or 0.14 µs processing time, to get one arctan value.

The error signal $\delta(r_r)$ is the output from the AM/PM-LUT and contains the phase error, which is used to drive the phase shifter. The transfer function of this shifter must be linear in phase over the range of 0–2π. The mathematical representation of a complex signal with a phase offset of φ is:

$$X(t) = A \cdot e^{\,j\omega t + \varphi} = \cos(\omega t + \varphi) + j\,\sin(\omega t + \varphi) = \cos(\omega t) \cdot \cos(\varphi) - \sin(\omega t) \cdot \sin(\varphi)$$
$$+ \; j\,[\sin(\omega t) \cdot \cos(\varphi) + \cos(\omega t) \cdot \sin(\varphi)] \tag{9.43a}$$

and with $I = \cos(\omega t)$, $Q = \sin(\omega t)$, *we can write* $I_\varphi = I \cdot \cos(\varphi) - Q \cdot \sin(\varphi)$,
$$Q_\varphi = Q \cdot \cos(\varphi) + I \cdot \sin(\varphi) \tag{9.43b}$$

Figure 9.50a and b present the output spectrum of the 50W PA when the transmitter is SSB-modulated with white noise (0 dB is equivalent to 50 Wpep, fo = 14.1 MHz, B = 2.7 kHz, channel spacing = 5kHz):

While the second-order distortion can easily be cancelled by a harmonic filter, the third-order distortion remains the dominant distortion in a PA. When the second-order products are disregarded, we can assume the transfer function from a PA to:

$$V_{out}(t) = k_1 \cdot V_{in}(t) - k_3 \cdot V_{in}^3(t) = k_1 \cdot V_{in} \cdot \sin(\omega t) - k_3 \cdot V_{in}^3 \sin^3(\omega t) \tag{9.44}$$

k_1 is the small signal voltage gain and k_3 represents the third-order product ($k_3 > 0$).

ATT: 60.00 dB
BW VIDEO: 0.10 kHz
BW RES: 0.07 kHz

MARKER: −89.84 dBm, 14090.0222 kHz

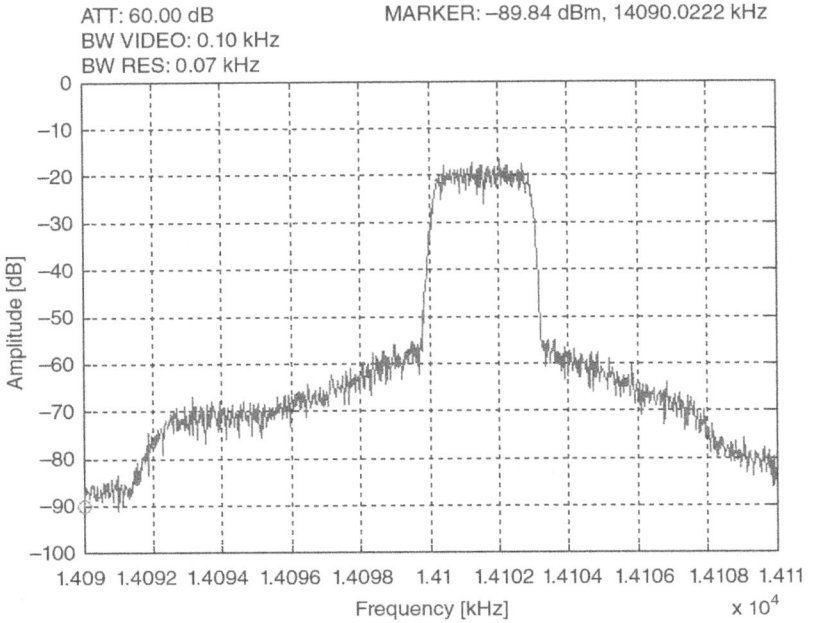

(a)

ATT: 40.00 dB
BW VIDEO: 0.10 kHz
BW RES: 0.07 kHz

MARKER: −91.79 dBm, 14090.0222 kHz

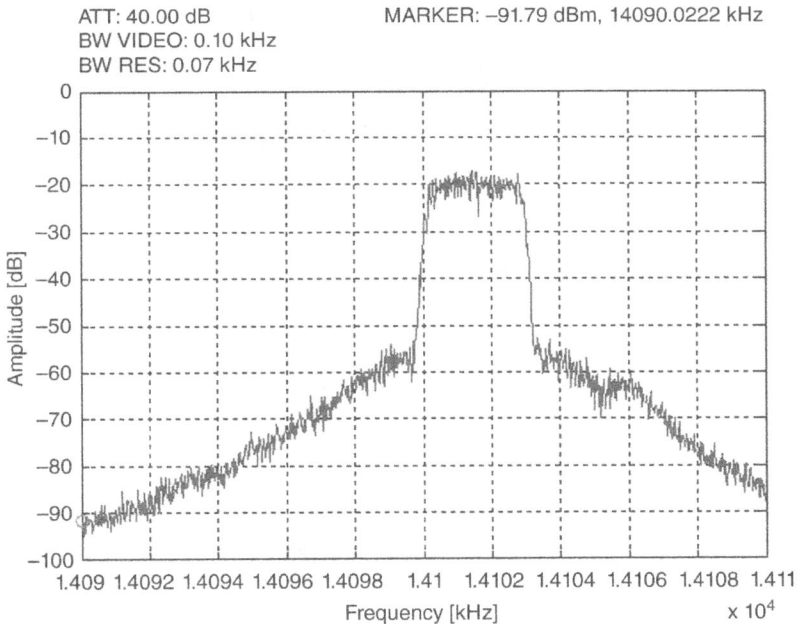

(b)

FIGURE 9.50 (a) TX spectrum with low-pass filter (LPF) = 7.5 kHz and adjacent channel power ratio (ACPR) = 47 dB. (b) TX spectrum with LPF = 5 kHz and ACPR = 50 dB.

With $\sin 3(x) = \dfrac{1}{4} \cdot [3\sin(x) - \sin(3x)]$ we get:

$$V_{out}(t) = k_1 \cdot V_{in} \cdot \sin(\omega t) - 0.75 \cdot k_3 \cdot V_{in}^3 \sin(\omega t) + 0.25 \cdot k_3 \cdot V_{in}^3 \cdot \sin(3\omega t) \quad (9.45)$$

The third-order intercept point OIP3 is defined as when the extrapolated fundamental and third-order distortion product appears with the same level on the output.

$$k_1 \cdot V_{in} - 0.75 \cdot k_3 \cdot V_{in}^3 = 0.25 \cdot k_3 \cdot V_{in}^3 \rightarrow V_{in\_IP3} = \sqrt{\dfrac{k_1}{k_3}} \quad (9.46)$$

The formula for $V_{out}(t)$ shows further that the fundamental signal is reduced with increasing input signal V_{in}. An important point in the transfer function is the 1 dB compression point. It can be calculated by using the relation:

$$V_{in1dB} = 0.3808 \cdot \sqrt{\dfrac{k_1}{k_3}} \quad (9.47)$$

Figure 9.51 shows an overdriven amplifier and its third harmonic.

The dashed line represents a linear gain. It is important to note that the output level begins to decrease at a certain drive level, even when the input signal is further increased. This situation leads to instability within the predistortion loop and must be avoided under all possible load conditions. Secure operation is maintained when the amplifier is driven slightly higher than the 1 dB compression point.

The result of a linearized SSB transmitter by means of the adaptive predistortion are presented in Figure 9.52. The third- and fifth-order intermodulation products are reduced by -64 dBc or -70 dB$_{PEP}$. The marker points to the carrier frequency. It is clearly visible, that the intermodulation products within the

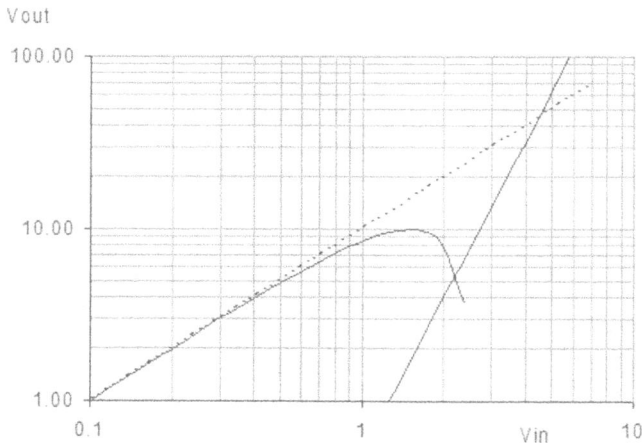

FIGURE 9.51 Gain and third harmonic with $k_1 = 10$ and $k_3 = 2$.

ATT: 61.00 dB   MARKER: −85.29 dBm, 14098.8056 kHz
BW VIDEO: 0.10 kHz
BW RES: 0.08 kHz

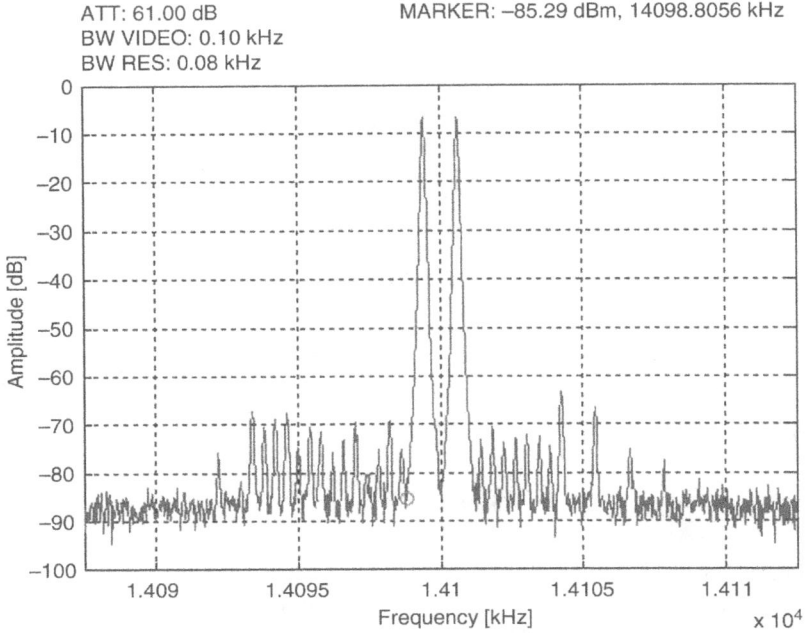

FIGURE 9.52   Transmit spectrum with f1 = 700 Hz, f2 = 1,900 Hz, and P = 50 W.

passband of the 5 kHz LPF are cancelled, whereas the seventh- and ninth-order products are even higher without predistortion (see Figure 9.53). A possible reason is that these products are further distorted by the rapidly changing phase on the filter slope of the LPF.

It looks unusual in Figure 9.52 that the intermodulation products are spaced by 400 Hz instead of by f2 − f1 = 1,200 Hz. The reason is that 1,200 Hz is not harmonic to the sample rate of 32 kHz. The lowest common multiple is 400 Hz. The use of 700 Hz and 1,500 Hz tones (difference of 800 Hz) is harmonic and will prove this assumption.

These presented results (Figure 9.54) neglect the memory effect of which there are different sources. They can be roughly classified in narrow band and wide band, often also referred to as long-term and short-term memory effects.

Possible causes for narrowband memory effects are:

- Thermal effects inside the transistor.
- Influences of the bias circuit, when it follows the envelope due to a too-small time constant.
- A bad alignment of the delay in the reference path (see Figure 9.54).

Wideband memory effects are mainly caused by bandwidth-dependent distortions. These distortions can be found by a two-tone modulation, where the distance in frequency is varied over the full bandwidth of the baseband.

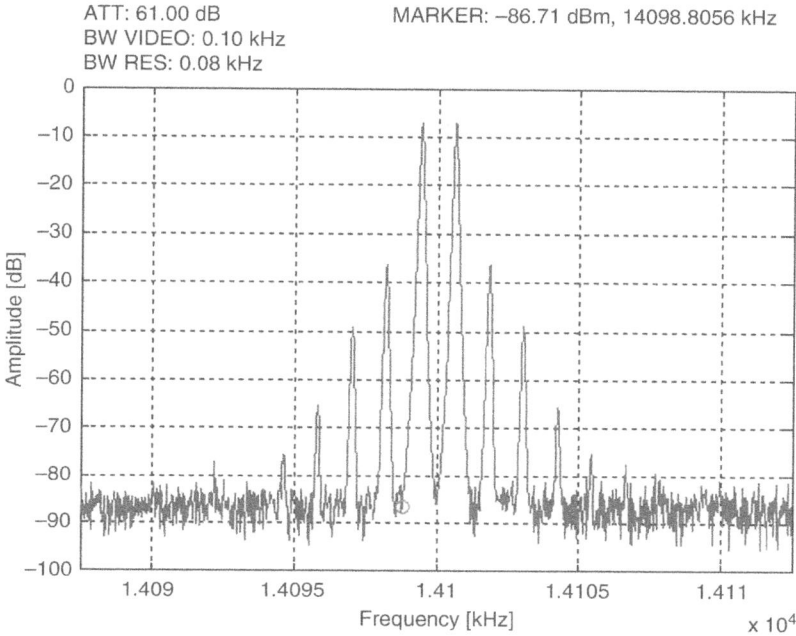

FIGURE 9.53 Transmit spectrum without predistortion (same settings as Figure 9.45).

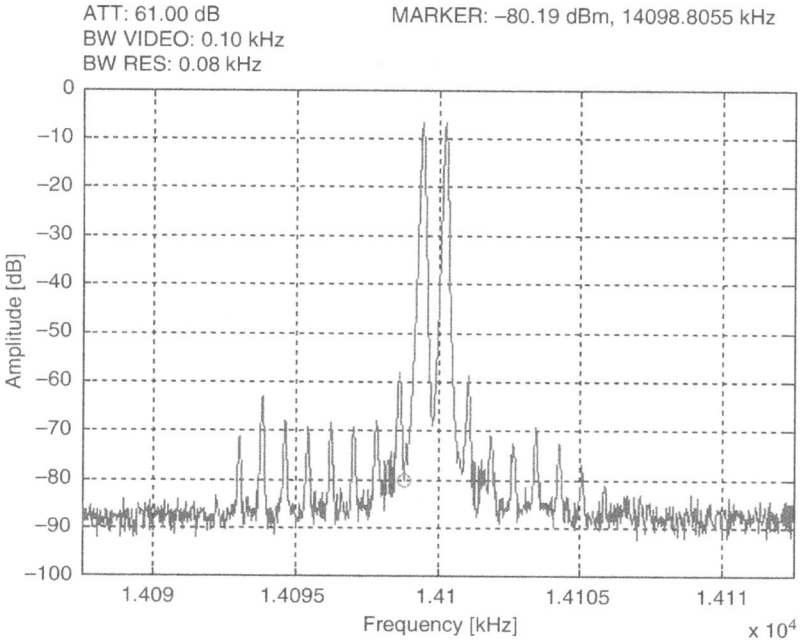

FIGURE 9.54 Spectrum with f2 − f1 = 800 Hz is harmonic to the sample rate.

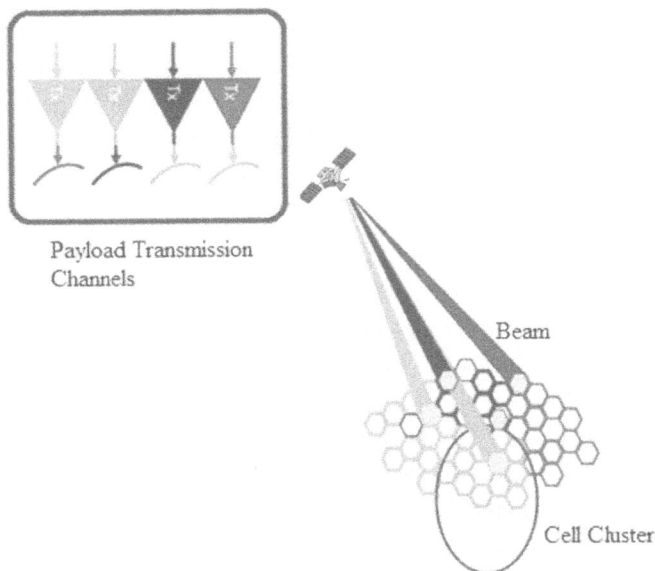

FIGURE 9.55   Typical block diagram of software-defined radio monitoring receiver [16].

To overcome memory effects is a complicated task and needs a model of all significant influences. Memory effects can be simulated using the Volterra series model. An interesting effect is that the lower and upper IM3 products are different when the two tones are shifted within the baseband [15].

An example of a radio monitoring receiver Rohde and Schwarz Wideband Monitoring Receiver model (R&S ESMD) (Figure 9.55) provides the UWB technology implementation for applications such as detecting unknown signals, identifying interference, spectrum monitoring, spectrum clearance, and signal search over wide frequency ranges, producing signal content and direction finding of identified signals.

The dynamic range of active antennas plays an important role and is an interesting parameter for further investigation for UWB technology implementation on SDR platforms for next-generation communication networks [16].

9.12   SDR-ENABLED PHASED ARRAYS AND BEAM HOPPING

With the rapid advancement of SDR, real-world networks at radio and microwave frequencies are undergoing a rapid change in design with the availability of easy-to-use tunable and reconfigurable hardware and software tools. SDR provides an opportunity to define the waveform features in software and implement them on easy-to-use hardware, thus, enabling easy alteration of waveforms in the actual system. Hence, the SDRs are equipped with highly reconfigurable pre-assembled hardware components with an easy-to-use software interface that provides accessibility to the hardware reconfigurability functions.

This concept of using a large number of software-defined waveform features in phased arrays is implemented in software-defined phased array radio (SDPAR) operating around the 5G new radio (NR) frequency band of 28 GHz [17]. The SDPAR hardware provided extensive beam reconfigurability functions, which are accessed directly from the software interface. The Python-based software controlled the easily reconfigurable phased array radio features, which were used for beam shaping/steering control as well as data TX/RX function control. On-chip digital circuits with millimeter wave (mmWave) circuit blocks enabled agile beam-steering and beam-shaping functions. Beams are formed using up to 128 independently controlled phased array elements using RF beamforming. The SDPAR system platform in mmWave phased array systems is targeted toward 5G communications [17].

The concept of tailored and customized software-controlled waveform features using SDR has enabled newer technologies to be explored in terms of power resource management. A maturing technology being explored for satellite applications is known as beam hopping. Beam hopping is used to adapt satellite resources depending on its demand, and is particularly appealing for future low Earth orbit (LEO) constellations. It reduces on-board satellite power consumption, and it allows the implementation of a half-duplex waveform to reduce user terminal costs.

Most modern satellites are set to be equipped with this technology, which will help satellites to dynamically adapt resources, hence, enabling flexible allocation of satellite resources according to demand as shown in Figure 9.56. It increases aggregate capacity by 15% and reduces unmet and excess capacity by 20% and lowers DC power consumption greater than 50% [18].

Another SDR application is used in eliminating blind spots in implanted sensing elements. A common cause of blind spots is the spatial and angular misalignments between the transmitter (TX) and receiver (RX) antennas. In Ref. [19], an SDR-based scalable data acquisition system is used, which is programmed to provide coverage over customized experimental areas. The incoming RF signal with the highest power among SDRs is selected in real-time to prevent data loss.

9.13   MULTIFUNCTIONAL PRINTED ANTENNAS FOR SDR APPLICATIONS

Design of a multifunctional and reconfigurable antenna for application in cognitive radio (CR) systems of SDR, which demand wideband and narrowband antennas and multiservice/multiband/multistandard radio (MSR) is presented in [20].

The design is realized using the concepts presented in Refs. [21] and [22]. A pair of split ring resonators (SRRs) and PIN diodes loaded in the coplanar waveguide (CPW) feed line constitute a dual state filter section, exhibiting a frequency-notched/narrow bandpass response depending on the status (ON/OFF) of the PIN diode. This filter section can be integrated into the feed section of any CPW-fed antenna. The designed antenna, due to this integrated dual-state filter section, exhibits a unique frequency reconfiguration capability that transforms a frequency-notched wideband antenna into a narrowband antenna, where the narrowband frequency complements the notch frequency. A single printed monopole antenna loaded with SRRs and PIN diodes on the feed section provides dual functionality. The frequency notch in the

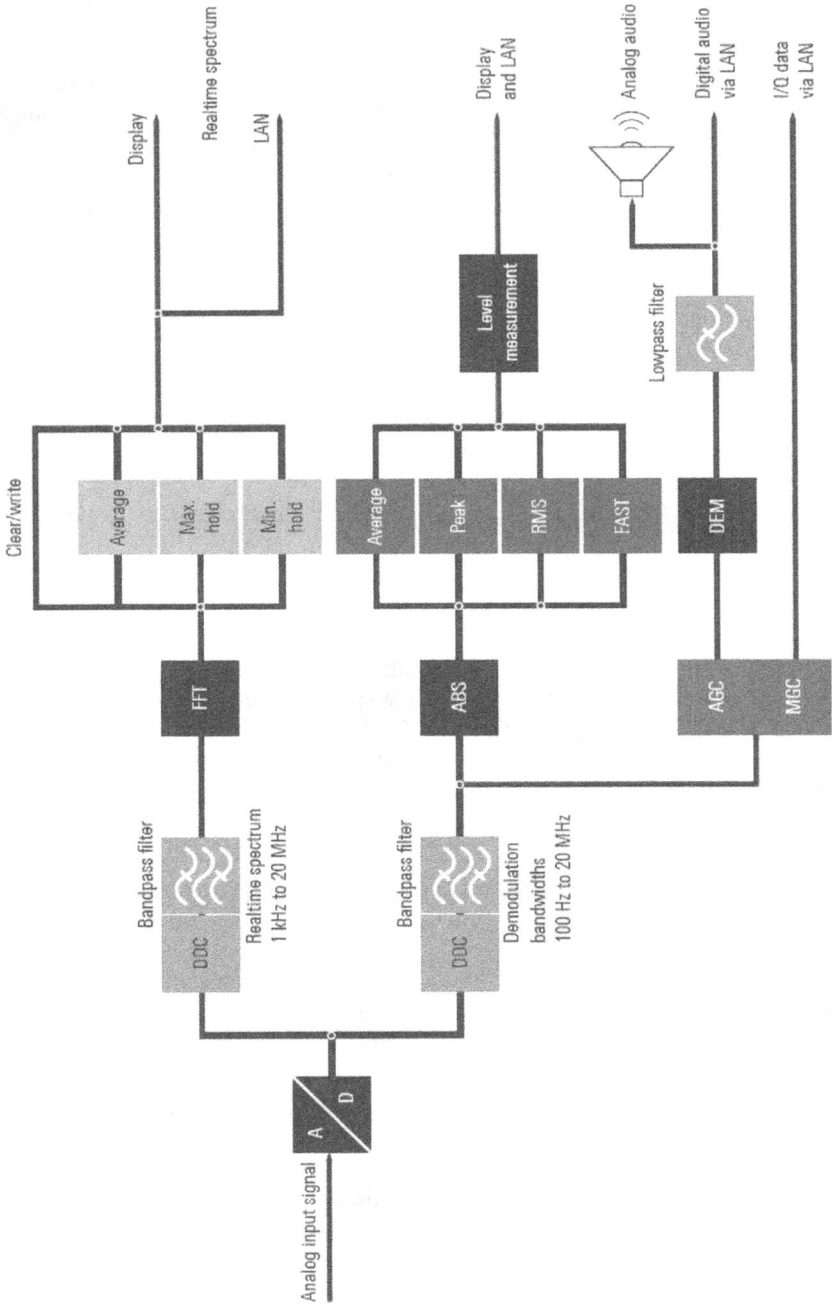

FIGURE 9.56   Schematic of beam hopping using a low Earth orbit satellite [18].

wideband printed monopole antenna is caused by the SRRs that are magnetically coupled with feeding CPW line and inhibit signal propagation around the SRRs' resonance frequency [22–24]. With the loading of PIN diodes on the CPW and appropriate biasing, the frequency-notched wideband antenna can be reconfigured into a narrowband antenna operating at the SRR's resonance. The design is realized using a printed circular monopole antenna with a wide bandwidth. The printed monopole is fed by a CPW transmission line loaded with a pair of SRRs and PIN diodes and is highly useful for SDR applications.

The schematic of the antenna loaded with SRRs and PIN diodes on the CPW feed section is shown in Figure 9.57. A circular monopole having radius R is fed by a CPW line consisting of ground planes having width W_1 and W_2, length Ls and a signal line of width S and length $Ls + t$. The signal line and the ground planes are separated by a symmetric pair of slot gaps, s_g. The antenna is printed on a substrate having thickness h and dielectric constant ε_r. Two square-shaped split-ring resonators having a dimension, a_{ext}, which is half the dimension of the side-length of the SRR, conductor thickness c, separation between rings d and split gaps g_1 and g_2, as shown in Figure 9.57, are printed on the other side of the substrate with their centers coinciding with the slot lines of the CPW feed. A pair of PIN diodes are placed on

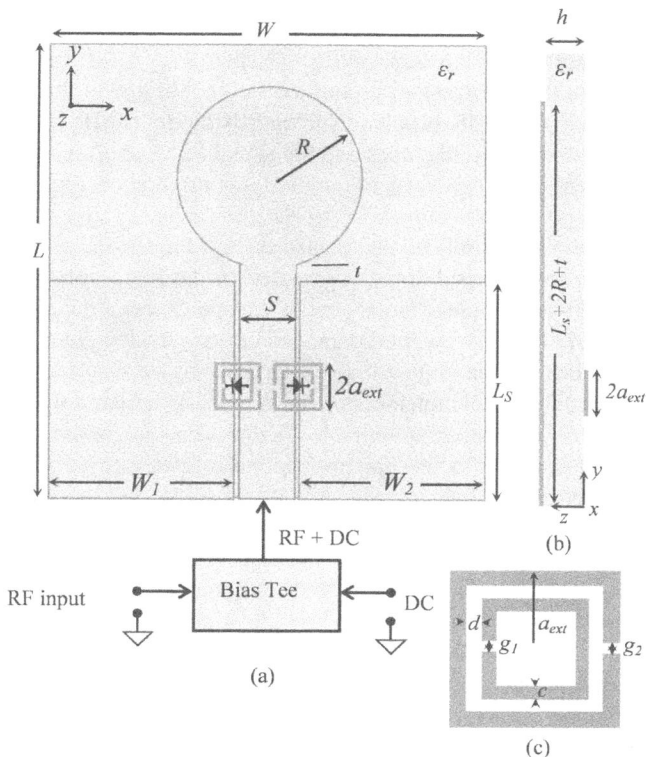

FIGURE 9.57   Schematic of SRR-coupled PIN-diode-loaded active antenna [20].

the slots of the CPW with their positions coinciding with the axes passing through the center of the SRRs.

The CPW is inductively coupled to a pair of SRRs, having narrow split gaps g_1 and g_2, which are symmetrically placed on the back side of the substrate. The propagating electric field vector is polarized along the plane of the SRRs and the magnetic field vector is polarized along the SRRs' axes.

The propagating signal excites the SRR, which prohibits transmission around its resonance frequency determined by the SRRs' dimensions and the constitutive parameters of the host substrate yielding in frequency a notch [24]. The symmetric position of the SRRs to optimize efficient magnetic coupling has been shown to yield a notch in the antenna impedance and radiation characteristics [24]. The CPW transmission line is also loaded with PIN diodes, which, on different biasing conditions (reverse and forward), would effectively open and short the signal line with the ground planes. This, in turn, provides a frequency-notched response and its complementary narrow band response.

The prototype was fabricated on Taconic substrate having $\varepsilon_r = 2.33$, $tan\ \delta = 0.0009$, and thickness h =1.575 mm. The circular monopole having radius $R = 12.5$ mm and fed with a CPW having ground plane length L_s =22.5 mm, width $W = 50$ mm, signal line width $S = 6$ mm, slot gap $S_g = 0.3$mm, and feed gap $t = 0.2$ mm was etched on one side of the substrate. The slot gaps and the signal line width were optimized to yield a line impedance close to 50 Ω. A pair of SRRs having dimensions $a_{ext} = 2.5$ mm, $c = 0.35$ mm, $d = 0.6$ mm, and split gaps $g_1 = g_2 = 0.5$ mm, were printed on the other side of the substrate with their axes coinciding with the slot line in the CPW, as shown in Figure 9.58. A pair of silicon PIN diodes (SMP 1145) were placed on the slots between the ground planes and the signal line and aligned to the position of SRRs axes. The required DC bias to the diodes is provided using a Mini-Circuits 15542 bias tee.

Figure 9.59 shows the simulated and measured S_{11} of the prototype active antenna with the diode in the ON condition. As revealed in the plot, with the diode in the ON condition, the antenna yields a narrowband response centered at 6.12 GHz and 6.02 GHz for simulated and measured data, and effectively complements the impedance behavior of the antenna in the diode OFF condition.

Figure 9.60 reveals the complementary nature of the antenna under diode OFF and ON conditions. The active antenna with the diode ON condition provides a narrow band response due to the narrow bandpass filter formed with the combination of SRRs and PIN diodes on the slots of the CPW. On the other hand, with diode in the OFF condition, the band-notch filtering of the SRRs contributes to a frequency-notched UWB response of the antenna. The biasing of the diodes is conducted through 5V DC supply connected to the bias tee. The current flowing through the diodes in the forward bias condition provides the DC short between the signal and ground planes of the CPW, which, in turn, reconfigures the notched wideband antenna into a narrowband antenna.

Figure 9.61 shows the measured maximum realized peak gain plotted against the frequency of the prototype active antenna in diode OFF and ON conditions. In the diode OFF condition, the gain drops sharply at the notch frequency of 6.01 GHz, prohibiting radiation, whereas for the rest of the frequencies, the gain remains above

(a)

(b)

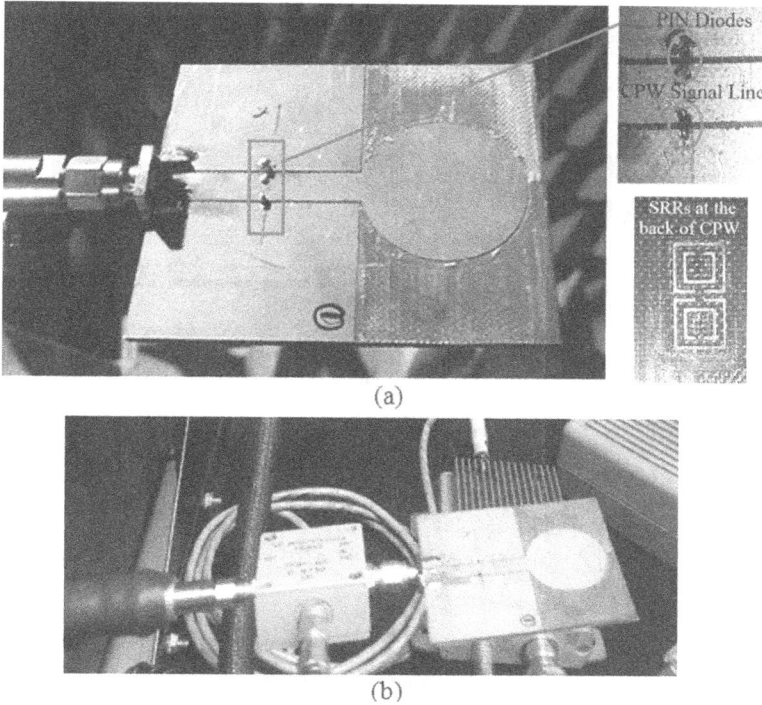

FIGURE 9.58   A printed monopole active antenna with PIN diodes accommodated between the signal line and ground planes of the coplanar waveguide (CPW). SRRs are printed on the opposite side of the CPW [20].

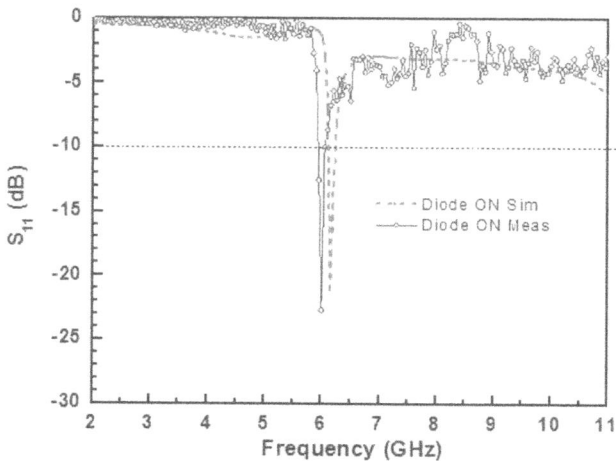

FIGURE 9.59   Measured and simulated S_{11} response of the SRR coupled PIN diode loaded coplanar waveguide-fed printed circular monopole active antenna with diodes in ON conditions [20].

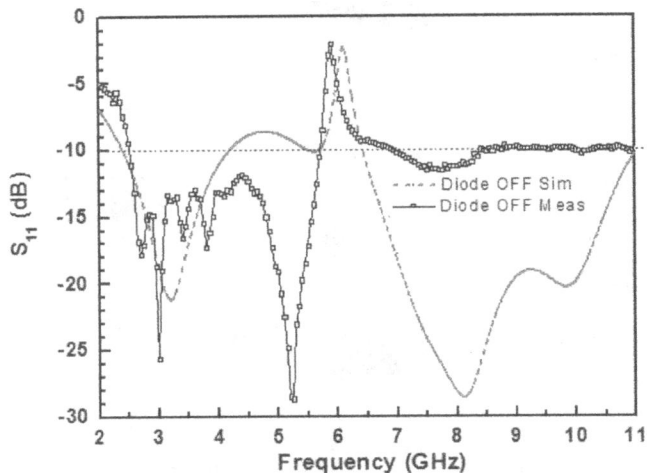

FIGURE 9.60 Measured and simulated S_{11} response of the SRR-coupled PIN-diode-loaded coplanar waveguide-fed printed circular monopole active antenna with diodes in OFF conditions [20].

0 dBi. A complementary gain profile is yielded when the diode is switched ON and the gain rises sharply at 6.02 GHz and drops off at either side of the radiating frequency. The measured gain value in the diode ON condition at the narrowband frequency 6.02 GHz was obtained as 3.59 dBi.

The measured normalized radiation patterns in two principal planes, the x-y plane (E-plane) and x-z plane (H-plane), for the prototype in diodes OFF condition

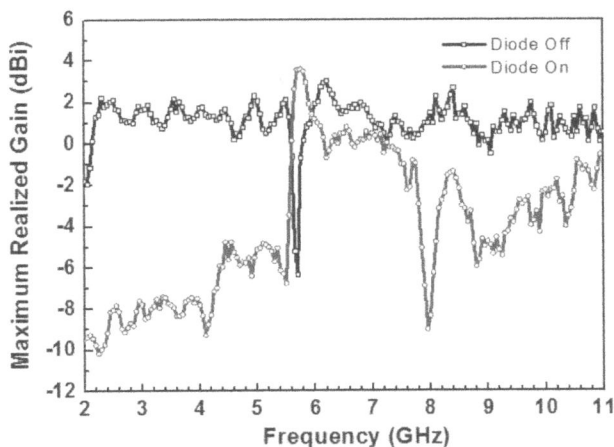

FIGURE 9.61 Measured realized peak gain of the SRR-coupled PIN-diode-loaded coplanar waveguide-fed printed circular monopole active antenna with diodes in OFF and ON conditions [20].

are shown in Figure 9.62a and b, respectively, for 3.8 GHz, 8 GHz, and 9.4 GHz. The E-plane and H-plane at 6.02 GHz for the diode ON condition are shown in Figure 9.62c and d. The radiation patterns indicate axial null along the y-axis, ensuring monopole type radiation for the x-y plane with high cross-polar discrimination. The H-plane radiation yields an omnidirectional pattern for the x-z plane. The simulated efficiency was obtained at 92% and 76% for diode OFF and diode ON conditions, respectively.

Figure 9.63 shows the simulated reflection coefficient of the proposed antenna for different a_{ext} dimensions in diode ON and diode OFF conditions. Three different a_{ext} dimensions of 2.3 mm, 2.5 mm, and 2.7 mm yielded notch frequencies in

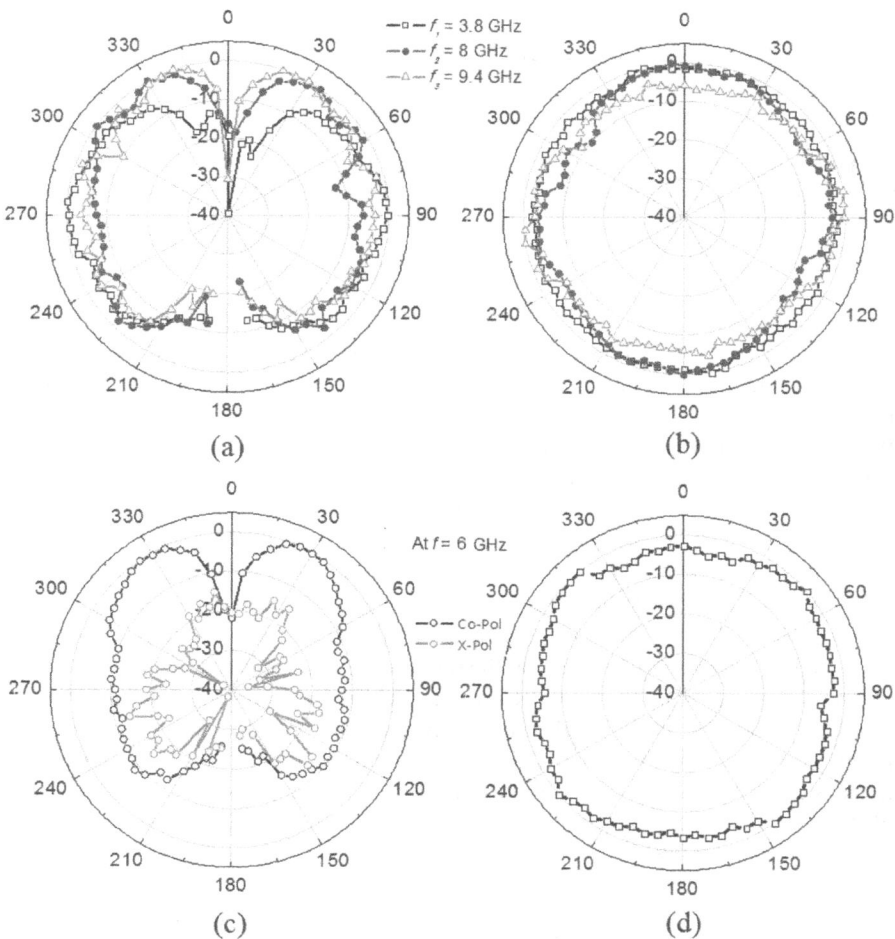

FIGURE 9.62 Measured normalized E- and H-plane radiation patterns of the fabricated active antenna. (a) and (c) The x-y plane (E-plane) in diode OFF and ON conditions, (b) and (d) x-z plane (H-plane) in diodes OFF and ON condition [20].

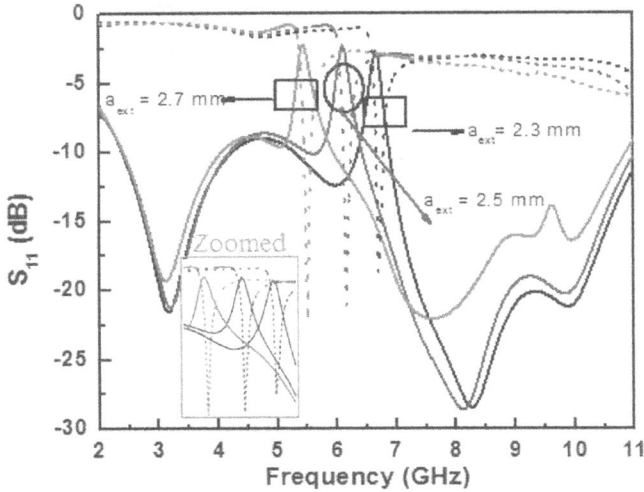

FIGURE 9.63 Simulated S_{11} of the SRR-coupled PIN-diode-loaded coplanar waveguide-fed printed circular monopole active antenna with diodes in the OFF and ON conditions for various a_{ext} values. Solid curves are for the diode OFF case and dotted curves are for the diode ON case [20].

the diode's OFF condition at 5.45 GHz, 6.1 GHz, and 6.65 GHz while with diode in diode's ON condition's narrowband response centered at 5.5 GHz, 6.15 GHz, and 6.7 GHz is obtained. The wide variation of the notch and narrowband frequency using the proposed concept by changing the SRR parameter and without altering the basic radiator dimension is evident from the figure. Similar variation can also be achieved by changing the other parameters of the SRRs like c, d, and g.

It is further demonstrated that the notch frequency and the narrowband frequency can be tuned by changing the physical dimension of the SRR geometry as well as shape of the SRRs without perturbing the radiator dimension. The various shapes of SRRs that can be utilized are circular, square, hexagonal, and triangular. Moreover, the inclusion of the SRRs and the PIN diodes on the feed section of the antenna does not have any adverse impact on the radiation performance of the reconfigurable antenna, making it suitable for SDR applications. Obtaining both wideband and narrowband responses from a single multifunctional antenna alleviates the problems of port coupling from multiple antennas as well as size constraints in compact SDR systems. Such antennas can be tailored for use in 5G SDR applications and the Internet of Things (IoT) by changing the dimensions of the SRRs and without altering the radiator dimensions [8].

9.14 ACTIVE PHASED ARRAY ANTENNA FOR 5G APPLICATIONS

The RF environment is often polluted by noise with interference signal falling in the band of interest with a multipath fading effect on the desired frequency. An active phased array antenna is an array of antenna elements designed to adapt and

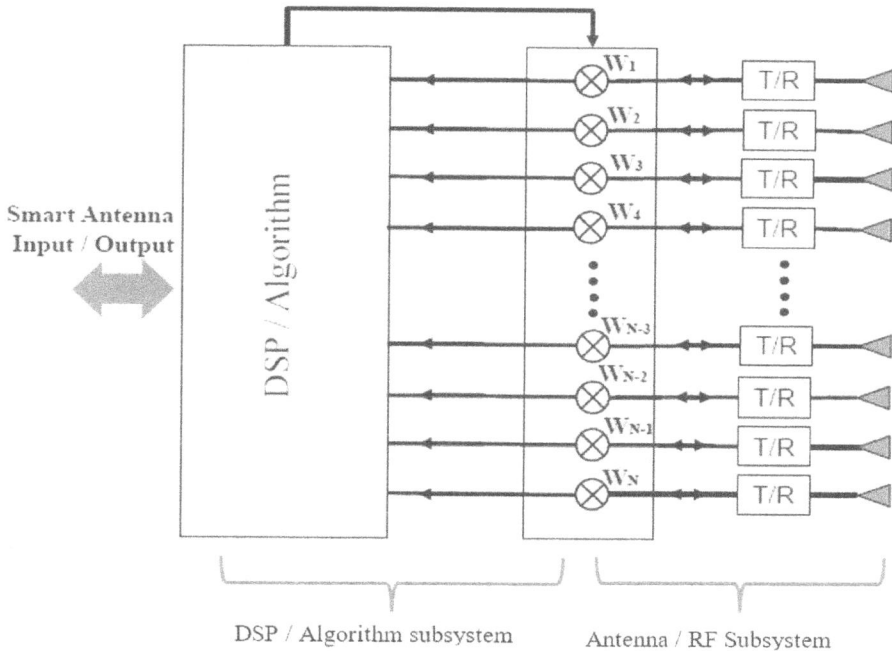

FIGURE 9.64 Block diagram of an active antenna array [25].

change the antenna radiation pattern to adjust the RF environment. The adaptations are realized by performing electrical beam-tilting and beam width adjustments. The antennas can direct beams toward particular users and track user movement and also steer nulls, reduce side lobes, and self-heal in case one of the array elements stop functioning. An active phased array antenna system helps to reduce signal fading and phase cancellation due to multipath and co-channel interference. A combination of antenna array and DSP running algorithms make it possible for the antenna to transmit and receive signals, adapt, and, hence, perform smart beamforming measures. Every active array comes equipped with a passive radiating element connected to each transmit/receive module in the RF subsystem as shown in Figure 9.64.

Each of these passive radiated elements are fed with different antenna-feeding weights. The weight is a function of amplitude and/or phase. Each of these weights is varied depending on the specific relationship with all other elements, which are configured in rows and columns to achieve the desired beam manipulation.

An LUT consisting of different antenna weights and the corresponding beam manipulation (beamforming and beamsteering) can be calculated or measured for each individual antenna array.

Active phased array antennas fall under two basic categories:

• Switched-beam antennas.
• Adaptive antenna arrays.

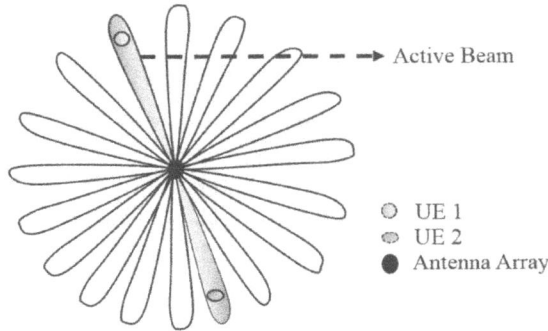

FIGURE 9.65   Switched-beam antenna radiation pattern [25].

9.14.1   SWITCHED-BEAM ANTENNA

A switched-beam antenna array is a system typically intended for a cellular base transceiver station (BTS), which has multiple predefined beam patterns designed to enhance the received signal power of the user equipment (UE). The arrangement of antennas at a BTS is designed to have a triangular structure. Each side of the triangle covers a 120° sector with multiple beams in each sector. Depending on the exact location of the user, the relevant beam is switched on and handed over to another relevant predefined beam with better signal strength when the user changes location. One major drawback of this technique arises when the user is not at the center of the allocated main predefined beam, because the signal quality drops. Likewise, if an interference signal falls close to the center of the main beam, it is unintentionally amplified more than the intended user signal.

9.14.2   ADAPTIVE ARRAY ANTENNAS

Adaptive antenna array systems can adjust and adapt their radiation pattern(s) almost in real time based on the movement of each individual user terminal. In principle, beamsteering is also useful in non-line-of-sight (NLOS) channels. Simultaneously, the interferers are rejected by performing a technique called side-lobe nulling and, thus, making the interferers fall intentionally into a direction of weak received gain. Figure 9.66 shows the radiation pattern of an adaptive array antenna.

For every type of active phased array antenna, stable phase-coherent signals are essential.

9.14.3   BEAMSTEERING AND BEAMFORMING

Beamsteering is the change of direction of the main lobe of a radiation beam. The phase of the element feed signals is varied in a controlled manner to achieve beamsteering.

Beamforming is used to steer and to "form" the shape of the radiation pattern of the array antenna for either signal transmission or signal reception. There are many

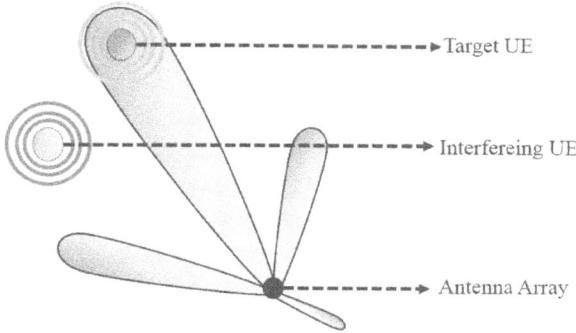

FIGURE 9.66   Adaptive array antenna radiation pattern [25]. (UE = user equipment).

techniques applied in the design of the antenna-feeding network to perform beamforming, but, in essence, the antenna-feeding amplitude and phase of the signal are varied. The design of the feeding network is optimized based on the application requirements.

The phase of the antenna feeds needs to be precisely controlled. Thus, phase coherent signaling is a key requirement for phased array antennas. Phase coherence of two RF signals means that there is a defined and stable phase relationship between two (or more) RF carriers, i.e. there is a fixed delta phase between the carriers. Phase coherence is defined for carriers derived from the same source. Fig. 9.67 depicts the Principles of beamsteering and beamforming [26].

9.14.4   TESTING FOR PASSIVE ARRAY ANTENNA

Figure 9.68 shows the test setup for characterizing smart active antenna arrays. An RX antenna is connected to a signal and spectrum analyzer. Active phased array antenna algorithm testing, beamforming algorithm testing, modulated signal analysis, and electric field pattern measurement can be performed using the signal and spectrum analyzer.

Single beam measurement applications for algorithm testing are maximum radiation testing, side-lobe reduction testing, side-lobe reduction testing with null filling, RF fair beam, nulling beam, and self-healing measurement. For these measurements, the signal and spectrum analyzer is connected to the RX antenna. The measurements are carried out in spectrum analyzer mode.

9.14.5   2D RADIATION PATTERN MEASUREMENT OF PASSIVE ARRAY ANTENNA

The measurement setup consists of signal generators, RX antenna, TX antenna, a power meter, and a turntable as shown in Figure 9.69. In this setup, the TX antenna is always used as the device under test (DUT). In the case where a high measurement dynamic range is required, the power meter can be replaced with a spectrum analyzer.

The phase coherence calibration setup is performed using a master/reference generator that provides the reference and the local oscillator (LO) signal to all other signal generators in the measurement setup.

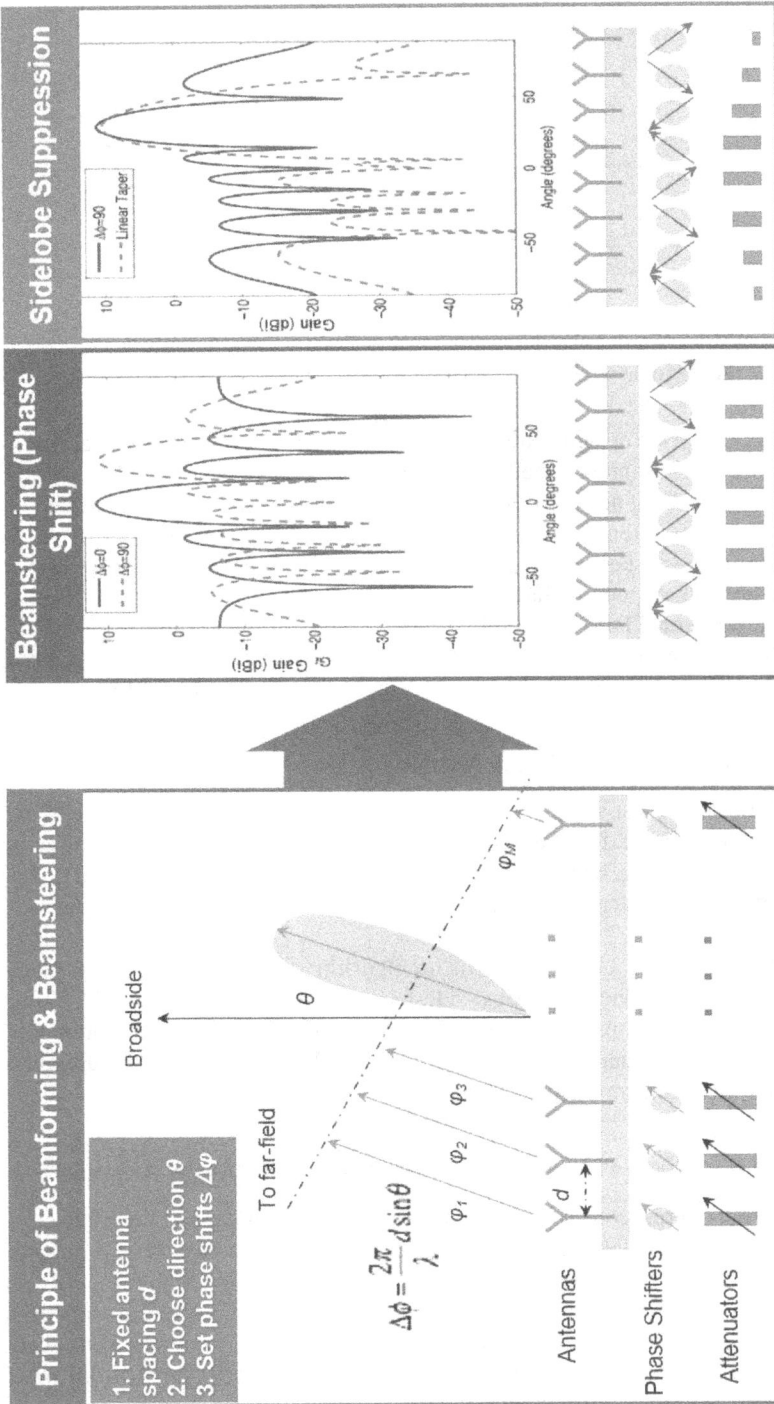

FIGURE 9.67   Principles of beamsteering and beamforming [26].

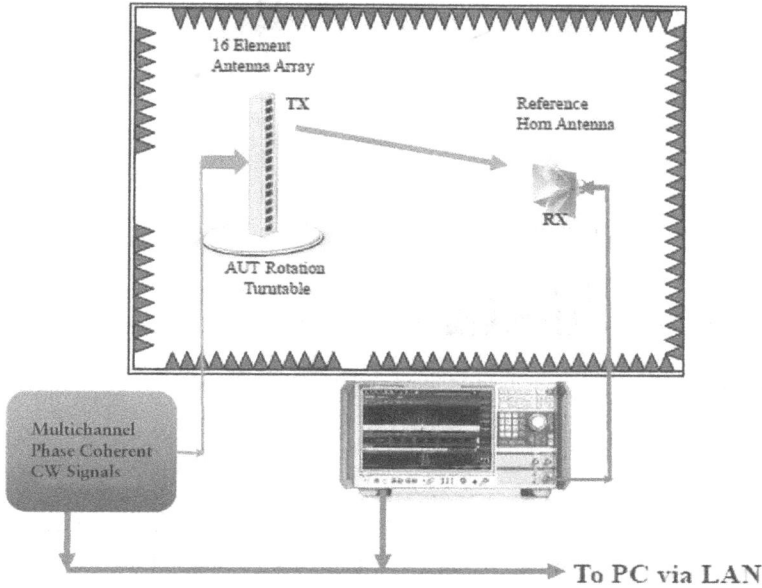

FIGURE 9.68 Test setup for active phased array antenna measurement in TX mode [25].

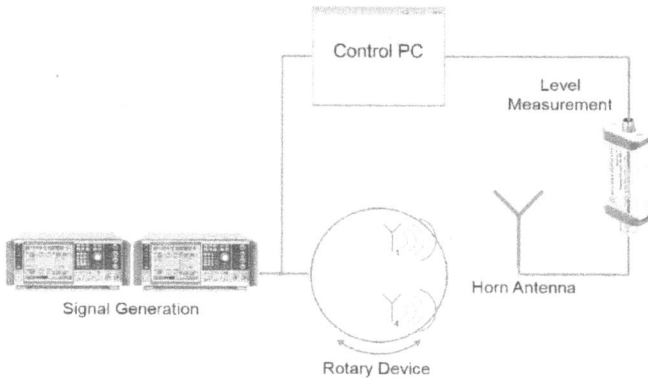

FIGURE 9.69 Test setup for 2D radiation pattern measurement. On the measurement side, a spectrum analyzer may also be used if an increased dynamic range is required measurement in TX mode [25].

Figure 9.70 shows a direct far-field measurement system where the DUT is placed inside an anechoic chamber. Types of radiation pattern measurements that can be performed are:

Far-field measurements (passive): These are the traditional over-the-air (OTA) measurements for legacy 2G/3G/4G antennas, where the antenna is fed via a cable from the vector network analyzer (VNA) and the

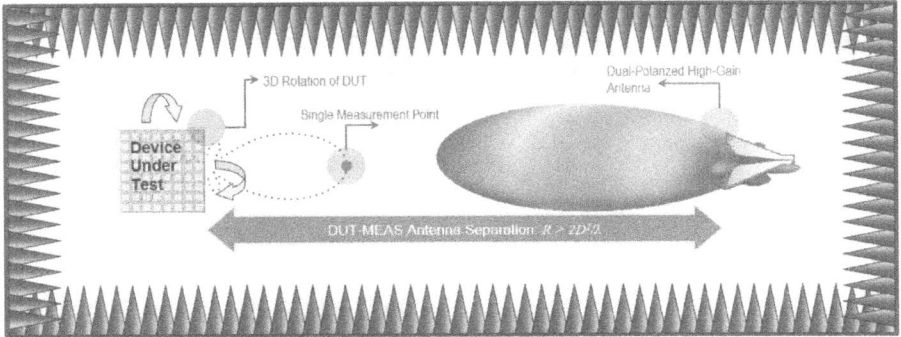

FIGURE 9.70   Direct far-field measurement system. (From Rohde & Schwarz.)

measurement antenna is fed via the second port on the VNA. The VNA then measures the S_{12} between the DUT and measurement antenna. The far-field depends on two parameters: Device size (D) and wavelength with the general equation: Far-field distance = $2D^2/\lambda$

Far-field measurements (active): For this chamber size, active far-field is only valid for small DUTs at mmWave frequencies. In this case, the DUT has its own transceivers that can transmit and receive signals. Either a power meter for magnitude only measurements or a spectrum analyzer to measure modulated signals are used.

Near-field: In the near-field of the DUT, measurement of both the phase and the magnitude of the signals are conducted, and then near-field to far-field Fourier Transform is performed to calculate far-field properties.

9.14.6 DIGITAL BEAMFORMING

In a hybrid beamforming architecture, analog beamforming is added to address a higher number of antenna elements. This narrows the beamwidth in a cost-efficient way and enables the RF signal to achieve a longer range. Modern highly integrated circuits (ICs) offer digitally controlled beamforming. They help overcome the space limitations at higher frequencies where the antenna elements become rather small. Digital beamforming with appropriate phase and level weighting takes place in the baseband to generate individual streams. Many beams can be overlayed to enable multiple links via one RF system. Complex digital chips such as FPGAs and ASICs have strict timings between the different sections of their booting sequences. Digital beamforming requires separate paths per stream with individual data converters. Time alignment between the different channels is essential, since jitter and time variances between channels degrade beamforming accuracy. Figure 9.71 shows various beamforming architectures.

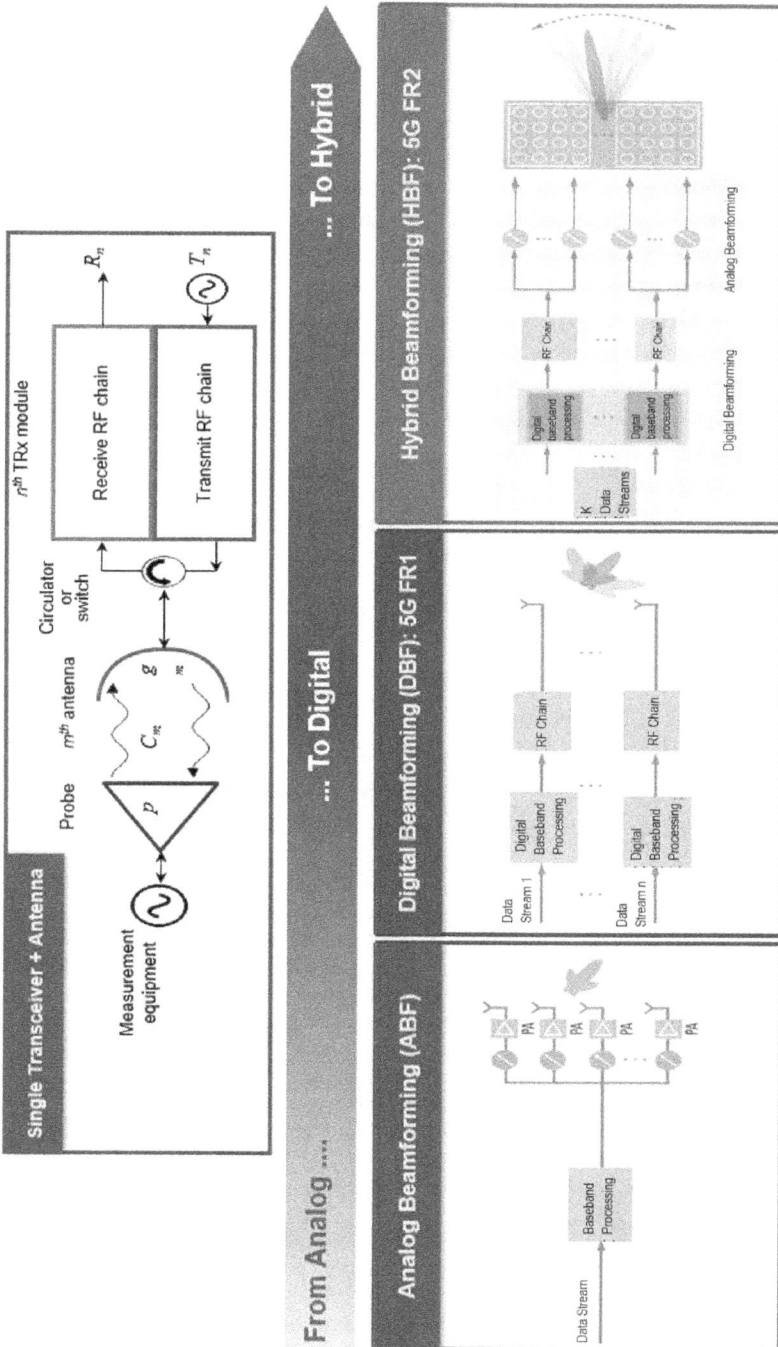

FIGURE 9.71 Beamforming architectures. (From Rohde & Schwarz.)

9.15 CONCLUSION

A fundamental challenge with SDR is how to achieve sufficient computational capacity, in particular for processing UWB high bit-rate waveforms, within acceptable size, power consumption, cost, and weight factors. SDR offers the flexibility of varying bandwidth and range, the ability to adapt to environmental parameters and employ optimal broadband pulse characteristics for channel equalization and robustness, and, finally, the capability to easily adapt to current- and later-generation communication infrastructures. Multifunctional antenna with real-time actuation to toggle between wideband and narrowband response will accelerate the development of compact SDR systems.

REFERENCES

1. Rohde, U. L. (1985). "Digital HF Radio: A Sampling of Techniques." Third International Conference on HF Communication Systems and Techniques (London, England).
2. Rohde, U. L. "Digital HF Radio: A Sampling of Techniques." Ham Radio Magazine (April, 1985).
3. https://th.bing.com/th?id=OIP.mVVHlxthM2RmGBdiXNdkeAHaFj&w=100&h=100 &c=8&rs=1&qlt=80&o=6&dpr=1.3&pid=3.1
4. Rohde, U. L., Poddar, A. K. & Sarkar, T. K. "Next generation networks: Software defined radio, emerging trends." *2016 International Conference on Electromagnetics in Advanced Applications (ICEAA)*, Cairns, QLD, Australia, 2016, pp. 784–788, doi: 10.1109/ICEAA.2016.7731516.
5. Rohde, U. L., Whitaker, J. C. & Zhand, H., *Communications Receivers Principle and Design*, 4th Edition, McGraw Hill, 2017.
6. Kong-Pang Pun, da Franca, José Epifanio, Azeredo-Leme, Carlo. *Circuit Design for Wireless Communications: Improved Techniques for Image Rejection in Wideband Quadrature Receivers*, Kluwer Academic Publishers, 2003.
7. Pace, P. E. (Ed.) *Advanced Techniques for Digital Receivers*, London, Artech House Boston, 2000.
8. Kester W. (Ed.) *The Data Conversion Handbook, 2005*, Analog Devices Inc, 2004.
9. Hogenauer, E. B. "An economical class of digital filters for decimation and interpolation." *IEEE Transactions on Acoustics, Speech and Signal Processing*, vol. 29, 1981.
10. Lyons, R. G., *Understanding Digital Signal Processing*, 2nd Edition, Prentice Hall, 2004, chapter 13.24.
11. Gerald Youngblood, "A Software Defined Radio for the Masses," QEX 2002/2, Part 1. https://www.arrl.org/files/file/Technology/tis/info/pdf/020708qex013.pdf
12. Jeffreys, H. & Jeffreys, B. S. "The gibbs phenomenon." *Methods of Mathematical Physics*, 3rd ed. Cambridge, England: Cambridge University Press, pp. 445–446, 1988.
13. Ralf Rudersdorfer. "Funkempfängerkompendium" chapter III-14; Elektor Verlag Aachen (in german).
14. Kenington, P. B. "High-linearity RF amplifier design," Artech House Microwave Library, 2000.
15. Brinkhoff, J. "Bandwidth-Dependant Intermodulation Distortion in FET Amplifiers," Dissertation, Department of Electronics, Macquarie University, Sydney (2004).
16. Rohde, U. L., Poddar, A. K. & Marius, S. A. "Next generation radios: SDR and SDN." *2017 IEEE-APS Topical Conference on Antennas and Propagation in Wireless Communications (APWC)*, Verona, Italy, 2017, pp. 296–299, doi: 10.1109/APWC.2017.8062305.

17. Sadhu, B., Paidimarri, A., Ferriss, M., Yeck, M., Gu, X. & Valdes-Garcia, A. "A Software-Defined Phased Array Radio with mmWave to Software Vertical Stack Integration for 5G Experimentation." *IEEE/MTT-S International Microwave Symposium - IMS 2018*, 2018, doi: 10.1109/MWSYM.2018.8439278.

18. Demonstrate wideband beam hopping (Rohde & Schwarz) https://www.rohde-schwarz.com/us/about/news-press/all-news/rohde-schwarz-and-satixfy-to-demonstrate-wideband-dvb-s2x-beam-hopping-and-dvb-rcs2-at-satellite-show-2023-press-release-detailpage_229356-1338368.html

19. Jia, Y. et al. "A software-defined radio receiver for wireless recording from freely behaving animals." *IEEE Transactions on Biomedical Circuits and Systems*, vol. 13, no. 6, pp. 1645–1654, Dec. 2019, doi: 10.1109/TBCAS.2019.2949233.

20. Saha, C., Siddiqui, J. Y., Freundorfer, A. P., Shaik, L. A. & Antar, Y. M. M., "Active reconfigurable ultra-wideband antenna with complementary frequency notched and narrowband response." *IEEE Access*, vol. 8, pp. 100802–100809, 2020, doi: 10.1109/ACCESS.2020.2997933.

21. Horestami, A. K., Shaterian, Z., Naqui, J., Martin, F. & Fumeaux, C. "Reconfigurable and tunable S-shaped split-ring resonators and application in band-notched UWB antennas." *IEEE Transactions on Antennas Propagation*, vol. 64, no. 9, pp. 3766–3775, 2016.

22. Siddiqui, J. Y., Saha, C. & Antar, Y. M. M. "A novel ultrawideband (UWB) printed antenna with a dual complementary characteristic." *IEEE Antennas and Wireless Propagation Letters*, vol. 14, pp. 974–977, 2015.

23. Martín, F., Bonache, J., Falcone, F., Sorolla, M. & Marqués, R. "Split ring resonator-based left-handed coplanar waveguide." *Applied Physics Letters*, vol. 83, no. 22, pp. 4652–4654, 2003.

24. Siddiqui, J. Y., Saha, C. & Antar, Y. M. M. "Compact SRR loaded UWB circular monopole antenna with frequency notch characteristics." *IEEE Transactions on Antennas and Propagation*, vol. 62, no. 8, pp. 4015–4020, 2014.

25. Characterizing Phased Array Antennas - Application Note (rohde-schwarz.com), https://www.rohde-schwarz.com/appnote/1MA248

26. Improving Transmit/Receive Module Test Accuracy and Throughput - Mobility Engineering Technology.

BIBLIOGRAPHY AND FURTHER READING

1. Ulrich L. Rohde, "How Many Signals Does a Receiver See?" Ham Radio Magazine June, 1977.

2. Ulrich L. Rohde, "Recent Developments in Communication Receiver Design to Increase the Dynamic Range," ELECTRO/80, Boston, May 1980.

3. Ulrich L. Rohde, "A Comparison of Solid State and Tube Based Receiver Systems Using CAD," QST June, 1993, pp. 24–28.

4. U. Rohde, "Modern Receiver Design Including Digital Signal Processing," UHF-VHF Conf., Germany, March 9–10, 1996.

5. U. L. Rohde and A. K. Poddar, "Reconfigurable wideband VCOs." *2005 IEEE 16th International Symposium on Personal, Indoor and Mobile Radio Communications*, Berlin, Germany, 2005, pp. 1250–1253 Vol. 2, doi: 10.1109/PIMRC.2005.1651641.

6. U. L. Rohde, "Front-end modern communication system-software defined radio: Recent & emerging trends." *2014 IEEE Benjamin Franklin Symposium on Microwave and Antenna Sub-systems for Radar, Telecommunications, and Biomedical Applications (BenMAS)*, Philadelphia, PA, USA, 2014, pp. 1–3, doi: 10.1109/BenMAS.2014.7529450.

7. Ulrich L. Rohde, A. K. Poddar, M. A. Silaghi, "Front-end receiver: Recent and Emerging Trend," *2015 Joint conference of the IEEE International Frequency Control Symposium & the European Frequency and Time Forum*, pp. 326–331, 2015.

8. U. L. Rohde, A. K. Poddar and S. K. Koul, "Modern radios: 5G and SDR emerging trends." *2016 Asia-Pacific Microwave Conference (APMC)*, New Delhi, India, 2016, pp. 1–4, doi: 10.1109/APMC.2016.7931428.

9. U. L. Rohde, A. K. Poddar and A. M. Silaghi, "Monitoring Radios: Dynamics and Emerging Trends," *2018 IEEE International Symposium on Circuits and Systems (ISCAS)*, Florence, Italy, 2018, pp. 1–5, doi: 10.1109/ISCAS.2018.8351835.

10. Ulrich L. Rohde, Universität der Bundeswehr Thomas Boegl, Rohde & Schwarz, Munich, Germany; The perfect HF Receiver. What would it look Like Today? Microwave Journal, May 13, 2022.

11. A. Apte, "A new analytical design method of ultra-low-noise voltage-controlled VHF crystal oscillators and its validation," March 2020, Thesis for: Doctoral, Anisha M Apte, University of Cottbus, Germany. https://opus4.kobv.de/opus4-btu/frontdoor/index/index/docId/5138

12. A. M. Apte, A. K. Poddar, M. Rudolph and U. L. Rohde, "A Novel Low Phase Noise X-Band Oscillator." *IEEE Microwave Magazine*, vol. 16, no. 1, pp. 127–135, Feb. 2015, doi: 10.1109/MMM.2014.2367957.

13. C. -Y. -D. Sim et al., "A PIFA Design with WLAN and Wi-Fi 6E Band for Laptop Computer Applications." *2022 IEEE International Symposium on Antennas and Propagation and USNC-URSI Radio Science Meeting (AP-S/URSI)*, Denver, CO, USA, 2022, pp. 1808–1809, doi: 10.1109/AP-S/USNC-URSI47032.2022.9886623.

14. A. K. Poddar, U. L. Rohde, A. M. Apte, V. Madhavan and T. Itoh, "Phase noise measurement: Uncertainty & limitations." *2014 IEEE Benjamin Franklin Symposium on Microwave and Antenna Sub-systems for Radar, Telecommunications, and Biomedical Applications (BenMAS)*, Philadelphia, PA, USA, 2014, pp. 1–3, doi: 10.1109/BenMAS.2014.7529478.

15. A. K. Poddar, U. L. Rohde, V. Madhavan, A. M. Apte and S. K. Koul, "Ka-Band metamaterial Möbius Oscillator (MMO) circuit." *2016 IEEE MTT-S International Microwave Symposium (IMS)*, San Francisco, CA, USA, 2016, pp. 1–4, doi: 10.1109/MWSYM.2016.7539970.

16. A. M. Apte, V. Madhavan, A. K. Poddar, U. L. Rohde and M. Rudolph, "A novel low phase noise X-band oscillator." *2014 IEEE Benjamin Franklin Symposium on Microwave and Antenna Sub-systems for Radar, Telecommunications, and Biomedical Applications (BenMAS)*, Philadelphia, PA, USA, 2014, pp. 1–3, doi: 10.1109/BenMAS.2014.7529471.

17. C. -Y. -D. Sim et al., "MIMO Antenna with Hexagonal Split Ring Resonator for LTE, Wireless LAN, and Wi-Fi 6 Services." *2022 IEEE International Symposium on Antennas and Propagation and USNC-URSI Radio Science Meeting (AP-S/URSI)*, Denver, CO, USA, 2022, pp. 1810–1811, doi: 10.1109/AP-S/USNC-URSI47032.2022.9886886.

18. U. L. Rohde, A. K. Poddar and A. M. Apte, "Getting Its Measure: Oscillator Phase Noise Measurement Techniques and Limitations." *IEEE Microwave Magazine*, vol. 14, no. 6, pp. 73–86, Sept.-Oct. 2013, doi: 10.1109/MMM.2013.2269860.

19. A. K. Poddar, U. L. Rohde and A. M. Apte, "How Low Can They Go?: Oscillator Phase Noise Model, Theoretical, Experimental Validation, and Phase Noise Measurements." *IEEE Microwave Magazine*, vol. 14, no. 6, pp. 50–72, Sept.-Oct. 2013, doi: 10.1109/MMM.2013.2269859.

20. U. L. Rohde and A. M. Apte, "Everything You Always Wanted to Know About Colpitts Oscillators [Applications Note]." *IEEE Microwave Magazine*, vol. 17, no. 8, pp. 59–76, Aug. 2016, doi: 10.1109/MMM.2016.2561498.

21. A. Apte, U. L. Rohde, A. Poddar, and M. Rudolph, "Optimizing Phase-Noise Performance: Theory and Design Techniques for a Crystal Oscillator." *IEEE Microwave Magazine*, vol. 18, no. 4, pp. 108–123, June 2017, doi: 10.1109/MMM.2017.2679998.

22. A. M. Apte, A. K. Poddar, U. L. Rohde, and E. Rubiola, "Colpitts oscillator: A new criterion of energy saving for high-performance signal sources." *2016 IEEE International Frequency Control Symposium (IFCS)*, New Orleans, LA, USA, 2016, pp. 1–7, doi: 10.1109/FCS.2016.7546729.
23. M. Rudolph and A. M. Apte, "Nonlinear Noise Modeling: Using Nonlinear Circuit Simulators to Simulate Noise in the Nonlinear Domain." *IEEE Microwave Magazine*, vol. 22, no. 7, pp. 47–64, July 2021, doi: 10.1109/MMM.2021.3069538.

Index

Note: Locators in *italics* represent figures and **bold** indicate tables in the text.

For Product Safety Concerns and Information please contact our EU
representative GPSR@taylorandfrancis.com
Taylor & Francis Verlag GmbH, Kaufingerstraße 24, 80331 München, Germany

www.ingramcontent.com/pod-product-compliance
Lightning Source LLC
Chambersburg PA
CBHW060754220326
41598CB00022B/2430